Advances in Intelligent Systems and Computing

Volume 1275

The series "Advances in Intelligent Systems and Computing" contains publications on theory, applications, and design methods of Intelligent Systems and Intelligent Computing. Virtually all disciplines such as engineering, natural sciences, computer and information science, ICT, economics, business, e-commerce, environment, healthcare, life science are covered. The list of topics spans all the areas of modern intelligent systems and computing such as: computational intelligence, soft computing including neural networks, fuzzy systems, evolutionary computing and the fusion of these paradigms, social intelligence, ambient intelligence, computational neuroscience, artificial life, virtual worlds and society, cognitive science and systems, Perception and Vision, DNA and immune based systems, self-organizing and adaptive systems, e-Learning and teaching, human-centered and human-centric computing, recommender systems, intelligent control, robotics and mechatronics including human-machine teaming, knowledge-based paradigms, learning paradigms, machine ethics, intelligent data analysis, knowledge management, intelligent agents, intelligent decision making and support, intelligent network security, trust management, interactive entertainment, Web intelligence and multimedia.

The publications within "Advances in Intelligent Systems and Computing" are primarily proceedings of important conferences, symposia and congresses. They cover significant recent developments in the field, both of a foundational and applicable character. An important characteristic feature of the series is the short publication time and world-wide distribution. This permits a rapid and broad dissemination of research results.

Indexed by SCOPUS, DBLP, EI Compendex, INSPEC, WTI Frankfurt eG, zbMATH, Japanese Science and Technology Agency (JST), SCImago.

More information about this series at http://www.springer.com/series/11156

Sinan Melih Nigdeli · Joong Hoon Kim ·
Gebrail Bekdaş · Anupam Yadav
Editors

Proceedings of 6th International Conference on Harmony Search, Soft Computing and Applications

ICHSA 2020, Istanbul

Springer

Editors
Sinan Melih Nigdeli
Istanbul University – Cerrahpaşa
Istanbul, Turkey

Gebrail Bekdaş
Istanbul University – Cerrahpaşa
Istanbul, Turkey

Joong Hoon Kim
Korea University
Seoul, Korea (Republic of)

Anupam Yadav
Dr. BR Ambedkar National Institute
of Technology
Jalandhar, India

ISSN 2194-5357 ISSN 2194-5365 (electronic)
Advances in Intelligent Systems and Computing
ISBN 978-981-15-8602-6 ISBN 978-981-15-8603-3 (eBook)
https://doi.org/10.1007/978-981-15-8603-3

This Springer imprint is published by the registered company Springer Nature Singapore Pte Ltd.
The registered company address is: 152 Beach Road, #21-01/04 Gateway East, Singapore 189721, Singapore

Preface

International Conference on Harmony Search, Soft Computing and Applications (ICHSA) is a prestigious academic event, where both potential and young researchers meet to discuss their knowledge and study. In 2020, the venue of the sixth edition of the event was Istanbul, Turkey. Unfortunately, the event was organized online due to Covid-19 outbreak because of travel restrictions and health care. The conference was organized with the support of Istanbul University—Cerrahpaşa. The earlier editions of the conference were organized at South Korea, Spain, India and China.

This book is a curated collection of the articles which were presented during the conference. The book focuses on the current and recent developments in the harmony search algorithm and their engineering applications. It demonstrates the new variants of harmony search algorithms for water distribution system operation, neural networks for predicting the drought index, prediction of soil plasticity using ANN, optimum designs of reinforced concrete retaining walls under static and dynamic loads, total protein optimization using metaheuristics, harmony search for extreme learning machines, ML-based pedotransfer function for estimating the soil–water characteristic curve, optimum transportation path assignment within airports in Turkey, defect detection in fruits and vegetables using soft computing techniques, potentials of AI in Military, and hybrid harmony search algorithm for optimum design of vibration absorber system.

In conclusion, the edited book comprises papers on diverse aspects of harmony search and other metaheuristic techniques with their application in areas, such as engineering optimization, water distribution system operation, drought indexing, landslide monitoring, mechanical engineering problems, and machine learning predicting models.

Turkey, China
July 2020

Sinan Melih Nigdeli
Joong Hoon Kim
Gebrail Bekdaş
Anupam Yadav

Contents

Performance Quantification of Search Operators in Hybrid Harmony Search Algorithms 1
Taewook Kim, Young Hwan Choi, and Joong Hoon Kim

Harmony Search Algorithms for Optimizing Extreme Learning Machines 11
Abobakr Khalil Al-Shamiri, Ali Sadollah, and Joong Hoon Kim

Freestyle Rap Harmony Search (FRHS) for Engineering Problem Optimization 21
Kyoung Won Min, Donghwi Jung, and Joong Hoon Kim

Improvement of Voltage Profile and Loss Reduction Based on Optimal Placement and Sizing of Renewable Distributed Generations Using 4-Rule Harmony Search Algorithm 33
Ali Sadollah, Mohammad Nasir, Abobakr Khalil Al-Shamiri, and Joong Hoon Kim

Control of a Jacket Platform Under Wave Load Using ATMD and Optimization by HSA 45
Maziar Fahimi Farzam, Babak Alinejad, Gabriel Bekdaş, and Seyyed Ali Mousavi Gavgani

Robustness of Structures with Active Tuned Mass Dampers Optimized via Modified Harmony Search for Time Delay 53
Aylin Ece Kayabekir, Sinan Melih Nigdeli, and Gebrail Bekdaş

Optimum Active Controlled SDOF Structures Under Pulse-Type Ground Motions via Harmony Search 61
Serdar Ulusoy, Sinan Melih Nigdeli, and Gebrail Bekdaş

Hybrid Harmony Search Algorithm for Optimum Design of Vibration Absorber System for Adjacent Buildings 73
Sinan Melih Nigdeli and Gebrail Bekdaş

The Effect of Initial Values on Metaheuristic-Based Optimum Design of Tuned Mass Dampers 81
Aylin Ece Kayabekir, Gebrail Bekdaş, and Sinan Melih Nigdeli

The Comparison of Classical and Artificial Neural Network-Based Formulations for Tuned Mass Damper Optimization 93
Melda Yücel, Sinan Melih Nigdeli, and Gebrail Bekdaş

Passive Control of Frame Structures by Optimum Tuned Mass Dampers 111
Apaer Mubuli, Sinan Melih Nigdeli, and Gebrail Bekdaş

Total Potential Optimization Using Metaheuristics: Analysis of Cantilever Beam via Plane-Stress Members 127
Yusuf Cengiz Toklu, Gebrail Bekdaş, Aylin Ece Kayabekir, Sinan Melih Nigdeli, and Melda Yücel

Performance of the Whale Optimization Algorithm in Space Steel Frame Optimization Problems 139
Bahar Neslişah Kale, İbrahim Aydoğdu, and Emre Demir

Evaluation of Metaheuristic Algorithm on Optimum Design of T-Beams 155
Aylin Ece Kayabekir, Gebrail Bekdaş, and Sinan Melih Nigdeli

Multi-objective Optimization of the Reinforced Concrete Beam 171
Zhi-Yu Zhang, Zulkarnaen Gifari, Young-Kyu Ju, and Joong Hoon Kim

Optimal Cost Design of Single-Story Reinforced Concrete Frames Using Jaya Algorithm 179
Elmas Rakıcı, Gebrail Bekdaş, and Sinan Melih Nigdeli

Optimum Design of Reinforced Concrete Retaining Walls Under Static and Dynamic Loads Using Jaya Algorithm 187
Nur Yılmaz, Sena Aral, Sinan Melih Nigdeli, and Gebrail Bekdaş

Jaya Optimization for the Design of Cantilever Retaining Walls with Toe Projection Restriction 197
Sena Aral, Nur Yılmaz, Gebrail Bekdaş, and Sinan Melih Nigdeli

Transportation Path Assignment Within the Airports in Turkey 207
Emre Demir and İbrahim Aydoğdu

Structural Strengthening of RC Buildings for Enhanced Seismic Resistance 219
Baris Gunes, Turgay Cosgun, Atakan Mangir, and Baris Sayin

Seismic Performance of Existing RC Framed Buildings Using Pushover Analysis Method 235
Baris Gunes, Atakan Mangir, Turgay Cosgun, and Baris Sayin

Computation of Axial Symmetric Cylindrical Reinforced Concrete Walls with Domes 249
Aylin Ece Kayabekir

A Review on the Mechanical Properties of Natural Fiber-Reinforced Composites 261
Celal Cakiroglu and Gebrail Bekdaş

Choosing the Appropriate Branch for Participation Banks Through Machine Learning 269
Bulut Karadağ and Eda Arıkan

Prediction of Soil Plasticity Index with the Use of Regression Analysis and Artificial Neural Networks: A Specific Case for Bakırköy District 281
Zülal Akbay Arama, Melda Yücel, Muhammed Selahaddin Akın, Said Enes Nuray, and Oğuzhan Alten

The Applicability of Regression Analysis and Artificial Neural Networks to the Prediction Process of Consistency and Compaction Properties of High Plastic Clays 295
Zülal Akbay Arama, Hazal Berrak Gençdal, Said Enes Nuray, and Melda Yücel

A Hybrid Recurrent Model Based on LSTM and Statistical Evaluation to Predict Soil Displacement for Landslide Monitoring 307
Yilin Li, Xiaojun Pu, Ying Qiao, and Hongan Wang

Prediction of Optimum 3-Bar Truss Model Parameters with an ANN Model 317
Melda Yücel, Gebrail Bekdaş, and Sinan Melih Nigdeli

Defects Detection in Fruits and Vegetables Using Image Processing and Soft Computing Techniques 325
V. G. Narendra and Ancilla J. Pinto

A Framework for Quality Evaluation of Edible Nuts Using Computer Vision and Soft Computing Techniques 339
V. G. Narendra, G. Shiva Prasad, and Ancilla J. Pinto

Discovery of Spatial Patterns of Types of Cooking Fuels Used in the Districts of India Using Spatial Data Mining 349
B. M. Ahamed Shafeeq and Zahid Ahmed Ansari

An Interactive Approach of Rule Mining and Anomaly Detection for Internal Risks 365
Kun Liu, Yunkun Wu, Wenting Wei, Zhonghui Wang, Jiaqi Zhu, and Hongan Wang

Wear Particles Classification Using Shape Features 377
Mohammad Shakeel Laghari, Ahmed Hassan, and Mubashir Noman

Smart Academic Guidance for Institutional Students 387
Mohammad Shakeel Laghari, Ahmed Dirir, and Mubashir Noman

Grey Wolf Optimizer with Crossover and Opposition-Based Learning 401
Shitu Singh and Jagdish Chand Bansal

Development of Discrete Artificial Electric Field Algorithm for Quadratic Assignment Problems 411
Anita, Anupam Yadav, Nitin Kumar, and Joong Hoon Kim

Fuzzy-Based Kernelized Clustering Algorithms for Handling Big Data Using Apache Spark 423
Preeti Jha, Aruna Tiwari, Neha Bharill, Milind Ratnaparkhe, Neha Nagendra, and Mukkamalla Mounika

Author Index 437

About the Authors

Dr. Sinan Melih Nigdeli Associative Professor, is Researcher in Structural Control and Optimization at Istanbul University—Cerrahpaşa. He obtained his D.Phil. in Structural Engineering from Istanbul Technical University with a thesis subject of active control. He co-organized the 15th EU-ME Workshop: Metaheuristic and Engineering in Istanbul. In optimization, he organized several mini-symposiums or special sections in prestigious international events such as 12th International Congress on Mechanics (HSTAM2019), 9th International Congress on Computational Mechanics (GRACM 2018), the Biennial International Conference on Engineering Vibration (ICoEV-2015), 3rd International Conference on Multiple Criteria Decision Making (MCDM 2015), 11th World Congress on Computational Mechanics (WCCM2014), International Conference on Engineering and Applied Sciences Optimization (OPTI 2014), 11th Biennial International Conference on Vibration Problems (ICOVP2013), 3rd European Conference of Civil Engineering and 10th, 11th and 15th International Conference of Numerical Analysis and Applied Mathematics (ICNAAM). He co-edited Metaheuristics and Optimization in Civil Engineering published by Springer, and he is one of the guest editors in 2017 special issue of KSCE Journal of Civil Engineering. He has authored about 250 papers for journals and scientific events.

Prof. Joong Hoon Kim Dean of Engineering College of Korea University, obtained his Ph.D. degree from the University of Texas at Austin in 1992 with the thesis title "Optimal replacement/rehabilitation model for water distribution systems". Prof. Kim's major areas of interest include optimal design and management of water distribution systems, application of optimization techniques to various engineering problems and development and application of evolutionary algorithms. He has been the Faculty of the School of Civil, Environmental and Architectural Engineering at Korea University since 1993 and is now serving as the Dean of Engineering College. He has hosted international conferences including APHW 2013, ICHSA 2014 & 2015 and HIC 2016 and has given keynote speeches at many international conferences including AOGS 2013, GCIS 2013, SocPros 2014 & 2015, SWGIC 2017 and RTORS 2017. He is a member of National Academy of Engineering of Korea since 2017.

Dr. Gebrail Bekdaş Associative Professor, is Researcher in Structural Control and Optimization at Istanbul University—Cerrahpaşa. He obtained his D.Phil. in Structural Engineering from Istanbul University with a thesis subject of design of cylindrical walls. He co-organized the 15th EU-ME Workshop: Metaheuristic and Engineering in Istanbul. In optimization, he organized several mini-symposiums or special sections in prestigious international events such as 12th International Congress on Mechanics (HSTAM2019), 9th International Congress on Computational Mechanics (GRACM 2018), the Biennial International Conference on Engineering Vibration (ICoEV-2015), 3rd International Conference on Multiple Criteria Decision Making (MCDM 2015), 11th World Congress on Computational Mechanics (WCCM2014), International Conference on Engineering and Applied Sciences Optimization (OPTI 2014), 11th Biennial International Conference on Vibration Problems (ICOVP2013), 3rd European Conference of Civil Engineering and 10th, 11th and 15th International Conference of Numerical Analysis and Applied Mathematics (ICNAAM). He co-edited Metaheuristics and Optimization in Civil Engineering published by Springer, and he is one of the guest editors in 2017 special issue of KSCE Journal of Civil Engineering. He has authored about 250 papers for journals and scientific events.

Dr. Anupam Yadav is an Assistant Professor, Department of Mathematics, Dr. BR Ambedkar National Institute of Technology Jalandhar, India. His research area includes numerical optimization, soft computing and artificial intelligence, and he has more than ten years research experience in the areas of soft computing and optimization. Dr. Yadav has done Ph.D. in soft computing from Indian Institute of Technology Roorkee, and he worked as a Research Professor at Korea University. He has published more than twenty-five research articles in journals of international repute and has published more than fifteen research articles in conference proceedings. Dr. Yadav, has authored a textbook entitled "An introduction to neural network methods for differential equations". He has edited three books which are published by AISC and Springer Series. Dr. Yadav was General Chair, Convener and member of steering committee of several international conferences. He is an Associate Editor in the journal of the experimental and theoretical artificial intelligence. Dr. Yadav is a member of various research societies.

Performance Quantification of Search Operators in Hybrid Harmony Search Algorithms

Taewook Kim, Young Hwan Choi, and Joong Hoon Kim

Abstract Meta-heuristic algorithms have been developed to solve various mathematical and engineering optimization problems. However, meta-heuristic algorithms show different performances depending on the characteristics of each problem. Therefore, there have been many kinds of research to decrease the performance gap for the different optimization problems by developing new algorithms, improving the search operators, and considering self-adaptive parameters setting on their algorithms. However, the previous studies only focused on improving the performance of each problem category (e.g., mathematical problem, engineering problem) without the quantitative evaluation for the operator performance. Therefore, this study proposes a framework for the quantitative evaluation to solve the no free lunch problem using the operators of the representative meta-heuristic algorithms (such as genetic algorithm and harmony search algorithm). Moreover, based on the quantitative analysis results for each operator, there are several types of hybrid optimization algorithms, which combined the operator of harmony search algorithm (HSA), genetic algorithm (GA), and particle swarm optimization (PSO). The optimization process to find the optimal solution is divided into five sections based on the number of function evaluations to see the performance of the search operator according to the section. Representative mathematical problems were applied to quantify the performance and operators. None of the five evaluated applied to mathematical benchmark problems were the best algorithms. Hybrid HSAs showed advanced performance for

T. Kim
Dept. of Civil, Environmental and Architectural Engineering, Korea University, Seoul, South Korea
e-mail: jaykim@korea.ac.kr

Y. H. Choi
Department of Civil Engineering, Gyeongnam National University of Science and Technology, Jinju, South Korea

J. H. Kim (✉)
School of Civil, Environmental and Architectural Engineering, Korea University, Seoul, South Korea
e-mail: jaykim@korea.ac.kr

S. M. Nigdeli et al. (eds.), *Proceedings of 6th International Conference on Harmony Search, Soft Computing and Applications*, Advances in Intelligent Systems and Computing 1275, https://doi.org/10.1007/978-981-15-8603-3_1

problems where traditional HSA did not show good performance. However, it still has not escaped the No Free Lunch theorem.

Keywords Operator · Meta-heuristic algorithm · Hybrid algorithm · Harmony search algorithm · Performance quantification

1 Introduction

Optimization problem has long been a problem that many researchers have tried to solve. Meta-heuristic algorithms are developed and have emerged as an effective way to solve optimization problem. Meta-heuristic algorithms are applied to various fields such as mathematics, economics, and engineering to solve optimization problems. Research continues to increase the performance of algorithms in various ways to solve the optimization problem that early meta-heuristic algorithm has not solved for structural limitation or other reasons [4–8]. However, the developed algorithm showed excellent performance in a limited field. These problems can be explained by no free lunch theorem. Wolpert and Macready [9] showed that there was no outstanding algorithm for all problems through geometric interpretation. This may allow algorithms that did not apply to a certain problem to perform better than algorithms that solved certain optimization problems. In other words, the reliability of the existing optimal solution could be low.

The reason why algorithms showed different performance in solving the problem is that the operator that the algorithms has is different. The operator is the rule that the algorithms have, and it explores the feasible solution area of the optimization problem to find an optimal solution. Genetic algorithm (GA; [3]) is an algorithm that imitate Charles Darwin's theory of natural selection and utilizes selection, crossover, and mutation as an operator. Harmony search algorithm (HSA; [2]) is an algorithm that imitates the process of jazz players finding the best harmony and utilizes harmony memory consideration (HMC), pitch adjusting (PA), and random search (RS). Particle swarm optimization (PSO, [1]) is an algorithm that mimics the swarm movement of fish or birds, and finds the optimal solution by calculating velocity and position vector. In this study, by analyzing the characteristics of the operator of three meta-heuristic algorithm, GA, HSA, and PSO, two new hybrid algorithms were developed. By combining the operatosr of other algorithms with HSA, it was expected to solve the no free lunch theorem. The performance of five algorithms was evaluated by applying them to mathematical benchmark problems. Also, the performance of search operators in two meta-heuristic algorithms, GA and HSA, and two newly developed hybrids algorithm based on HSA, HSA with crossover and HSA with the modified formula from PSO was quantified. As a result, the GA confirmed that the use of crossover and mutation operator decreased, and for HSA, HMC operator was increased and PA and RS converged to zero. HSA with crossover had increased the usage of HMC and crossover operator, and PA had decreased. In HSA with PSO, HMC operator tended to increase, but does not converge to 100% like HMC in basic HSA. PSO operator in HSA with PSO tended to decrease but does not converge to zero. In both hybrid algorithms, RS converged to zero after finding the optimal solution.

2 Meta-Heuristic Algorithm

Meta-heuristic algorithms have been developed to solve optimization problems by mimicking the nature or artificial phenomena. Each algorithm has operators expressing the phenomena. The meta-heuristic algorithm is developed by properly combining operators of two rules: the global search, a tendency to explore the entire feasible solution area; and the local search, a tendency to intensively explore a specific area of feasible solution area near the optimal solution.

2.1 Genetic Algorithm

GA is the algorithm that expresses the processes of natural selection. GA first creates chromosomes that consist of a row of genes in binary form. The population consist of these chromosomes, and the original population is called parent generation. For the next generation, GA then selects chromosomes that have better fitness values by roulette selection, tournament selection, etc. Selected chromosomes do crossover with a high probability. Finally, the algorithm does mutation to find the optimal solution.

2.2 Harmony Search Algorithm

HSA was developed by Geem et al. [2]; it is an algorithm that is being used not only in engineering but also in various optimization problems. HSA inspired by the observation that jazz players play notes (decision variables) to find the best harmony (optimal solution). In this process, performers can play notes that produce good harmony (harmony memory consideration, HMC), adjust pitch of good harmony (pitch adjusting, PA) or play random sounds (random search).

2.3 Particle Swarm Optimization

PSO finds optimal solution by moving particles that search feasible solution area. At first, randomly scattered particles are generated in the feasible solution area. In the position of particles with the best fitness value, all other particles are collected. PSO expresses this process in the algorithm to find the optimal solution.

Each particle has its position, x_i, and saves the moment that was closest to the optimal solution as personal best, p_B. Of all the particles, the one with the best fitness value is called the global best, g_B. Position of particles moves to g_B using velocity vector, v_i, which is

$$v_i^{\text{new}} = v_i^{\text{old}} + \alpha \bullet c_1 \bullet (p_B - x_i) + \beta \bullet c_2 \bullet (g_B - x_i) \tag{1}$$

where v_i^{new} is updated velocity vector, v_i^{old} is original velocity vector, α and β are coefficients that have value about two, c_1 and c_2 are random values between zero and one. After, v_i^{new} is updated, the new positions of particles are updated:

$$x_i^{\text{new}} = x_i^{\text{old}} + v_i^{\text{new}} \tag{2}$$

3 Hybrid Algorithm

It was assumed that the reason why the performance of the algorithm was different depended on the problems applied was the difference of operator that algorithm had. When combining operators of different algorithms into the HSA, two new hybrid algorithms were developed to determine how the performance of the algorithms varies depending on the operator. One is the combination of crossover operators in GA to the HSA, and the other is combined with PSO and HSA. The two hybrid algorithms are called HSA with crossover, and HSA with PSO, respectively. The HSA with crossover mixed crossover operator in the phase of generating a solution of HS. After the HMC progresses, it will go through crossover with 50% probability. After crossover operator, the PA process, finally a new solution is decided. The HSA with PSO was generated using the velocity vector equation of the PSO during the PA process of the HSA to move the newly generated solutions to the location of the optimal solution.

4 Quantifying Operator of Algorithm

In order to quantify the performance of each search operator a new index was needed. In the course of the algorithm's progress, the moments that generate good solutions and search operators used at the moment were identified. Therefore, the number of finding the optimal solution (NFOS) is increased when the new solution is outstood among the previous solutions. When NFOS is increased which means local best

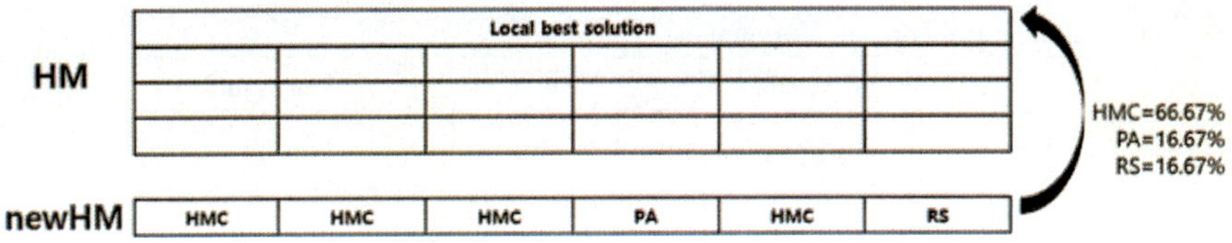

Fig. 1 Example of quantifying performance of search operator

solution is updated, the operators that have worked to find the local best solution are defined in percentage. Figure 1 shows an example of this method:

If there is an existing HM, and the new HM has generated better local best solution the solution in HM, NFOS increases by one. At this moment search operator utilized 66.67% for HMC, 16.67% for PA, and 16.67% for RS, respectively. These indicators showed the algorithm's search for solutions divided into five sections, which helped to determine the results of which search operator worked well on which sections.

5 Application Results

Three meta-heuristic algorithms and two hybrid algorithms were applied to mathematical benchmark functions, and the performance was evaluated. Since the operator of PSO is not stochastics by parameters, operators of the remaining algorithms except PSO have been quantified.

5.1 Mathematical Benchmark Function

In this study, five mathematical benchmark problems were used to evaluate the performance of algorithms. All functions are shown in Table 1.

5.2 Performance Evaluation

All parameters had been set as parameters to find the optimal solution for all algorithms through sensitivity analysis. The population of GA and PSO and the harmony memory size (HMS) of HSA and the two hybrid algorithms were set to 30. The number of function evaluations in this study was set as 500,000.

Result for algorithms are shown in Table 2. Algorithms showed different performances depending on the problems applied. HSA showed good result in Rastrigin function and Rosenbrock function. PSO showed great performance in Schwefel 2.22 problem. HSA with crossover and HSA with PSO showed good result in Schwefel

Table 1 Mathematical benchmark function

Problem	Formula	Constraints
Ackley function	$\min f(x, y) = -20\exp\left[-0.2\sqrt{\frac{1}{N}\sum_{i=1}^{N} x_i^2}\right] - \exp\left[\frac{1}{N}\sum_{i=1}^{N}\cos(2\pi x_i)\right] + e + 20$	$-32.768 \le x \le 32.768$
Rastrigin function	$\min f(x) = An + \sum_{i=1}^{n}[x_i^n - Acos(2\pi x_i)]whereA = 10$	$-5.12 \le x \le 5.12$
Rosenbrock fucntion	$\min f(x) = \sum_{i=1}^{n-1}[100(x_{i+1} - x_i^n)^2 + (1 - x_i)^2]$	$-2.048 \le x \le 2.048$
Schwefel fcuntion 2.22	$\min f(x) = \sum_{i=1}^{n}\lvert x_i^n\rvert + \prod_{i=1}^{n}\lvert x_i^n\rvert$	$-10 \le x \le 10$
Schwefel function 2.26	$\min f(x) = \sum_{i=1}^{n} x_i \sin(\sqrt{\lvert x_i^n\rvert}]$	$-512 \le x \le 512$

Table 2 Optimization results

Algorithm	Ackley function	Rastrigin function	Rosenbrock function	Schwefel 2.22 problem	Schwefel 2.26 problem
GA	2.438E−02	7.467E−03	1.584E+01	2.605E-02	2.595E+03
HSA	1.212E−03	**5.257E−06**	**3.844E−05**	1.475E−02	2.594E−01
PSO	2.764E−06	4.923E+01	1.440E+01	**1.748E−87**	2.483E+03
HSA with CO	3.063E−01	4.925E−01	4.083E−05	1.444E−02	**3.606E−03**
HSA with PSO	**1.865E−14**	1.244E−03	8.438E+00	1.173E−01	1.588E+01

2.26 problem and Ackely function, respectively. In this result, the overall performance of the PSO and HSA with PSO was reduced slightly (two algorithms were not converging to optimal solution in Schwefel 2.26 problem), but there was a strong tendency to converge when the optimum solution was found.

Next, quantifying performance of search operator was analyzed. The value in tables is the mean values for the NFOS after quantifying each section of the 10 times run on the mathematical benchmark problem.

In Table 3, NFOS of GA showed decrease, with the largest decrease in Sect. 2. Crossover operator also showed decreasing result as optimization proceeded, but there was no significant change. The usage of mutation operator decreased in Sect. 2, but increased by about 3% in the final section. The results showed that GA initially used crossover operator to find the optimal solution, thereby reducing the use of the mutation operator. However, it could be seen that the algorithm wanted to develop the solution through mutation operator in order to come closer from the local best solution to the global best solution.

In HSA, NFOS also showed decreasing results in Table 4. However, the value of

Table 3 GA—NFOS

Section	Ackley function	Rastrigin function	Rosenbrock function	Schwefel 2.22 problem	Schwefel 2.26 problem
1	170.50	158.20	1015.90	155.30	147.60
2	32.90	48.50	426.00	30.20	37.40
3	13.60	24.50	440.30	15.10	21.70
4	9.80	11.90	396.30	10.80	10.78
5	7.80	10.20	326.40	9.10	7.22

Table 4 HSA-NFOS

Section	Ackley function	Rastrigin function	Rosenbrock function	Schwefel 2.22 problem	Schwefel 2.26 problem
1	492.60	2556.10	1202.90	1353.10	597.20
2	185.50	1947.10	567.70	454.70	466.60
3	280.80	1784.10	578.20	365.50	509.20
4	396.50	1130.10	586.10	367.20	503.60
5	424.50	1081.90	585.40	376.30	513.70

NOFS is about 100 times larger than the NFOS of GA. This is because of the structure of GA and HSA. GA continuously conducted high probability to do crossover, looking for optimal solution. However, HS used the existing good solution through HMC. The results of HMC, the algorithm used HMC operator nearly 100% when updating optimal solution, which means HSA generate local best solution using HMC operator. Results of PA and RS operators showed that they finally converged to 0, which meant the algorithm converged to the one optimal solution and could not find more optimal solution better than the solution in existing HM.

In HSA with crossover, NFOS also decreased as optimization proceeded shown as Table 5. This algorithm organized by dividing the HMC operator into the HMC and crossover operator and thus the aspects of the two values were similar. The sum of two values showed similar results with the HMC in the original HSA which converged 100% at the last section. PA and RS operator showed same result with the

Table 5 HSA with CO-NFOS

Section	Ackley function	Rastrigin function	Rosenbrock function	Schwefel 2.22 problem	Schwefel 2.26 problem
1	850.50	1064.60	1011.00	879.90	879.90
2	517.40	1076.30	908.90	431.00	431.00
3	710.60	1084.30	866.60	749.20	749.20
4	703.20	1055.30	855.70	769.40	769.40
5	736.70	1007.70	823.50	776.60	776.60

Table 6 HSA with PSO-NFOS

Section	Ackley function	Rastrigin function	Rosenbrock function	Schwefel 2.22 problem	Schwefel 2.26 problem
1	3237.40	573.50	352.20	5193.80	153.70
2	7424.40	707.10	383.40	4712.80	24.90
3	9177.10	761.43	395.50	3682.10	14.60
4	10,385.90	6421.67	389.20	4031.90	12.10
5	10,739.00	6718.67	392.00	3559.30	10.80

original HSA, but some differences were shown in Rastrigin function and Rosenbrock function that did not converge to zero. These results showed that this two benchmark problems could be converged to optimal solution if there were more computation.

HSA with PSO showed good performance in Ackley function and Rastrigin function. In Table 6, NFOS of HSA with PSO showed high value at Ackley function and Ratrigin function, which meant that a strong propensity for PSO to converge to optimal solution. As the PSO operator showed good performance in converging to optimal solution, the performance of PSO operator did not decrease as the optimization proceeded. This result could be compared with PA operator in the original HSA, because the structure of original HSA and HSA with PSO are same. PSO operator in original HSA did not work after the optimal solution was found, but PSO operator worked more to converge to optimal solution. RS operator showed the same result with original HSA.

6 Conclusion

In this study, the performance of the search operator of GA, HSA, HSA with crossover, and HSA with PSO was quantified by applying to mathematical benchmark problems. As GA and HSA had different structures and different operators, there was a big difference in NFOS value. Performance of crossover and mutation operator showed the same tendency, even though it was applied to different functions. In original HSA, the performance of HMC operator was almost 100% at the final section, which means HSA reinforce the optimal solution using HMC operator. HSA with crossover showed bad performance in Rastrigin function and Rosenbrock function, which did not converge to optimal solution after 500,000 function evaluation had done. However, the tendency of HMC and crossover operator results in HSA with crossover showed similar results with HMC in original HSA. In HSA with PSO, NFOS values showed largest values than any other results in Rastrigin function and Schwefel 2.26 problem, which means PSO operator showed strong convergence to optimal solution.

The study found that PA, RS, and mutation operators had to be used in the early stages of the optimization process to explore feasible solution areas, and that their impact was reduced in the mid-term, and the impact of HMC increased. For the last section, using PSO operators had been identified to play a major role in converging into one optimal solution.

In future study, by using machine learning techniques, new algorithm could be developed by combining all the operators of well-known meta-heuristic algorithm. It will allow using the appropriate operator depending on the problem applied or how much optimization has progressed. So, to make future algorithms more practical for all problems, more operators should be quantified.

Acknowledgment This work was supported by the National Research Foundation of Korea (NRF) grant funded by the Korea government (MSIT). (No. 2019R1A2B5B03069810).

References

1. R. Eberhart, J. Kennedy, A new optimizer using particle swarm theory, in *MHS'95. Proceedings of the Sixth International Symposium on Micro Machine and Human Science* (IEEE, 1995), pp. 39–43
2. Z.W. Geem, J.H. Kim, G.V. Loganathan, A new heuristic optimization algorithm: harmony search. Simulation **76**(2), 60–68 (2001)
3. Goldberg, D. E., Holland, J. H.: Genetic algorithms and machine learning (1988).
4. S.H. Jun, Y.H. Choi, D. Jung, J.H. Kim, Copycat harmony search: considering poor music player's followship toward good player, in *Harmony Search and Nature Inspired Optimization Algorithms* (Springer, Singapore, 2019), pp. 113–118
5. A. Kaveh, S. Talatahari, Particle swarm optimizer, ant colony strategy and harmony search scheme hybridized for optimization of truss structures. Comput. Struct. **87**(5–6), 267–283 (2009)
6. M. Mahdavi, M. Fesanghary, E. Damangir, An improved harmony search algorithm for solving optimization problems. Appl. Math. Comput. **188**(2), 1567–1579 (2007)
7. M.G. Omran, M. Mahdavi, Global-best harmony search. Appl. Math. Comput. **198**(2), 643–656 (2008)
8. K. Premalatha, A.M. Natarajan, Hybrid PSO and GA for global maximization. Int. J. Open Problems Compt. Math **2**(4), 597–608 (2009)
9. D.H. Wolpert, W.G. Macready, No free lunch theorems for optimization. IEEE Trans. Evol. Comput. **1**(1), 67–82 (1997)

Harmony Search Algorithms for Optimizing Extreme Learning Machines

Abobakr Khalil Al-Shamiri, Ali Sadollah, and Joong Hoon Kim

Abstract Extreme learning machine (ELM) is a non-iterative algorithm for training single-hidden layer feedforward neural network (SLFN). ELM has been shown to have good generalization performance and faster learning speed than conventional gradient-based learning algorithms. However, due to the random determination of the hidden neuron parameters (i.e., input weights and biases) ELM may require a large number of neurons in the hidden layer. In this paper, the original harmony search (HS) and its variants, namely, improved harmony search (IHS), global-best harmony search (GHS), and intelligent tuned harmony search (ITHS) are used to optimize the input weights and hidden biases of ELM. The output weights are analytically determined using the Moore–Penrose (MP) generalized inverse. The performance of the hybrid approaches is tested on several benchmark classification problems. The simulation results show that the integration of HS algorithms with ELM has obtained compact network architectures with good generalization performance.

Keywords Harmony Search · Extreme Learning Machine · Classification

1 Introduction

Single-hidden-layer feedforward neural networks (SLFN) have been shown to have universal approximation and classification capability, and are usually trained by traditional gradient-based learning methods, such as back-propagation (BP) algorithm. The gradient-based algorithms suffer from slow convergence and may get stuck in

A. K. Al-Shamiri
Research Institute for Mega Construction, Korea University, Seoul, South Korea

A. Sadollah
Department of Mechanical Engineering, University of Science and Culture, Tehran, Iran

J. H. Kim (✉)
School of Civil, Environmental and Architectural Engineering,
Korea University, Seoul, South Korea
e-mail: Jaykim@korea.ac.kr

S. M. Nigdeli et al. (eds.), *Proceedings of 6th International Conference on Harmony Search, Soft Computing and Applications*, Advances in Intelligent Systems and Computing 1275, https://doi.org/10.1007/978-981-15-8603-3_2

local minima. Extreme learning machine (ELM) is a relatively new method for training SLFN [1]. In ELM, the input weights and hidden biases are randomly initialized and remain fixed during the learning process. The output weights are determined analytically by using the Moore–Penrose (MP) generalized inverse. Therefore, ELM has been shown to have good generalization performance and fast learning speed. Since the input weights and hidden biases are initialized with random values and remain fixed during the training, ELM generally requires more hidden neurons than gradient-based algorithms. To obtain a compact network architecture with good generalization performance, ELM has been integrated with evolutionary and swarm intelligence techniques. Zhu et al. [2] proposed an evolutionary ELM (E-ELM) algorithm which uses the differential evolution (DE) algorithm to optimize the hidden layer parameters (i.e., the input weights and hidden biases), and MP generalized inverse to compute the output weights analytically. Cao et al. [3] proposed a hybrid learning algorithm which uses the self-adaptive DE algorithm to select the input weights and hidden biases of ELM. ELM has been combined with harmony search for time-series forcasting in [4, 5]. To enhance the performance of ELM, Alshamiri et al. [6] integrated ELM with artificial bee colony (ABC) and invasive weed optimization algorithms.

Optimizing the hidden layer parameters of ELM is a real-parameter optimization problem. In this paper, the original harmony search (HS) and its variants, namely, improved harmony search (IHS), global-best harmony search (GHS), and intelligent tuned harmony search (ITHS) are used to optimize the input weights and hidden biases of ELM. The proposed hybrid approaches are refereed to as HS-ELM, IHS-ELM, GHS-ELM and ITHS-ELM. In these hybrid approaches, the output weights are analytically determined using the MP generalized inverse.

The rest of this paper is organized as follows: Sect. 2 presents an overview of ELM. Section 3 introduces HS algorithm and its variants. Section 4 presents the proposed hybrid approaches. Experimental results are presented in Sect. 5. Finally, Sect. 6 concludes the paper.

2 Extreme Learning Machine

Traditional algorithms for training feedforward neural networks are usually based on a gradient-based approach in which the network weights and biases are tuned in an iterative manner. Gradient-based learning methods such as BP often get stuck in local minima or converge slowly. To overcome these limitations, Huang et al. [1] proposed an efficient method for training SLFNs, called ELM. ELM significantly reduces the computational time required for training the feedforward neural network. In ELM, the hidden layer parameters are randomly initialized and fixed. The output weights are analytically computed using the MP generalized inverse [7–9].

Consider N training samples $(\mathbf{x}_i, \mathbf{t}_i)_{i=1}^{N}$, where $\mathbf{x}_i = [x_{i1}, x_{i2}, \ldots, x_{id}] \in \mathbf{R}^d$ and $\mathbf{t}_i = [t_{i1}, t_{i2}, \ldots, t_{im}] \in \mathbf{R}^m$. Let L denote the number of neurons in the hidden layer of an SLFN. If this SLFN can approximate these N samples with zero error, then the output of SLFN is as follows:

$$\sum_{i=1}^{L} \boldsymbol{\beta}_i h_i(\mathbf{x}_j) = \mathbf{t}_j = \mathbf{h}(\mathbf{x}_j)\boldsymbol{\beta}, \quad j = 1, \ldots, N, \tag{1}$$

where $\boldsymbol{\beta}_i = [\beta_{i1}, \beta_{i2}, \ldots, \beta_{im}]$ is the weight vector connecting the ith hidden neuron to m output neurons. $\boldsymbol{\beta} = [\boldsymbol{\beta}_1, \boldsymbol{\beta}_2, \ldots, \boldsymbol{\beta}_L]^T$ is the output weight matrix, and $\mathbf{h}(\mathbf{x}_j) = [h_1(\mathbf{x}_j), h_2(\mathbf{x}_j), \ldots, h_L(\mathbf{x}_j)]$ is the hidden layer output vector corresponding to the input $\mathbf{x}_j$. $h_i(\mathbf{x}_j) = g(\mathbf{w}_i, b_i, \mathbf{x}_j))$ is the output of the ith neuron in the hidden layer, where $\mathbf{w}_i \in \mathbf{R}^d$ and $b_i \in R$ are the input weights and bias of the ith hidden neuron, respectively. $g(\cdot)$ is the hidden neuron activation function. Equation (1) can be written compactly as follows [7]:

$$\mathbf{H}\boldsymbol{\beta} = \mathbf{T}, \tag{2}$$

where $\mathbf{H}$ is the output matrix of the ELM hidden layer:

$$\mathbf{H} = \begin{bmatrix} \mathbf{h}(\mathbf{x}_1) \\ \vdots \\ \mathbf{h}(\mathbf{x}_N) \end{bmatrix} = \begin{bmatrix} g(\mathbf{w}_1, b_1, \mathbf{x}_1) & \ldots & g(\mathbf{w}_L, b_L, \mathbf{x}_1) \\ \vdots & \ldots & \vdots \\ g(\mathbf{w}_1, b_1, \mathbf{x}_N) & \ldots & g(\mathbf{w}_L, b_L, \mathbf{x}_N) \end{bmatrix}, \tag{3}$$

and $\mathbf{T}$ is the target matrix of the training data:

$$\mathbf{T} = \begin{bmatrix} \mathbf{t}_1 \\ \vdots \\ \mathbf{t}_N \end{bmatrix} = \begin{bmatrix} t_{11} & \ldots & t_{1m} \\ \vdots & \ldots & \vdots \\ t_{N1} & \ldots & t_{Nm} \end{bmatrix}. \tag{4}$$

The output weights $\boldsymbol{\beta}$ can be computed as follows [1]:

$$\boldsymbol{\beta} = \mathbf{H}^{\dagger}\mathbf{T}, \tag{5}$$

where $\mathbf{H}^{\dagger}$ is the Moore–Penrose generalized inverse of $\mathbf{H}$.

3 Harmony Search and other Variants

In this section, a brief overview of the standard HS algorithm and other variants is presented.

3.1 Harmony Search

Harmony search (HS) is a meta-heuristic optimization algorithm, proposed by Geem et. al [10], which imitates the improvisation process of music players. In the standard HS algorithm, each solution called "harmony" is represented by D-dimension real vector. The initial solutions are randomly generated and stored in a harmony memory (HM). Then three rules, namely, memory consideration rule, pitch adjustment rule and random selection rule are used to generate a new solution from all the solutions in the HM. Finally, the new solution is compared with the worst solution (i.e., harmony vector) in the HM. The HM is updated if the new solution is better than the worst harmony vector in the HM. The process of generating a new solution and replacing the worst harmony vector in HM is repeated until a specified termination criterion is met or a maximum number of iterations is reached. There are five parameters in the standard HS algorithm. These parameters include harmony memory size (HMS), harmony memory consideration rate ($HMCR$), pitch adjustment rate (PAR), distance bandwidth (BW), and the number of improvisations/iterations (NI). HMS is the number of harmony vectors or solutions in HM. In HS, the balance between exploration and exploitation is controlled by the parameter $HMCR$ which takes a value between 0 and 1. If the memory consideration rule is performed, PAR determines whether further adjustment is required according to BW parameter. The pseudo-code of HS is shown in Algorithm 1. In Algorithm 1, r, r_1, r_2, and r_3 are uniform random numbers in the range $[0, 1]$. UB and LB are the lower and upper bounds for each element/design variable of the solution.

3.2 Improved Harmony Search

Unlike the HS algorithm, which uses fixed values of PAR and BW parameters, Mahdavi et al. [11] proposed an improved harmony search (IHS) algorithm which dynamically adjusts the values of PAR and BW with respect to the number of iterations. The IHS algorithm is similar to the HS algorithm with the exception that the values of PAR and BW are dynamically updated as follows:

$$PAR(it) = PAR_{min} + \frac{PAR_{max} - PAR_{min}}{NI} \times it \tag{6}$$

$$BW(it) = BW_{max} exp(c \times it), c = \frac{\ln\left(\frac{BW_{min}}{BW_{max}}\right)}{NI} \tag{7}$$

where it is the current iteration, PAR_{min} is the minimum pitch adjustment rate and PAR_{max} is the maximum pitch adjustment rate. BW_{max} and BW_{min} are the maximum and the minimum bandwidths, respectively. The value of BW is exponentially decreased with the number of iterations.

Algorithm 1: The pseudo-code of the HS algorithm

Set the HS parameters: HMS, $HMCR$, PAR, BW and NI
Initialize the population of HMS solutions $S_{i,j}$ where $i = 1, 2, \ldots, HMS$, $j = 1, 2, \ldots, D$ and compute the objective function value of each solution $f(S_i)$
repeat
  Improvise a new harmony S_{new} as follows:
  for $j = 1$ *to* D **do**
    if $r_1 <= HMCR$ **then**
      $S_{new,j} = S_{a,j}$ where $a \in (1, 2, ..., HMS)$
      if $r_2 <= PAR$ **then**
        $S_{new,j} = S_{new,j} \pm r_3 \times BW$
      end
    else
      $S_{new,j} = LB_j + r \times (UB_j - LB_j)$
    end
  end
  Update the HM as follows:
  if $f(S_{new})$ *is better than* $f(S_{worst})$ **then**
    $S_{worst} = S_{new}$
  end
until *Maximum number of iterations* (NI) *is reached*
Return the best solution S_{best} in HM

3.3 Global-best Harmony Search

To enhance the performance of HS algorithm, Omran and Mahadavi [12] adopted concepts from particle swarm optimization (PSO) and proposed a new variant of HS, called global-best HS (GHS) algorithm. In the GHS algorithm, the best harmony in HM is considered when constructing a new harmony vector. The GHS algorithm has the same steps as the HS algorithm except that it dynamically updates the value of parameter PAR by using Eq. (6) and modifies the pitch adjustment step as follows:

$$S_{new,j} = S_{best,k} \tag{8}$$

where k is a random integer between 1 and D, and S_{best} is the best harmony vector in HM.

3.4 Intelligent Tuned Harmony Search

Inspired by the concept of the decision-making theory, Yadav et. al [13] proposed a new variant of the HS algorithm called intelligent tuned harmony search (ITHS) algorithm which maintains a proper balance between exploitation and exploration during the search process. In ITHS, the HM is divided into two groups. The first group consists of all the harmony vectors whose objective function value is better than the mean of the objective function values (f_{mean}) of the whole HM. The second group consists of the rest. While the first group is responsible for both exploitation and exploration, the second group is responsible only for exploration. The procedure for improvising a new harmony in ITHS is shown in Algorithm 2.

Algorithm 2: Improvisation scheme of ITHS algorithm

$PAR(it) = PAR_{max} - \frac{PAR_{max} - PAR_{min}}{NI} \times it$
for $j = 1$ *to* D **do**
  if $r_1 <= HMCR$ **then**
    $S_{new,j} = S_{a,j}$, where $a \in (1, 2, ..., HMS)$
    if $r_2 <= PAR(it)$ **then**
      if $f(S_a)$ *is better than* f_{mean} **then**
        /* First group */
        if $r_3 <= 0.5$ **then**
          $S_{new,j} = S_{best,j} - r_4 \times (S_{best,j} - S_{new,j})$
        else
          $S_{new,j} = S_{best,j} + r_4 \times (S_{worst,j} - S_{new,j})$
        end
      else
        /* Second group */
        $S_{new,j} = S_{new,j} + r_5 \times (S_{best,k} - S_{new,j})$, where $k \in (1, 2, ..., D)$ and $r_1, r_2, r_3, r_4, r_5 \in (0, 1)$
      end
    end
  else
    $S_{new,j} = LB_j + r \times (UB_j - LB_j)$, where $r \in (0, 1)$
  end
end

4 Proposed Hybrid Approaches (HS-ELM, IHS-ELM, GHS-ELM, and ITHS-ELM)

In ELM, the hidden layer parameters are initialized with random values and remain fixed during the training process. Since the output weights are calculated based on random input weights and hidden biases, ELM tends to require more hidden neurons than the gradient-based algorithms. However, there might be a set of nonoptimal input weights and hidden biases, and for some applications, compact network architectures are preferred. In this paper, hybrid approaches named HS-ELM, IHS-ELM, GHS-ELM, and ITHS-ELM are proposed for classification problems. In these approaches, HS, IHS, GHS and ITHS are independently used to optimize the input weights and hidden biases, and MP generalized inverse is used to compute the output weights.

First, the initial solutions are randomly generated and stored in HM. Each solution S in HM consists of all the input weights and hidden biases:

$$S = [w_{11}, w_{12}, \ldots, w_{1L}, w_{21}, w_{22}, \ldots, w_{2L}, \ldots, w_{d1}, w_{d2}, \ldots, w_{dL}, b_1, b_2, \ldots, b_L] \quad (9)$$

where the dimension of each solution S is $D = (d + 1) \times L$ and all w_{ij} and b_j are randomly generated within a specific range $[LB, UB]$. Here, $LB = -1$ and $UB = 1$. Second, after initializing all solutions, the corresponding output weights for each solution are calculated by using Eq. (5). Then the fitness (i.e., the objective function value) of each solution is evaluated. Suppose that the root mean squared error (RMSE) is adopted as an objective function, then the fitness of the solution S is computed as follows:

$$f(S) = \sqrt{\frac{\sum_{j=1}^{N} \left\| \sum_{i=1}^{L} \boldsymbol{\beta}_i g(\mathbf{w}_i \cdot \mathbf{x}_j + b_i) - \mathbf{t}_j \right\|_2^2}{m \times N}}. \quad (10)$$

Here, the fitness is computed as the RMSE on the validation set only instead of the whole training set [2]. After the fitness of all solutions in HM is calculated, a new solution is improvised using harmony search (i.e., HS, IHS, GHS, and ITHS). The new solution is compared with the worst solution in HM. The comparison is based on the fitness (RMSE on the validation set) of the solutions and the norm of the output weights $\|\boldsymbol{\beta}\|$ [2]. Therefore, the HM is updated as follows:

$$S_{worst} = \begin{cases} S_{new} & \text{if } f(S_{worst}) - f(S_{new}) > \eta f(S_{worst}) \\ S_{new} & \text{if } |f(S_{worst}) - f(S_{new})| < \eta f(S_{worst}) \\ & \quad \text{and } \|\boldsymbol{\beta}_{new}\| < \|\boldsymbol{\beta}_{worst}\| \\ S_{worst} & else, \end{cases} \quad (11)$$

where $\eta > 0$ is a tolerance rate. The above optimization process is repeated until the termination criterion is met. Thus, the best solution in HM represents the optimal input weights and hidden biases of ELM. In GHS-ELM, the pitch adjustment step is modified such that $S_{new,j} = S_{best,j}$ where $j = 1, 2, ..., D$ [14].

Table 1 Specifications of classification data sets

Data sets	Number of observations			Attributes	Classes
	Training	Validation	Testing		
Balance	469	78	78	4	3
Cardiotocography-3	1488	319	319	21	3
Diabetes	538	115	115	8	2
Glass	150	32	32	9	6
Iris	100	25	25	4	3

5 Experimental results

In this paper, the performance of ELM and the proposed HS-ELM, IHS-ELM, GHS-ELM, and ITHS-ELM algorithms is tested on five benchmark classification data sets obtained from the UCI machine learning repository.[1] The specifications of the data sets are given in Table 1. All algorithms run 30 times on each of these data sets. For each run of simulations, the training, validation, and testing sets are randomly generated according to the specifications given in Table 1. The logistic sigmoid function is used as an activation function in the ELM hidden layer. In every simulation, the number of hidden neurons is gradually increased and the one with the best generalization performance is reported in Table 2. The input weights and hidden biases are randomly initialized within the range of $[-1, 1]$. For all hybrid algorithms, $HMS = 10$, $HMCR = 0.9$, and $NI = 3000$. For HS-ELM, $PAR = 0.3$, and $BW = 0.2$. For IHS-ELM, $BW_{min} = 0.0001$ and $BW_{max} = (UB - LB)/20$. For IHS-ELM, GHS-ELM, and ITHS-ELM, $PAR_{min} = 0.01, 0.3$, and 0, respectively, and $PAR_{max} = 0.99, 0.99$ and 1, respectively.

Table 2 shows the average classification performance of ELM and the hybrid approaches. From Table 2, we can observe that ELM requires more hidden neurons to obtain good testing accuracy. IHS-ELM with lowest number of hidden neurons outperforms the other algorithms on Balance data set. HS-ELM obtains the best accuracy on Cardiotocography-3 data set. For Diabetes data set, it can be observed that even though GHS-ELM obtains the best accuracy, IHS-ELM is able to achieve comparable performance with much less number of neurons. HS-ELM obtains the best performance on Glass data set, and IHS-ELM is the second best. For Iris data set, IHS-ELM, GHS-ELM, and ITHS-ELM obtain similar performance. It can be observed that IHS-ELM obtains the best performance on two data sets (i.e., Balance and Iris data sets) and the second best on other two data sets (i.e., Diabetes and Glass data sets). Overall, the proposed hybrid approaches are able to obtain good performance with more compact network architectures.

[1] https://archive.ics.uci.edu/ml/.

Table 2 Classification performance of ELM and the proposed hybrid algorithms

Data sets	Methods	Testing (%)		Neurons
		Accuracy	Dev	
Balance	ELM	90.812	2.5879	40
	HS-ELM	90.8974	2.8038	18
	IHS-ELM	90.9402	2.4726	17
	GHS-ELM	90.8547	2.6667	18
	ITHS-ELM	90.812	2.6097	19
Cardiotocography-3	ELM	91.4838	1.4379	310
	HS-ELM	91.5569	1.4885	180
	IHS-ELM	91.3062	1.1842	175
	GHS-ELM	91.4316	1.4195	180
	ITHS-ELM	91.5152	1.4702	179
Diabetes	ELM	77.7681	3.7571	20
	HS-ELM	77.8261	3.7436	11
	IHS-ELM	78.2899	3.9519	11
	GHS-ELM	78.6667	3.3071	17
	ITHS-ELM	77.8551	3.9929	15
Glass	ELM	64.4792	6.641	45
	HS-ELM	65.7292	8.5477	25
	IHS-ELM	65.3125	7.5817	22
	GHS-ELM	64.8958	8.2754	21
	ITHS-ELM	65.1042	7.1583	24
Iris	ELM	97.0667	3.1397	20
	HS-ELM	97.0667	2.5587	7
	IHS-ELM	97.2	2.605	8
	GHS-ELM	97.2	3.1775	8
	ITHS-ELM	97.2	2.8089	7

6 Conclusions

In this paper, hybrid approaches based on ELM and harmony search were proposed. In the proposed approaches, the standard HS algorithm and its other three variants were used to optimize the input weights and hidden biases, and the MP generalized inverse to determine the output weights. The proposed approaches considered the RMSE on the validation data set and the norm of the output weights during the search for optimal input weights and hidden biases. The simulation results show that the proposed approaches are able to obtain compact network architectures with high generalization performance.

Acknowledgements This work was supported by the National Research Foundation of Korea (NRF) grant funded by the Korean government (MSIT) (No. 2019R1A2B5B03069810).

References

1. G.-B. Huang, Q.-Y. Zhu, C.-K. Siew, Extreme learning machine: a new learning scheme of feedforward neural networks, in *IEEE IJCNN, Budapest, Hungary* (2004), pp. 985–990
2. Q.Y. Zhu, A.K. Qin, P.N. Suganthan, G.-B. Huang, Evolutionary extreme learning machine. Pattern Recogn. **38**, 1759–1763 (2005)
3. J. Cao, Z. Lin, G.-B. Huang, Self-adaptive evolutionary extreme learning machine. Neural Process. **36**, 285–305 (2012)
4. W.K. Wong, Z.X. Guo, A hybrid intelligent model for medium-term sales forecasting in fashion retail supply chains using extreme learning machine and harmony search algorithm. Int. J. Prod. Econ. **128**, 614–624 (2010)
5. R. Dash, P.K. Dash, R. Bisoi, A self adaptive differential harmony search based optimized extreme learning machine for financial time series prediction. Swarm Evol. Comput. **19**, 25–42 (2014)
6. A.K. Alshamiri, A. Singh, B.R. Surampudi, Two swarm intelligence approaches for tuning extreme learning machine. Int. J. of Mach. Learn. Cyb. **9**, 1271–1283 (2018)
7. G.-B. Huang, Q.-Y. Zhu, C.-K. Siew, Extreme learning machine: theory and applications. Neurocomputing **70**, 489–501 (2006)
8. G.-B. Huang, L. Chen, C.-K. Siew, Universal approximation using incremental constructive feedforward networks with random hidden nodes. IEEE T. Neural Networ. **17**(4), 879–892 (2006)
9. Y. Lan, Y. Soh, G.-B. Huang, Constructive hidden nodes selection of extreme learning machine for regression. Neurocomputing **73**, 3191–3199 (2010)
10. Z.W. Geem, J.H. Kim, G.V. Loganathan, A new heuristic optimization algorithm: harmony search. Simulation **76**(2), 60–68 (2001)
11. M. Mahdavi, M. Fesanghary, E. Damangir, An improved harmony search algorithm for solving optimization problems. Appl. Math. Comput. **188**(2), 1567–1579 (2007)
12. M.G.H. Omran, M. Mahdavi, Global-best harmony search. Appl. Math. Comput. **198**(2), 643–656 (2008)
13. P. Yadav, R. Kumar, S.K. Panda, C.S. Chang, An intelligent tuned harmony search algorithm for optimization. Inf. Sci. **196**, 47–72 (2012)
14. Q.-K. Pan, P.N. Suganthan, M.F. Tasgetiren, J.J. Liang, A self-adaptive global best harmony search algorithm for continuous optimization problems. Appl. Math. Comput. **216**(3), 830–848 (2010)

Freestyle Rap Harmony Search (FRHS) for Engineering Problem Optimization

Kyoung Won Min, Donghwi Jung, and Joong Hoon Kim

Abstract A maximum benefit with minimum cost has been pursued in the optimal design of water distribution systems (WDSs). For this problem, trial-and-error methods, mathematical approaches, dynamic programming, and meta-heuristic techniques have been proposed while satisfying all the hydraulic constraints such as minimum pressure and demand. Recently, various variants of HSA mostly with multiple harmony memories and hierarchical memories have drawn a lot of attention because of their promising search. In this study, we propose a new version of HSA, freestyle rap harmony search (FRHS), with multiple memories among which different types of solutions are shuffled. The core properties of Freestyle rap are Rhyme, Flow, and Punch line. In FRHS algorithm, multiple Harmony Memories (HMs) are created as flow property in freestyle rap. When the value of the number of solution vector generations (NSVG) is above a certain value as a parameter, the bad solution of each HM is replaced by the good solution of the neighboring HM. The movement of the harmony memory vector is like a rhyme property with similar pronunciation repeated. When NSVG is above a certain value as a parameter, this condition is the same as the property of the punch line, the most shocking part of the freestyle rap. First, to check the performance of the algorithm, FRHS is applied to the least-cost design of WDSs. The performance of FRHS and HS is evaluated by how fast to find nearly global optimal solution. FRHS converges to the nearly global optimal solution faster than HS.

Keywords Freestyle · Rap · Harmony search · Optimization · Water distribution system

K. W. Min
Dept. of Civil, Environmental and Architectural Engineering, Korea University, Seoul, South Korea

D. Jung · J. H. Kim (✉)
School of Civil, Environmental and Architectural Engineering, Korea University, Seoul, South Korea
e-mail: jaykim@korea.ac.kr

S. M. Nigdeli et al. (eds.), *Proceedings of 6th International Conference on Harmony Search, Soft Computing and Applications*, Advances in Intelligent Systems and Computing 1275, https://doi.org/10.1007/978-981-15-8603-3_3

1 Introduction

Water Distribution Systems (WDSs) is to efficiently supply water from water sources to consumers. Water utility manager tends to minimize WDS design and operational costs while satisfying hydraulic constraints (e.g., conservation of mass and energy, minimum/maximum pressure and pipe flow constraints). Such characteristics of WDS make its design and operation problem non-linear non-convex complex optimization problem with high dimensionality.

During the last four decades, various optimization methods and models have been introduced to track WDS optimization problems. Alperovits and Shamir [1] proposed a linear programming model for least-cost WDS design whereas Schaake and Lai [7] applied a dynamic programming method. In order to overcome the aforementioned models' risk of finding local optimum, meta-heuristic optimization algorithms such as genetic algorithm [8], simulated annealing [2], and harmony search algorithm (HSA) [4] have been employed to WDS optimal design. Among many algorithms, HSA was one of the most popular algorithms which have been applied in the domain of water resources and water distribution system engineering.

Recently, various variants of HSA mostly with multiple harmony memories and hierarchical memories have drawn a lot of attention because of their promising search performance [5, 6].

In this study, we propose a new version of HSA, freestyle rap harmony search (FRHS), with multiple memories among which different types of solutions are shuffled. Three core properties of freestyle rap, Rhyme, Flow, and Punch line, are embedded in the proposed FRHS. Transferring a solution to neighboring memory (flow) is similar to rhyme's characteristic which repeats some words with similar pronunciation over phrases. The frequency of this shuffling is controlled by a punch line (strong wording with some interval) parameter. FRHS, designed to provide a rapid convergence ability to the global optimal solution, is demonstrated in the least-cost design of two benchmark WDSs, two-loop, and Hanoi networks. FRHS with different number of memories are compared with respect to the final solution's quality measures (best, worst, and mean solutions and their variation).

2 Methodology

FRHS is presented, embedding the nature of the freestyle rap into HS, traditional metaheuristic optimization algorithm would provide rapid convergence ability to the global optimal solution.

2.1 Freestyle Rap Harmony Search

FRHS combines the HS algorithm with Freestyle rap properties to quickly find a solution that is close to the global solution. The core properties of Freestyle rap are Rhyme, Flow, and Punch line. Rhyme refers to repeating the same or similar pronunciation. Flow is meant to include all beings that give rap a sense of rhythm beyond simple reading. Punch line is a story, especially the last part of the humor that is most shocking (Table 1, Figs. 1 and 2).

In FRHS algorithm, multiple HMs are created as flow property in freestyle rap, and the parameters (e.g., HMCR, PAR) in each HM are randomly set differently. When NSVG is above a certain value as a parameter, the bad solution of each HM is replaced by the good solution of the neighboring HM. The migration of the harmony vector is corresponding to including repetitive different words with similar pronunciation in phrases. When NSVG is above a certain value as a parameter, this condition is

Table 1 The FRHS operator corresponding to metaphor of freestyle rap

Freestyle rap operator	FRHS operator
Flow	The number of harmony memory
Rhyme	Migration of the best harmony vector
Punch line	The value of NI where the migration starts

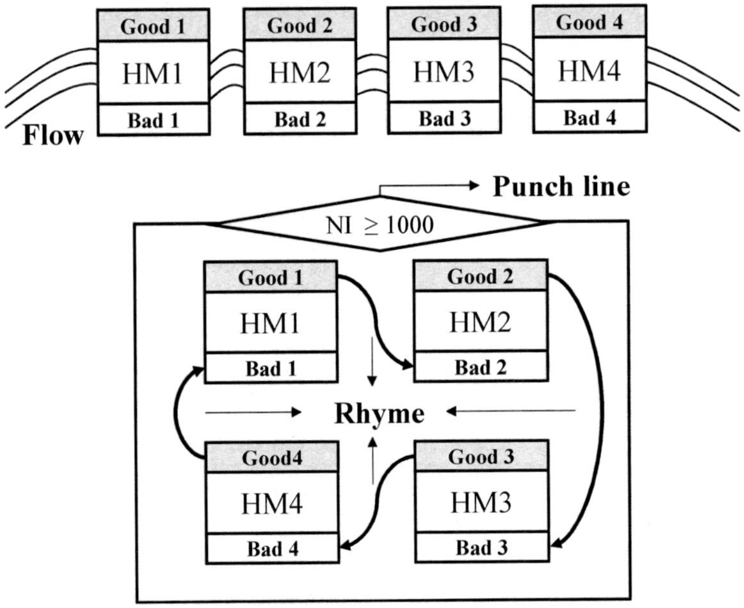

Fig. 1 The properties of Freestyle Rap Harmony Search

the same as property of the punch line, the most shocking part of the freestyle rap (Figs. 1 and 2).

2.1.1 Initialize Parameters and Initialize HM

The difference from the existing HS algorithm is that multiple HMs are created, and the parameters in each HM are randomly set differently within the range. The ranges of the number of HM, HMCR, PAR, etc. are all parameters.

2.1.2 Change Harmony Memory Vector

When the value of NI is 1000 or more, the worst solution of HM is replaced with a good solution of neighboring HM. Here, the value of 1000 is also a parameter.

3 Problem Formulation

3.1 Objective Function

The objective function is to minimize the cost of WDSs design and assumed to be a function of pipe diameters and lengths.

$$\text{Cost} = \sum_{i=1}^{N} C_c(D_i) \times L_i + Penalty \tag{1}$$

$C_c(D_i)$ is the design cost according to pipe diameter per unit length, D_i is the pipe diameter, L_i is the pipe length, $Penalty$ is provided when the following constraints are not satisfied.

3.2 Conservation of Mass Equation.

In WDSs, a conservation of mass equation for each node should be satisfied.

$$\sum Q_{in} - \sum Q_{out} = Q_e \tag{2}$$

Q_{in} is the flow rate to the node, Q_{out} is the flow rate out of the node and Q_e is the flow rate form the node to the outside or the demand for the node.

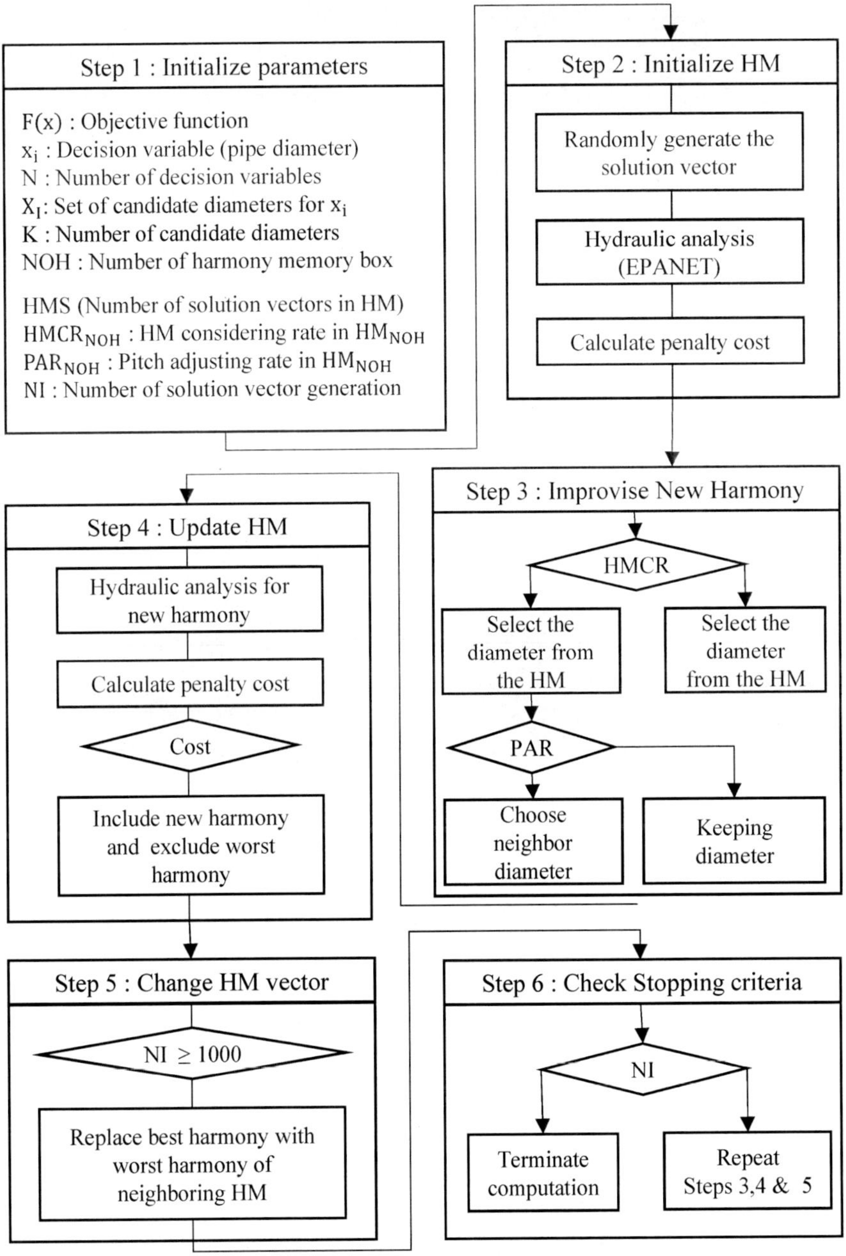

Fig. 2 Freestyle Rap Harmony Search for WDS design

3.3 Conservation of Energy Equation

In WDSs, a conservation of energy equation for each loop should be satisfied.

$$\sum h_f - \sum E_p = 0 \tag{3}$$

h_f is the head loss computed by Hazen–Williams or Darcy–Weisbach formula and E_p is the energy added by a pump.

3.4 Minimum Pressure Requirements

In WDSs, the minimum pressure requirement for each node is given to supply the demand of the node.

$$H_j \geq H_j^{min};\ j = 1, 2, \ldots, M \tag{4}$$

H_j is the pressure head at node j, H_j^{min} is the minimum required pressure head at node j, and M is the number of nodes.

The calculation and analysis of hydraulic constraints were done through the EPANET programming.

4 Application

4.1 Study Network

The HS and FRHS models were applied to the following two WDSs.

4.1.1 Two-Loop

The Two-loop WDS was first presented by Alperovits and Sharmir [1]. The Two-loop WDS consists of seven nodes and eight pipes with two loops and is fed by gravity from a constant reservoir which has 210 m fixed head. The length of all the pipes is 1000 m. The minimum head limitation is 30 m above ground level. Candidate diameter is [1, 2, 3, 4, 6, 8, 10, 12, 14, 16, 18, 20, 22, 24] in inches and corresponding cost is [2, 5, 8, 11, 16, 23, 32, 50, 60, 90, 130, 170, 300, 550] in dollar/meter.

4.1.2 Hanoi

The Hanoi WDS is a real WDS that formerly presented by Fujiwara and Khang [3] in Vietnam. The Hanoi WDS consists of 32 nodes and 34 piped with 3 loops and is fed by gravity from a constant reservoir which has 100 m fixed head. The minimum head limitation is 30 m above ground level. Candidate diameter is [12, 16, 20, 24, 30, 40] in inches and corresponding cost is [45.726, 70.4, 98.378, 129.333, 180.748, 278.28] in dollar/meter.

4.2 Application Results

In Case 1, the HS algorithm is used. In Case 2 to 7, the FRHS algorithm is used. Table 2 shows the parameter values and results of each case in a two-loop network. From the results, the average and worst values show that the more HM in the FRHS, the better the solution. The optimal solution in Best was found in all cases because Two-Loop WDS is a simple network. The value of standard deviation (Stdv) was found to be smaller as there were more HM. Figure 3 shows the average value of each iteration of NI 50 times. The NI value in FRHS is smaller than in HS when the nearly global solution is calculated. This means that FRHS find the solution that is close to global optimum faster than HS.

Table 3 shows the parameter values and results of each case in a Hanoi network. From the results, the average and worst values show that the more HM in the FRHS, the better the solution. The optimal solution in Best was not found in all cases because the value of NI is small for complex network. The value of Stdv was found to be smaller as there were more HM. Figure 4 shows the average value of each iteration of NI 50 times. The NI value in FRHS is smaller than in HS when the nearly global solution is calculated. This means that FRHS find the solution that is close to global optimum faster than HS.

5 Conclusions

Existing techniques were limited in focusing on finding the global optimum, but supplementing this point, FRHS was presented. FRHS combines the HS algorithm with Freestyle rap properties such as rhyme, flow, and punch line to quickly find a solution that is close to the global solution. A FRHS algorithm was applied to the least-cost design of WDSs. FRHS found a nearly global optimum earlier than HS. A nearly global optimal solution should be quickly sought in real-time operation problems in WDS. Given a limited time for optimization, water system operators would prefer and employ such solution rather than truly global solution which generally takes a lot of time for acquisition with low likelihood. In future works, a methodology to

Table 2 Two-loop results

	HM	HMS	HMCR	PAR	NI	Run = 50 × NI			
						Average	Best	Worst	Stdv
Case 1	1	100	0.95	0.05	5,000	435,160	419,000	476,000	14,975
Case 2	3					428,860	419,000	464,000	10,671
Case 3	5					425,880	419,000	447,000	7,182
Case 4	7					422,900	419,000	441,000	5,258
Case 5	3		Random [0.90, 0.99]	Random [0.01, 0.10]		429,300	419,000	465,000	12,075
Case 6	5					426,460	419,000	457,000	8,180
Case 7	7					423,500	419,000	441,000	5,300

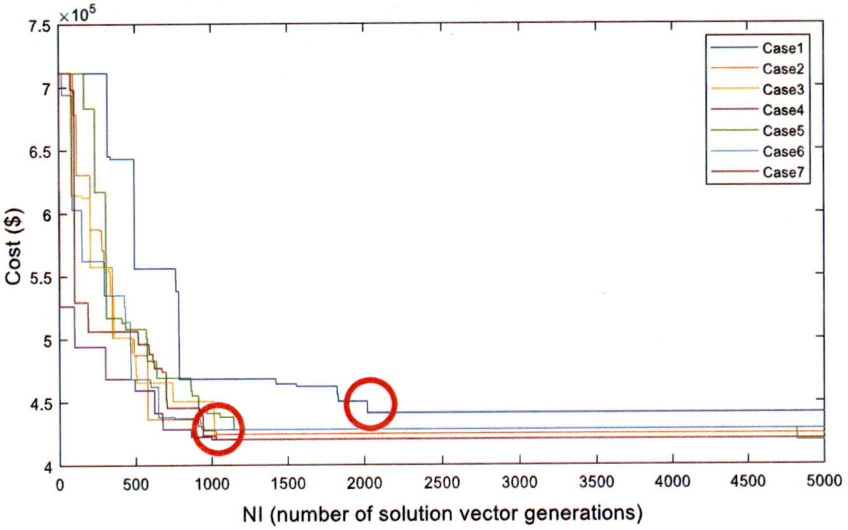

Fig. 3 Mean cost of each iteration of NI 50 times in two-loop WDS

help make faster decisions by applying FRHS to real-time WDS operation such as pump and valve scheduling will be presented.

Table 3 Hanoi results

	HM	HMS	HMCR	PAR	NI	Run = 50 × NI			
						Average	Best	Worst	Stdv
Case 1	1	50	0.93	0.18	200,000	6,522,600	6,200,600	6,875,500	149,430
Case 2	3					6,333,400	6,108,900	6,583,300	95,576
Case 3	5					6,302,100	6,125,200	6,550,600	84,119
Case 4	7					6,277,600	6,131,500	6,453,000	70,991
Case 5	3		Random [0.90, 0.99]	Random [0.01, 0.20]		6,476,000	6,110,700	6,892,100	121,521
Case 6	5					6,445,200	6,119,900	6,871,000	110,232
Case 7	7					6,326,000	6,127,700	6,802,900	101,360

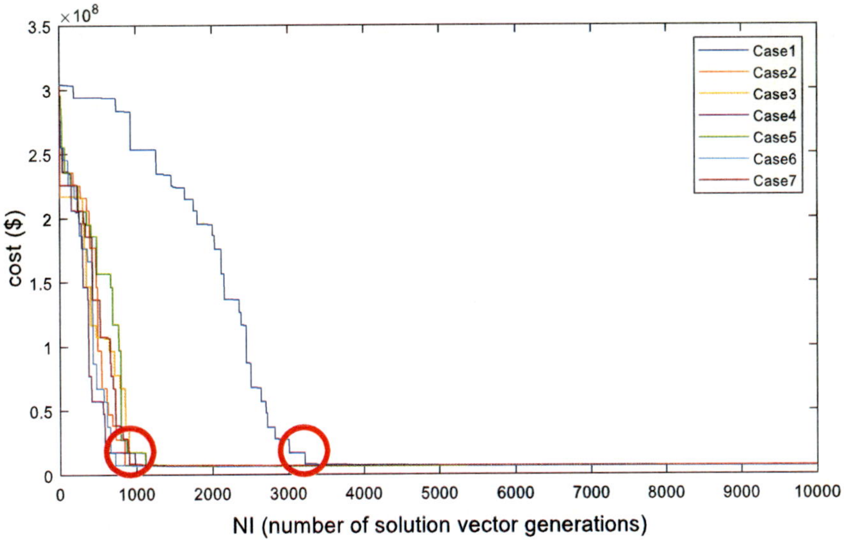

Fig. 4 Mean cost of each iteration of NI 50 times in Hanoi WDS

Acknowledgments This work was supported by the National Research Foundation of Korea (NRF) grant funded by the Korea government (MSIT) (No. 2019R1A2B5B03069810).

References

1. E. Alperovits, U. Shamir, Design of optimal water distribution systems. Water Resour. Res. **13**(6), 885–900 (1977)
2. M.D.C. Cunha, J. Sousa, Water distribution network design optimization: simulated annealing approach. J. Water Resour. Plann. Manage. **125**(4), 215–221 (1999)
3. O. Fujiwara, D.B. Khang, A two-phase decomposition method for optimal design of looped water distribution networks. Water Resour. Res. **26**(4), 539–549 (1990)
4. Z.W. Geem, J.H. Kim, G.V. Loganathan, A new heuristic optimization algorithm: harmony search. Simulation **76**(2), 60–68 (2001)
5. A. Hong, D. Jung, J. Choi, J.H. Kim, Sensitivity analysis on migration parameters of parallel harmony search, in *International Conference on Harmony Search Algorithm* (Springer, Singapore, 2017, February), pp. 3–7
6. H.M. Lee, E.H. Lee, Y.H. Choi, J.H. Kim, Multi-layered harmony search algorithm: introduction of a novel and efficient structure, in *International Conference on Harmony Search Algorithm* (Springer, Singapore, 2017, February), pp. 8–14
7. J.C. Schaake, F.H. Lai, *Linear Programming and Dynamic Programming Application to Water Distribution Network Design* (MIT Hydrodynamics Laboratory, 1969)
8. A.R. Simpson, G.C. Dandy, L.J. Murphy, Genetic algorithms compared to other techniques for pipe optimization. J. Water Resour. Plann. Manage. **120**(4), 423–443 (1994)

Improvement of Voltage Profile and Loss Reduction Based on Optimal Placement and Sizing of Renewable Distributed Generations Using 4-Rule Harmony Search Algorithm

Ali Sadollah, Mohammad Nasir, Abobakr Khalil Al-Shamiri, and Joong Hoon Kim

Abstract Distributed generation (DG) resources based on solar and wind energies are among the DGs that can optimize the voltage profile and reduce network losses by optimal locating and sizing in distribution systems. In this paper, 4-Rule Harmony Search (4-RHS) algorithm is presented for optimal placement and sizing of the renewable DG resources. In this regard, installation of the DG is performed by loss sensitivity factors first in the most candidate buses. Then, the 4-RHS is employed for placement and sizing of DGs from the selected buses. In 4-RHS, an extra rule is added to standard HS. The 4-RHS is studied on IEEE 69-bus distribution system. The results comparison of the 4-RHS algorithm with the standard HS algorithm and other optimization methods in the literature shows the efficiency and effectiveness of the 4-RHS in finding the optimal location and size of renewable DG resources and subsequently reducing losses and improving voltage profiles.

Keywords Distributed generation · 4-rule harmony search · Harmony search · Optimal placement · Voltage profile improvement · Loss reduction

A. Sadollah
Department of Mechanical Engineering, University of Science and Culture, Tehran, Iran

M. Nasir
Materials and Energy Research Center, Dezful Branch, Islamic Azad University, Dezful, Iran

A. K. Al-Shamiri
Research Institute for Mega Construction, Korea University, Seoul, South Korea

J. H. Kim (✉)
School of Civil, Environmental and Architectural Engineering, Korea University, Seoul, South Korea
e-mail: jaykim@korea.ac.kr

S. M. Nigdeli et al. (eds.), *Proceedings of 6th International Conference on Harmony Search, Soft Computing and Applications*, Advances in Intelligent Systems and Computing 1275, https://doi.org/10.1007/978-981-15-8603-3_4

1 Introduction

Distributed Generation (DG) Resource is a source of electrical power that is directly connected to the distribution system or on the customer's location. Since various DGs use renewable sources such as wind turbines and photovoltaic systems, they provide cleaner production energy.

Renewable-based DG sources are resources that use renewable energies such as solar and wind energy to respond to load demand. The most important advantages of using renewable-based DGs are line losses reduction, efficiency increment, improving power quality, increasing system reliability and reducing fuel costs and operating and maintenance costs [1]. However, inappropriate selection of DG location and size results in higher losses and costs compared to without DG mode. Therefore, the optimal placement and size determination of DGs in the distribution system is of significant importance.

Due to the importance of this topic, the optimal placement and sizing problem of DG has been solved using various techniques in the literature. Since metaheuristic optimization algorithms have a considerable ability in finding optimal solution in solving problem, they have been employed widely in tackling the problem of the DG placement and sizing.

In this regard, several metaheuristic optimization algorithms such as backtracking search optimization algorithm [1], flower pollination algorithm [2], symbiotic organisms search algorithm [3], and stud krill herd algorithm [4] have been applied for placing and sizing DGs resources in the distribution system.

Also, placement and sizing of multiple DG and battery swapping stations have been considered using Grasshopper Optimizer Algorithm (GOA) [5]. Furthermore, A hybrid approach combining the GOA and Cuckoo Search (CS) technique has been introduced for loss reduction based on optimal placement of DG units in the distribution system [6].

Kayalvizhi and Vinod Kumar [7] solved optimal operation of the distribution system using the HS in a framework of a multi-objective optimization using a renewable DG resource. Jha et al. [8] studied the HS to solve the problem of dynamic economic distribution of microgrids. In their research, the objective was to distribute the dynamic economic dispatch of the microgrid for minimizing the generation cost and to coordinate the different DGs with respect to the dynamic cost of the microgrid.

In Ref. [9], fuzzy logic and Harmony Search (HS) have been utilized to locate DG units to reduce power losses and improve voltage profiles along with optimal location and unit size. The obtained optimization results prove the effectiveness of the proposed hybrid method. Also, Nekooei and his colleagues [10] employed the improved HS algorithm for locating DG in radial distribution systems.

The HS algorithm, developed by Geem et al. [11], is derived from the concepts of musical improvisations and harmony knowledge. The HS algorithm and its improved variants have proved its advantages over other optimizers [12].

In this paper, a modified version of HS so-called 4-RHS has been utilized for optimal solving of placement and sizing the renewable DG resources for distribution systems by introducing an additional rule to three well-known rules of the standard HS algorithm [13].

The remainder of this paper is organized as follows. The next section describes the 4-RHS algorithm in detail. Objective function and loss sensitivity factors are taken into account as the problem formulation in Sect. 3, and simulation results and conclusions are provided in Sects. 4 and 5, respectively.

2 4-Rule Harmony Search Algorithm

Recently, the HS has been applied to various research areas and obtained considerable attention in different disciplines [14]. The HS intellectualizes the musical process of searching for a perfect state of harmony [11]. Musicians seek a fantastic harmony determined by aesthetic estimation. Similarly, optimization techniques seek a best state (global optimum) determined by an objective function value.

Aesthetic estimation depends on the set of the sounds played by a musical ensemble, whereas the objective function is evaluated by a set of adjustable variables. More aesthetic sounds can be produced by constant practice, and the optimization of the objective function can generally be improved by repeated iterations. Further details of HS can be found in the work of Geem et al. [11] and Kim et al. [15].

Standard HS consists of three general approaches. They are given as follows: (i) memory consideration phase, (ii) pitch adjusting phase, and (iii) uniform random search phase. The latter has been considered as exploration phase in HS similar to the concept of mutation in GAs. Accordingly, first and second phases (i.e., memory consideration and pitch adjusting) are related to the exploitation capability of HS. It is worth mentioning that memory consideration phase may act as exploration at early iterations due to the large diversity in the early populations.

However, looking at exploitation equations in standard HS and even IHS [16], we can see that they have suffered from having big guided jump at early iterations toward the best solution. As can be seen in Particle Swarm Optimization (PSO) [17], movements toward the best experiences (i.e., global and personal experiences) lead us finding the optimum solutions. Therefore, in the proposed improved HS, we considered one additional rule for overcoming this drawback observed in standard HS and IHS. We have named this rule as "information exchange in HM" [13].

One of the advantages of 4-RHS compared with the standard HS is the collaboration between exploration and exploitation phases caused by the suggested rule. A newly added rule, named as "information exchange in HM," performs more explorations rather than exploitation at early iterations due to large variation in quality solutions (i.e., harmonies) in HM. By continuous iteration, the added rule acts as exploitation operator by reason of small variation of solutions in HM.

At first, based on the random HM, a new solution vector x^{New} is improvised. When improvising a new value for the ith decision variable x^{New} in the new vector, 4-RHS algorithm utilizes one of the four operations (i.e., memory consideration, pitch adjusting, information exchange in HM, and random selection) as given [13]:

$$x_i^{\text{New}} \leftarrow \begin{cases} x_i \in \text{HM} = \left\{x_i^1, x_i^2, \ldots, x_i^{\text{HMS}}\right\}, & \text{w.p. HMCR} \\ x_i + \text{BW} \times (2\text{rand} - 1) \quad if \;\; x_i \in HM, \; or \;\; x_i(k+m) \;\; if \;\; x_i(k) \in \text{HM}, & \text{w.p. HMCR} \times \text{PAR} \\ x_i^{r(\text{HM})} + 1.5 \times (2\text{rand} - 1) \times (x_i^{B(\text{HM})} - x_i^{r(\text{HM})}), & \text{w.p. HMCR} \times (1 - \text{PAR}) \\ x \in \left[x_i^{\text{Lower}}, x_i^{\text{Upper}}\right] \;\; \text{or} \;\; \left\{x_i \in \left\{x_i(1), \ldots, x_i(k), \ldots., x_i(K)\right\}\right\}, & \text{w.p. } (1 - \text{HMCR}) \end{cases} \tag{1}$$

where

w.p. means with respect to;

m is neighboring index which has normally 1 or −1 for discrete problems;

$\{x_i \in \{x_i(1), \ldots, x_i(k), \ldots., x_i(K)\}\}$ is the complete value set for the ith design variable;

$\left\{x_i^1, x_i^2, \ldots, x_i^{\text{HMS}}\right\}$ is the value set stored in HM for the ith decision variable;

$x_i^{B(\text{HM})}$ is the best harmony in the HM for the ith design variable;

$x_i^{r(\text{HM})}$ is a random harmony in the HM for the ith decision variable;

And *rand* is uniformly distributed random number between zero and one.

In the proposed version of HS, if the harmony is selected from HM and rejected by the pitch adjusting rule, then the new rule will be applied on the current harmony. First, a random harmony is selected from the HM (i.e., storage of best harmonies), afterwards, the chosen harmony moves to the best harmony in the HM. Therefore, in addition of pitch adjusting rule, this assumption (i.e., information exchange in HM) is considered as another exploration–exploitation phase for the HS.

As it is obvious, using dynamic parameters are preferred than constant parameters. Therefore, in the 4-RHS, similar to IHS [2], probability value of PAR and bandwidth (BW) are changed adaptively at each iteration given as follows [13]:

$$\text{PAR}(t) = \text{PAR}_{\min} + \frac{(\text{PAR}_{\max} - \text{PAR}_{\min})}{\text{Max\_It}} \times t \tag{2}$$

$$\text{BW}(t) = \text{BW}_{\max} \exp\left(\frac{Ln\left(\frac{\text{BW}_{\min}}{\text{BW}_{\max}}\right)}{\text{Max\_It}} \times t\right) \tag{3}$$

where $\text{PAR}_{\min}$ and $\text{PAR}_{\max}$ are the minimum and maximum pitch adjusting rates, $\text{BW}_{\min}$ and $\text{BW}_{\max}$ are the minimum and maximum bandwidths, t and Max_It are iteration counter and maximum number of iterations, respectively. Large PAR values with small BW values usually cause the improvement of best solutions in final iterations. With higher value for the PAR, the chance for the added rule (i.e., 4th rule) reduces. It means that the new rule in this improved version of HS plays more important roles at early iterations when the probability value of PAR is small. Indeed, at early iterations, due to the high variance in HM, forcing harmonies to move to the best harmony can be assumed as guided exploitation inside the HM. By iteration goes, this chance will be reduced as the variation in HM decreases.

3 Mathematical Formulation

In the section, the formulation of the optimal placement and sizing of the renewable DG resources problem is presented. In this study, Photovoltaic (PV) and Wind Turbine (WT) are taken into account as DG resources.

The proposed objective function reduces power loss and improves voltage profile and Voltage Stability Index (VSI). In this regard, the summation weighted method converts the multi-objectives optimization problem into a single objective problem. Thus, DG locations and their sizes can be optimized by solving the following objective function:

$$F_t = w_1 f_1 + w_2 f_2 + w_3 f_3 \tag{4}$$

where f_1 denotes the reduction of active losses and it can be expressed by Eq. (5), f_2 indicates the improvement of voltage profile status and it can be defined as Eq. (6), and f_3 provides VSI improvement given in Eq. (7).

$$f_1 = \frac{\sum_{i=1}^{L} (P_{\text{Lineloss}}(i))_{\text{afterDG}}}{\sum_{i=1}^{L} (P_{\text{Lineloss}}(i))_{\text{beforeDG}}} \tag{5}$$

$$f_2 = \frac{\sum_{i=1}^{N} \left|V_i - V_{i,\text{ref}}\right|_{\text{afterDG}}}{\sum_{i=1}^{N} \left|V_i - V_{i,\text{ref}}\right|_{\text{beforeDG}}} \tag{6}$$

$$f_3 = \frac{1}{VSI(k)_{\text{afterDG}}} \tag{7}$$

where the VSI is given as follows [18]:

$$VSI(k) = |V_i|^4 - 4(P_k.X_{ik} - Q_k.R_{ik})^2 - 4(P_k.R_{ik} + Q_k.X_{ik}).|V_i|^2 \tag{8}$$

w_1, w_2, and w_3 are weight factors. In this study, 0.5 is considered for w_1, while 0.25 is considered for w_2 and w_3, respectively. As the studied problem is a constrained minimization objective problem, several constraints including equality and inequality constraints are imposed as given:

(a) *Power conservation constraint:* The algebraic sum of the inlet and outlet power flow on the distribution system must be equal [18], thus

$$P_{\text{Swing}} + \sum_{i=1}^{N_{\text{DG}}} P_{\text{DG}}(i) = \sum_{i=1}^{L} P_{\text{Lineloss}}(i) + \sum_{q=1}^{N} Pd(q) \tag{9}$$

$$Q_{\text{Swing}} + \sum_{i=1}^{N_{\text{DG}}} Q_{\text{DG}}(i) = \sum_{i=1}^{L} Q_{\text{Lineloss}}(i) + \sum_{q=1}^{N} Qd(q) \tag{10}$$

(b) *Voltage constraint*: The amount of voltage in each bus must be limited by Eq. (11):

$$V_{\min} \leq |V_i| \leq V_{\max} \tag{11}$$

where $V_{\min}$ and $V_{\max}$ are set to 0.95 and 1.05, respectively.

(c) *DG limits constraints:* The installed capacity of DG in the network is limited to prevent reverse power flow so as not to exceed the power provided by the substation [18]:

$$\sum_{i=1}^{N_{\text{DG}}} P_{\text{DG}}(i) \leq \frac{3}{4} \times \left[\sum_{i=1}^{L} P_{\text{Lineloss}}(i) + \sum_{q=1}^{N} Pd(q) \right] \tag{12}$$

$$\sum_{i=1}^{N_{\text{DG}}} Q_{\text{DG}}(i) \leq \frac{3}{4} \times \left[\sum_{i=1}^{L} Q_{\text{Lineloss}}(i) + \sum_{q=1}^{N} Qd(q) \right] \tag{13}$$

$$P_{DG}^{\min} \leq P_{\text{DG}}(i) \leq P_{\text{DG}}^{\max} \tag{14}$$

$$Q_{DG}^{\min} \leq Q_{\text{DG}}(i) \leq Q_{\text{DG}}^{\max} \tag{15}$$

(d) *Line capacity constraint:* The complex power through each line must be less than the rating value obtained as given:

$$S_{\text{Li}} \leq S_{\text{Li(rated)}} \tag{16}$$

In this paper, Loss Sensitivity Factors (LSFs) method is used to determine candidate buses for embedding into the DGs [19]. In the LSFs method, the search range is greatly reduced and as a result, the computational time in the optimization process is also decreased.

In this method, the voltage amplitudes of all the chains are first obtained by performing load flow calculations, and based on the grid base voltage, they are normalized and then subdivided to 0.95 (low voltage limit). Buses with a calculated number less than 1/01 can be considered as candidate bus for DG installation. The

LSFs then calculate the candidate buses and affect algorithm in descending order. The LSFs values used in the study are extracted from the literature [18].

4 Simulation Results

In this section, the IEEE 69-bus distribution system has been solved using the 4-RHS is examined. The system has a total load of 3800 kW and 2690 kW and a voltage level of 12.6 kV. System data are extracted from the literature [20]. It is worth mentioning that backward/forward sweep method is used in this paper for the load-flow analysis of the radial distribution system.

The task of optimization has been carried out for 30 independent runs. MATLAB programming language has been utilized for coding purpose. After sensitivity analysis, the parameter values of the proposed HS and standard HS are given as follow: HMS of 30, HMCR of 0.98, PAR_{min} of 0.3, PAR_{max} of 0.99, BW_{min} of 0.0001, and BW_{max} of 0.01. Table 1 summarizes the results for installing two DGs for both 4-RHS and standard HS algorithms.

As can be seen from Table 1, 4-RHS algorithm acquired better results than the HS algorithm in terms of power loss and power loss percentage. In addition, for further clarification, the effects of DG installation on the voltage profile and VSI are shown in Fig. 1. By observing Fig. 1, the designed WT type has better results than the PV type in terms of voltage profile and VSI. Therefore, through the use of reactive power storage, the overall power loss for WT is primarily reduced.

Table 1 Results for the IEEE 69-bus system using the 4-RHS and original HS

Item	Without DG	With 2DG (kVA/P.F)			
		HS		4-RHS	
		PV	WT	PV	WT
Total losses (kW)	224.94	54.25	19.522	52.23	18.453
Loss Reduction (%)	–	75.88%	91.32%	76.78%	91.80%
Min. voltage	0.9102	0.965	0.9721	0.989	0.9984
Max. voltage	1	1	1	1	1
DG size bus	–	960.2	1336.1	1336.1	568.82
		@bus 12	@bus 61	@bus 61	@bus 16
		1873.5	460.2	460.23	2001.06
		@bus 61	@bus 18	@bus 18	@bus 61
VSI	61.2379	65.502	66.102	66.852	67.201
Cost of losses ($)	1,18,228.46	28,513.80	10,260.76	27,452.08	9698.89
Saving ($/year)	–	89,714.66	1,07,967.69	90,776.37	1,08,529.56

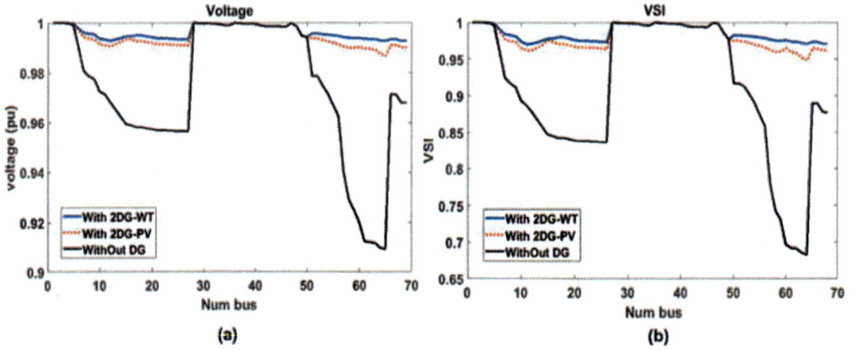

Fig. 1 The effect of installing 2 DGs in IEEE 69-bus system on (a) voltages, (b) VSI

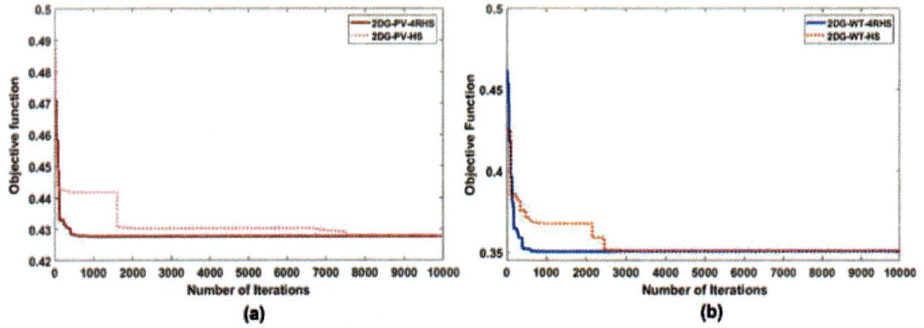

Fig. 2 Convergence history for IEEE 69-bus system for: (a) 2DG-PV, (b) 2DG-WT

The convergence history graph of the 4RHS algorithm after the DGs setup for the IEEE 69-bus system is shown in Fig. 2. It can be observed from Fig. 2, the 4-RHS outperforms the standard HS in terms of convergence rate.

Furthermore, for extending our comparison pool, Table 2 shows the comparison of other existing optimizers in the literature. Looking at Table 2, It can be seen that the 4-RHS surpassed other used algorithms for solving IEEE 69-bus systems in reducing losses. The best-obtained solutions have been highlighted in bold in Table 2.

5 Conclusions

In this paper, a recently enhanced HS algorithm so-called 4-Rule HS (4-RHS) algorithm along with loss sensitivity factors is presented to find an optimum location and determine the size of renewable DG resources in distribution system. The aim of this paper was to reduce power loss, improve voltage profile, and voltage stability index. In the 4-RHS, an extra rule is included in standard HS along with its three traditional

Table 2 Obtained optimization results for installing two DG in IEEE 69-bus system

DG type	Methods	DG installation		Power loss	
		Size (kVA/P.F)	Bus	Value (kW)	Percentage
Without DG		–	–	224.94	–
2DG-PV	4-RHS	**1336**	**61**	**52.23**	**76.78**
		460.2	**18**		
	HS	1164.2	11	53.68	75.88
		1766.7	61		
	GA1 [21]	1777	61	71.79	68.08
		555	11		
	GA2 [22]	6	1	84.23	62.55
		1794	62		
	CS [23]	600	22	76.4	66
		2100	61		
	SGA [23]	1000	17	82.9	63.1
		2400	61		
	PSO [23]	700	14	78.8	64.97
		2100	62		
	MTLBO [24]	519.71	17	71.77	68.09
		1732	61		
	ALOA [19]	538.78	17	70.75	68.54
		1700	61		
2DG-WT	**4-RHS**	**568.8**	**16**	**18.45**	**91.72**
		2001	**61**		
	HS	460.41	17	19.88	91.32
		1336.1	61		
	CS [23]	800	18	39.9	82.26
		2000	61		
	SGA [23]	600	18	44	80.4
		2300	62		
	PSO [23]	900	18	42.4	81.15
		1900	62		
	ALOA [19]	726.63	17	20.93	90.69

rules. The proposed improved HS was simulated by considering the objective function and problem constraints on the IEEE 69-bus distribution system. Comparison of attained optimization results with other reported methods shows the superiority of 4-RHS algorithm with respect to performance and efficiency in reducing power loss, improving voltage profile, and VSI.

Acknowledgments This work was supported by the National Research Foundation of Korea (NRF) grant funded by the Korea government (MSIT) (No. 2019R1A2B5B03069810).

References

1. A. El-Fergany, Optimal allocation of multi-type distributed generators using Backtracking search optimization algorithm. Int. J. Electr. Power Energy Syst. **64**, 1197–1205 (2015)
2. M.S. Jayasree, P. Sreejaya, G.R. Bindu, Multi-Objective Metaheuristic Algorithm for Optimal Distributed Generator Placement and Profit Analysis. Technol Econ Smart Grids Sustain Energy **4**, 11 (2019). https://doi.org/10.1007/s40866-019-0067-z
3. B. Das, V. Mukherjee, D. Das, DG placement in radial distribution network by symbiotic organisms search algorithm for real power loss minimization. Applied Soft Computing **49**, 920–936 (2016)
4. S.A. ChithraDevi, L. Lakshminarasimman, R. Balamurugan, Stud Krill herd Algorithm for multiple DG placement and sizing in a radial distribution system. Engineering Science and Technology, an International Journal **20**(2), 748–759 (2017)
5. Sultana U., Khairuddin A.B., Sultana B., Rasheed N., Hussain Qazi S., Riaz Malik N., Placement and sizing of multiple distributed generation and battery swapping stations using grasshopper optimizer algorithm, Energy, 2018, vol. 165, Part A, pp. 408–421.
6. Suresh M.C.V., Belwin Edward J., A hybrid algorithm based optimal placement of DG units for loss reduction in the distribution system, Applied Soft Computing, 2020, vol. 91, 106191.
7. S. Kayalvizhi, D.M. Vinod Kumar, 2015, "Dispatchable DG Planning in Distribution Networks considering costs", International Conference on Recent Developments in Control, Automation and Power Engineering (RDCAPE), pp. 320–325.
8. R. Jha, N. K. Meena, A. Swarnkar, N. Gupta,K. R. Niazi, 2015, "Dynamic Economic Dispatch of Micro-grid using Harmony Search Algorithm", Annual IEEE India Conference (INDICON), pp. 1–6.
9. P. Suresh Babu, R. Madhan Mohan, 2015, "Optimal performance enhancement of DG for loss reduction using fuzzy and harmony search algorithm", International Conference on Electrical, Electronics, Signals, Communication and Optimization (EESCO), pp. 1–5.
10. K. Nekooei, M.M. Farsangi, K.H. Nezamabadi-Pour, S.Y. Lee, An Improved Multi-Objective Harmony Search for Optimal Placement of DGs in Distribution Systems. IEEE Transactions On SMART GRID **4**(1), 557–567 (2013)
11. G.W. Geem, J.H. Kim, G.V. Loganathan, A new heuristic optimization algorithm: harmony search. Simul. **76**(2), 60–68 (2001)
12. Zhang, P., Ouyang, H., Gao, L.: Improved Harmony Search Algorithm with Perturbation Strategy, Control and Decision Conference (CCDC), 27th Chinese, DOI: https://doi.org/10.1109/CCDC.2015.7162873 (2015).
13. Sadollah, A., Kim, J.H., Choi, Y., Karamodin, N., "4-Rule harmony search algorithm for solving computationally expensive optimization test problems", The Fifth International Conference on Harmony Search, Soft Computing and Applications (ICHSA 2019), July 20–22, 2019, Kunming, China.
14. D.G. Yoo, J.H. Kim, Z.W. Geem, Overview of harmony search algorithm and its applications in civil engineering. Evol. Intel. **7**, 3–16 (2014)
15. J.H. Kim, Z.W. Geem, E. Kim, Parameter estimation of the nonlinear Muskingum model using harmony search. J. Am. Water Resour. Assoc. **37**(5), 1131–1138 (2001)
16. M. Mahdavi, M. Fesanghary, E. Damangir, An improved harmony search algorithm for solving optimization problems. Appl. Math. Comput. **188**(2), 1567–1579 (2007)
17. Kennedy, J., Eberhart, R.C.: Particle swarm optimization, in IEEE International Conference on Neural Networks, Piscataway, NJ, 1942–1948 (1995).

18. E.S. Ali, S.M. Abd Elazim, A.Y. Abdelaziz, Ant Lion Optimization Algorithm for optimal location and sizing of renewable distributed generations. Renewable Energy **101**, 1311–1324 (2017)
19. D. Hung, N. Mithulananthan, Multiple distributed generator placement in primary distribution networks for loss reduction. IEEE Trans. Ind. Electron. **60**(4), 1700–1708 (2013)
20. S.G. Naik, D.K. Khatod, M.P. Sharma, Planning and operation of distributed generation in distribution networks. Int. J. Emerg. Technol. Adv. Eng. **2**(9), 381–388 (2012)
21. Pisica. I, Bulac. C, Eremia. M, 2009, "Optimal distributed generation location and sizing using genetic algorithms", in: 15th Int. Conf. On Intelligent System Applications to Power Systems, ISAP '09, Curitiba, pp. 1–6.
22. T. Shukla, S. Singh, V. Srinivasaraob, K. Naik, Optimal sizing of distributed generation placed on radial distribution systems. Electr. Power Compon. Syst. **38**(3), 260–274 (2010)
23. Tan. W, Hassan. M, Majid. M, Rahman. H, 2012, "Allocation and sizing of DG using cuckoo search algorithm", IEEE Int. Conf. Power Energy, 133–138.
24. J. García, A. Mena, Optimal distributed generation location and size using a modified teaching-learning based optimization algorithm. Int. J. Electr. Power Energy Syst. **50**, 65–75 (2013)

Control of a Jacket Platform Under Wave Load Using ATMD and Optimization by HSA

Maziar Fahimi Farzam, Babak Alinejad, Gabriel Bekdaş, and Seyyed Ali Mousavi Gavgani

Abstract Jacket platforms are life bloods of oil-rich countries and one of the well-studied offshore structures subjected to environmental loads. In the Persian Gulf, wave loads are the dominant load in designing procedure. Vibration control of jacket platforms is always one of the important issues for demanding requirements such as production activity, safety, and serviceability. Tuned Mass Damper (TMD) is one of the well-established control systems in this context. The present study investigates the dynamic behavior of the Ressalat platform located in the Persian Gulf under the wave loads. An Active Tuned Mass Damper (ATMD) is employed for vibration control, while taking the actuator saturation into account. A fuzzy controller is applied to calculate the control force. Moreover, the Harmony Search Algorithm (HSA) is examined to optimize the power of the actuator in the control system. Results obtained through this study indicate superior performance of the proposed ATMD in comparison to the TMD, leading to significant reduction in vibration of the Ressalat platform under studied wave loads.

Keywords Jacket platform · Tuned mass damper · Harmony search algorithm · Fuzzy control · Wave load

1 Introduction

In recent decades, different control strategies (i.e., passive, active, semi-active or hybrid) have been extensively employed to suppress the response of offshore structures (e.g., jacket platforms, tension leg platforms, spar structures, floating production storage, offloading vessels, and riser structures) subjected to dynamic lateral loads

M. F. Farzam (✉) · B. Alinejad · S. A. M. Gavgani
Department of Civil Engineering, Faculty of Engineering, University of Maragheh, Maragheh, Iran
e-mail: m.farzam@maragheh.ac.ir

G. Bekdaş
Department of Civil Engineering, Faculty of Engineering, Istanbul University, Istanbul, Turkey

S. M. Nigdeli et al. (eds.), *Proceedings of 6th International Conference on Harmony Search, Soft Computing and Applications*, Advances in Intelligent Systems and Computing 1275, https://doi.org/10.1007/978-981-15-8603-3_5

[1, 2]. Jacket platforms are life bloods of oil-rich countries and one of the well-studied offshore structures subjected to environmental loads. Vandiver and Mitome are among the first researchers who evaluated the control of jacket platforms using tuned liquid storage tanks [3].

Several other researchers implemented passive or semi-active control methods to mitigate the dynamic response of jacket platforms [1, 2]. Enferadi et al. [4] used shape-memory alloy to control the vibrations of a steel jacket platform under sea waves. The authors used a multi-degree of freedom system with lumped mass [4]. Other dampers were also examined to control the vibration of jacket platforms, including frictional, viscous, and viscoelastic dampers, which were studied by Patil and Jangid [5]. Given the diverse variety of loads applied on platforms, passive and semi-active control devices may not be efficient in addressing all loading scenarios. This highlights the need for active control systems for improving the control performance [6]. Tuned Mass Damper (TMD) and Active Tuned Mass Damper (ATMD) are employed in this study to control the vibration of Ressalat platform located at a depth of 68 m from surface and exposed to the Persian Gulf's sea waves.

2 Mathematical Model of Ressalat Platform with ATMD and Optimization Procedure

A simplified lumped mass linear model is adopted for the Ressalat platform [7]. The lumped characteristics (i.e., mass and stiffness) were calculated according to the properties of the real structure such that the simplified model exhibited the same natural periods of the detailed model as reported by Mohajer Nasab et al. [8]. The equivalent seven degrees of freedom system is shown in Fig. 1, where the platform deck is located in the 7th level. The equivalent mass and stiffness values are presented in Table 1, and the damping matrix of the structure is considered as the Rayleigh damping while assuming 2% damping ratio for the first and second modes. The period of first mode is 2.34 s.

Random wave and constrained new-wave theories are utilized in generation of the wave records [7]. A wave with a 100-year return period was assumed, with significant wave height, peak spectral period, and peak level of 5.83 m, 7.10 s, and 4.77 m, respectively. Schematic overviews of the wave velocity and acceleration applied to the structural elevation levels are presented in Fig. 2.

Both the TMD and ATMD are assessed to control the vibration of the platform under defined wave loads. The assumed parameters for designing the TMD are presented in Table 2. A Fuzzy Logic Algorithm is applied for active control of the platform, with membership functions and relevant rules adopted from Shariatmadar et al. [9] as shown in Fig. 3 and Table 3, respectively.

Harmony Search (HS) Algorithm is also implemented to optimize the amount of actuator power in active control. Reducing the maximum displacement of the

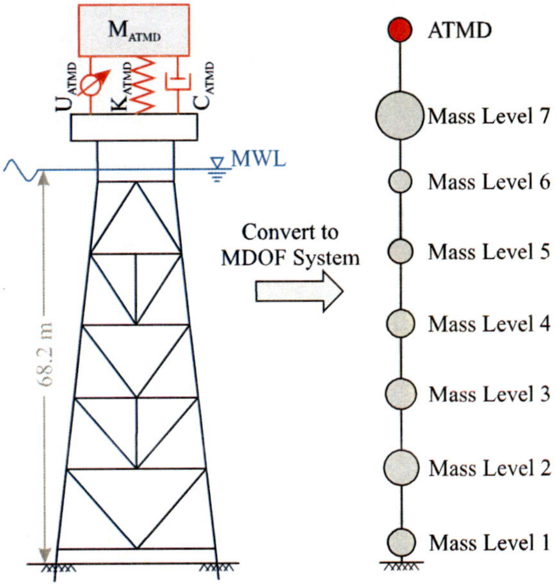

Fig. 1 Active controlled platform and its simplified lumped mass model

Table 1 Specifications of the Ressalat platform

Ressalat platform	Level 1	Level 2	Level 3	Level 4	Level 5	Level 6	Level 7
Mass (ton)	106	129	116	105	92	63	1790
Stiffness (MN/m)	179	146	146	121	106	90	38
Volume (m^3)	134	134	117	113	103	22	0
Cross-sectional area (m^2)	227	238	213	209	191	35	0

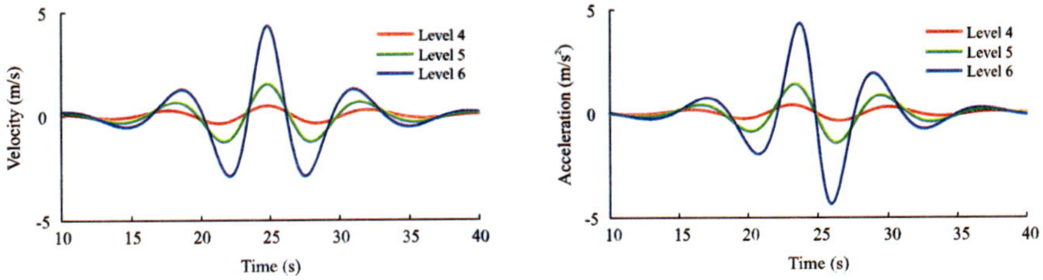

Fig. 2 The velocity and acceleration of the wave load at the different elevations of the structure

Table 2 Parameters for designing TMD

Parameters	Value
Structural damping ratio	2%
Mass ratio	3%
Structural circular frequency	2.68 rad/s
Optimal frequency ratio	0.9639
Optimal damping ratio	0.106

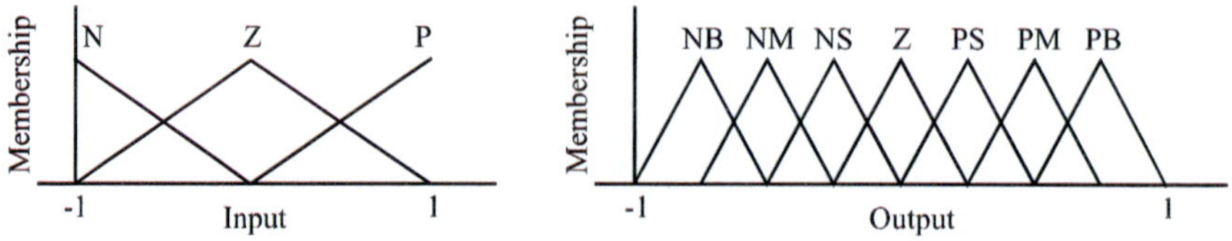

Fig. 3 Input and output membership functions

Table 3 Fuzzy rules

Velocity			
P (Positive)	Z (Zero)	N (Negative)	Displacement
PS (Positive Small)	PM (Positive Medium)	PB (Positive Big)	N (Negative)
NS (Negative Small)	Z (Zero)	PS (Positive Small)	Z (Zero)
NB (Negative Big)	NM (Negative Medium)	NS (Negative Small)	P (Positive)

deck during the vibration is considered as the objective function of the optimization problem.

Modeling and analyzing the uncontrolled and controlled structures are performed in MATLAB and SIMULINK software. The outline of the model prepared in SIMULINK is shown in Fig. 5. Additionally, the fluid–structure interaction, the effect of added mass due to accelerated motion of the body in the fluid, and the effect of actuator saturation are considered.

The platform equation of motion is defined by Eq. 1.

$$M\ddot{x} + C\dot{x} + Kx = F_I + F_D + F_C \tag{1}$$

Where M, C, and K are the mass, damping, and stiffness matrices, and x, $\dot{x}$ and $\ddot{x}$ are displacement, velocity, and acceleration vector of the structure, respectively. Moreover, inertia force (F_I) and drag force (F_D) are determined from the Morison equation (according to Eq. 2) and control force (F_C) is calculated by fuzzy controller.

$$F_I = \rho C_m V\dot{U} \ , \quad F_D = \rho C_d A'(U - \dot{X})\left|U - \dot{X}\right| \tag{2}$$

U and $\dot{U}$ represent the wave velocity and acceleration, respectively, and ρ is the water density assumed to be 1024 kg/m^3. According to API, the inertia and drag coefficients (C_m and C_d) are set at 1.2 and 1.05, respectively [11]. The volume (V) and the cross-sectional area (A') of the members were earlier presented in Table 1. The mass matrix M is the summation of the structural mass M_0 and added mass M_a presented in Eq. 3.

$$M = M_0 + M_a \ , \quad M_a = \rho(C_m - 1)V \tag{3}$$

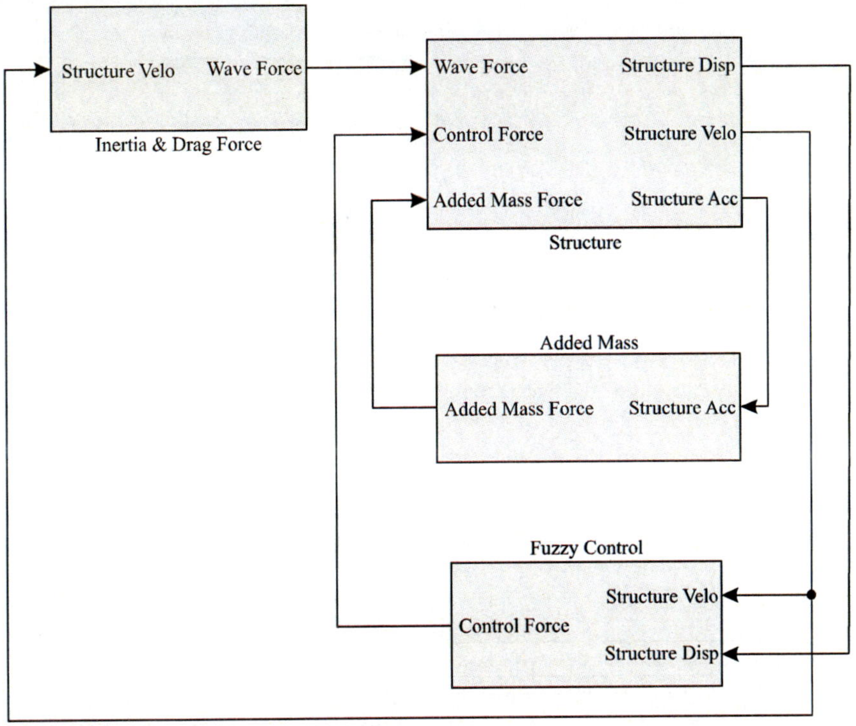

Fig. 5 SIMULINK model

3 Results and Discussion

The results presented in Fig. 6 indicate that the best response was obtained in less than 10 iterations by adjusting the parameters of the HS Algorithm appropriately.

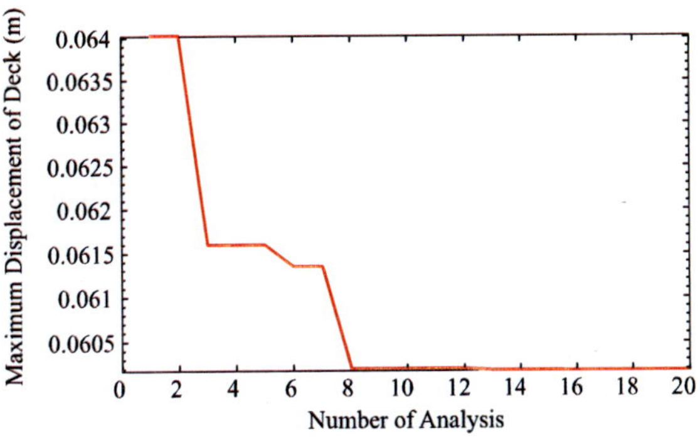

Fig. 6 Convergence curves for the jacket platform

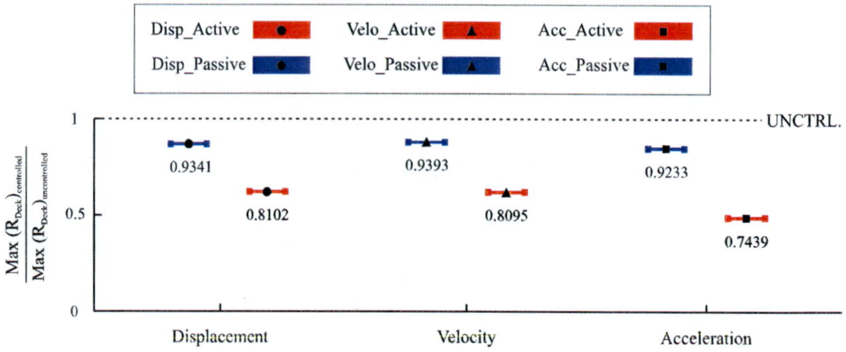

Fig. 7 Structural responses in different control modes

A summary of various calculated normalized performance criteria (response of controlled to uncontrolled structure) for passive- and active-controlled platform, including normalized maximum displacement, velocity and acceleration of the deck level are shown in Fig. 7. In this figure, the blue and red colors represent the normalized structural response of passive and active-controlled platform, respectively.

Passive control reduces the structural responses (displacement, velocity, and acceleration) by 7%, 7%, and 8%, respectively. Further reduction in structural response is observed with the active control, where the displacement, velocity, and acceleration values are reduced by 19%, 20%, and 26%, respectively. The results suggest that the vibration of the Ressalat platform is significantly reduced under studied wave loads, indicating superior performance of the proposed ATMD in comparison to the TMD.

References

1. R. Kandasamy, F. Cui, N. Townsend, C.C. Foo, J. Guo, A. Shenoi, Y. Xiong, A review of vibration control methods for marine offshore structures. Ocean Eng. **127**, 279–297 (2016)
2. B.-L. Zhang, Q.-L. Han, X.-M. Zhang, Recent advances in vibration control of offshore platforms. Nonlinear Dyn. **89**(2), 755–771 (2017)
3. J.K. Vandiver, S. Mitome, Effect of liquid storage tanks on the dynamic response of offshore platforms. Appl. Ocean Res. **1**(2), 67–74 (1979)
4. M.H. Enferadi, M.R. Ghasemi, N. Shabakhty, Vibration control of steel jacket platform through shape memory alloys damper. Modares Mech. Eng. **20**(1), 13–24 (2019)
5. K. Patil, R. Jangid, Passive control of offshore jacket platforms. Ocean Eng. **32**(16), 1933–1949 (2005)
6. B.-L. Zhang, Q.-L. Han, X.-M. Zhang, G.-Y. Tang, Active control of offshore steel jacket platforms (Springer, 2019)
7. M. Dastan Diznab, S. Mohajernassab, M. Seif, M. Tabeshpour, H. Mehdigholi, Assessment of offshore structures under extreme wave conditions by modified endurance wave analysis. Mar. Struct. **39**, 50–69 (2014)

8. S. Mohajernasab, M.A. Dastan Diznab, M.R. Tabeshpour, H. Mehdigholi, M.S. Seif, Application of New-wave theory in the Endurance Wave method to assess offshore structures under the Persian Gulf wave conditions. J. Mar. Eng. **9**(18), 71–82 (2014)
9. H. Shariatmadar, S. Golnargesi, M. Akbarzadeh Totonchi, Vibration control of buildings using ATMD against earthquake excitations through interval type-2 fuzzy logic controller. Asian J. Civil Eng. (BHRC) **15**(3), 321–338 (2014)
10. A. RP2A-WSD, American petroleum institute recommended practice for planning, designing and constructing fixed offshore platforms—working stress design (American Petroleum Institute, Washington, 2007)

Robustness of Structures with Active Tuned Mass Dampers Optimized via Modified Harmony Search for Time Delay

Aylin Ece Kayabekir, Sinan Melih Nigdeli, and Gebrail Bekdaş

Abstract By using structural control techniques, it is possible to reduce structural vibrations resulting due to seismic sources. The implemented control systems need a perfect tuning according to the behavior of the structure and seismic excitations. In this manner, optimization is a must for structural control. Also, the uncertainty and non-linearity of optimization problems prevent the usage of mathematical methods. In this respect, metaheuristic algorithms play an important role in finding the optimum parameters using numerical iterations. In the control systems, the robustness against the time delay of the system is also an important factor in realistic active control. In this study, active tuned mass dampers (ATMDs) optimized via modified harmony search (MHS) for reduction of responses of seismic structures is evaluated for time-delay systems. The investigations were done for different time delay values.

Keywords Active control · Harmony search · Time delay · Active tuned mass dampers · Robustness · Structural control · Optimization · Metaheuristic

1 Introduction

In order to reduce structural vibrations resulting from earthquakes, strong winds, vehicle, and human traffic, structural control systems with the types such as passive, active, semi-active, and hybrid are used. Passive systems contain only mechanical components while time-varying force is produced by active systems. The controller parameters for the production of control force and mechanical properties of all control systems need tuning with respect to structural properties.

Mass dampers are one of the structural control types, and it can be also passive, active, hybrid, or semi-active. For the tuning of passive tuned mass dampers (TMDs),

A. E. Kayabekir (✉) · S. M. Nigdeli · G. Bekdaş
Department of Civil Engineering, Istanbul University-Cerrahpaşa, 34320 Avcılar, Istanbul, Turkey
e-mail: ecekayabekir@gmail.com

S. M. Nigdeli
e-mail: melihnig@istanbul.edu.tr

S. M. Nigdeli et al. (eds.), *Proceedings of 6th International Conference on Harmony Search, Soft Computing and Applications*, Advances in Intelligent Systems and Computing 1275, https://doi.org/10.1007/978-981-15-8603-3_6

several metaheuristic methods such as particle swarm optimization [1], harmony search [2–3], artificial bee colony optimization [4], bat algorithm [2, 5], and flower pollination algorithm [6] had been employed. For active tuned mass dampers, several different control algorithms can be used in the development of active control force. The example control algorithms are sliding mode control [7–8], linear quadratic Gaussian control (LQG) [9–10], fuzzy logic controller (FLC) [11–12], hybrid type of FLC and LQR [13–16], proportional derivative (PD) type controllers [16], fuzzy proportional derivative integral (PID) type controllers [17] and PID controllers [18]. Kayabekir et al. proposed a modified harmony search method for optimization of mechanical parameters of ATMD and the gains of PID controllers to actively control seismic structures [19].

In the present study, the robustness of active control systems against time delay of the response of the control system is investigated. As the case study, the results of optimum PID controller and ATMD were taken from Kayabekir et al. [19] using modified harmony search algorithm.

2 The Equations of Motion for ATMD with PID Control

As seen in Fig. 1, an ATMD is positioned on the top of the structure. The three mechanical parameters of ATMD are mass (m_d), stiffness (k_d), and damping coefficient (c_d). Also, the structural properties of the story are shown as m_i, k_i, and c_i for mass, stiffness, and damping coefficient of ith story, respectively.

For a shear building, the equation of motion can be written as Eq. (1) in which the matrices of mass, stiffness, and damping are denoted by M, K, and C, respectively. The matrices were also given as Eqs. (2)–(4) and the response vector is shown as Eq. (5), which is multiplied with stiffness matrix, and the derivations of it respect to time are multiplied with the other matrices.

$$\mathbf{M}\ddot{x}(t) + \mathbf{C}\dot{x}(t) + \mathbf{K}x(t) = -\mathbf{M}\{\mathbf{1}\}\ddot{x}_g(t) + F \tag{1}$$

$$\mathbf{M} = \mathrm{diag}[m_1 m_2 \ldots\ m_N m_d] \tag{2}$$

$$\mathrm{C} = \begin{bmatrix} (c_1 + c_2) & -c_2 & & & & \\ -c_2 & (c_2 + c_3) & -c_3 & & & \\ & \cdot & \cdot & & & \\ & \cdot & \cdot & \cdot & & \\ & & \cdot & \cdot & \cdot & \\ & & & -c_N & (c_N + c_d) & -c_d \\ & & & & -c_d & c_d \end{bmatrix} \tag{3}$$

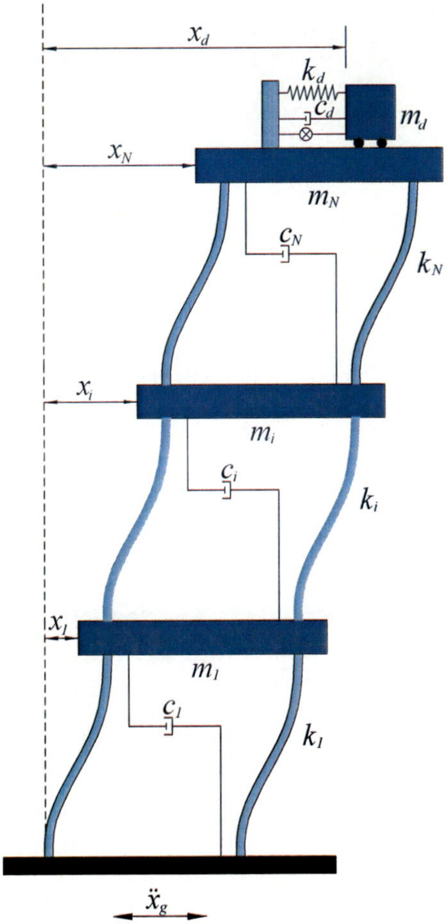

Fig. 1 A structural model with ATMD on the top

$$K = \begin{bmatrix} (k_1 + k_2) & -k_2 & & & & \\ -k_2 & (k_2 + k_3) & -k_3 & & & \\ & \cdot & \cdot & & & \\ & \cdot & \cdot & \cdot & & \\ & & \cdot & \cdot & \cdot & \\ & & & -k_N & (k_N + k_d) & -k_d \\ & & & & -k_d & k_d \end{bmatrix} \quad (4)$$

$$x(t) = \begin{Bmatrix} x_1 \\ x_2 \\ \vdots \\ x_N \\ x_d \end{Bmatrix} \quad (5)$$

The left side of the Eq. (1) results from the internal forces, and the right side is the derivation of external excitations like earthquake load and active force. The earthquake force is defined according to the ground excitation ($\ddot{x}_g(t)$) and it is multiplied with M and vector ones. As shown in Eq. (6), the active control force vector (F) has nonzero terms in two last rows which are corresponding to last story and ATMD. This control force shown as F_u can be written as Eq. (7), where K_f defines the thrust constant, and i_{ATMD} is the current of armature coil that can be found by solving Eq. (8). R, K_e, $\dot{x}_d$, $\dot{x}_N$, and u represent resistance value, induced voltage constant of the armature coil, the velocity of ATMD, the top story velocity respect to ground, and the control signal defined according to the PID controller, respectively.

$$F(t) = \begin{Bmatrix} 0 \\ 0 \\ \vdots \\ F_u \\ -F_u \end{Bmatrix} \quad (6)$$

$$F_u = K_f i_{ATMD} \quad (7)$$

$$Ri_{ATMD} + K_e(\dot{x}_d - \dot{x}_N) = u \quad (8)$$

The control signal is calculated by sensing the error signal (e(t)), which is taken as the top story velocity of the structure in the investigated case. The PID controller with parameters proportional gain (K_p), derivative time (T_d) and integral time (T_i) is formulated as Eq. (9). These parameters were also optimized by using modified harmony search.

$$u = K_p\left[e(t) + T_d\frac{de(t)}{dt} + \frac{1}{T_i}\int e(t)dt\right] \quad (9)$$

3 The Numerical Investigation

A ten-story shear structure with an ATMD optimized by Kayabekir et al. [19] is investigated for different amounts of time delay of the control signal produced by PID controller. The structure has 360t mass, 650MN/m stiffness, and 6.2MNs/m damping coefficient for each story [10]. The optimum parameters for different limit cases of the stroke capacity (st_max) are given as Table 1. Also, the objective function (f(x)) which is the maximum story displacement, and the normalized stroke value (g_1) formulated as Eq. (10) are given in Table 1.

The optimum results were found according to 22 earthquake records, and the objective function was considered for the critical one of far-fault ground motions given in FEMA 6–695: Quantification of Building Seismic Performance Factors [21]. The information about these records are given in Table 2, and the critical excitation for the numerical investigation is the BOL090 component of the Duzce (Turkey) record.

The ATMD parameters such as period (T_{atmd}) and damping (ξ_d) are presented and the parameters such as k_d and c_d can be found by using Eqs. (11)–(12).

$$g_1 = \frac{\max(|x_d - x_N|)_{\text{with ATMD}}}{\max(|x_N|)_{\text{without ATMD}}} < st\_max \quad (10)$$

$$T_{atmd} = 2\pi\sqrt{\frac{m_d}{k_d}} \quad (11)$$

$$\xi_d(\%) = \frac{c_d}{2m_d\sqrt{\frac{k_d}{m_d}}} x100 \quad (12)$$

By adding a time delay to control signal, the following results in Tables 3–5 were found for different amounts of time delay (td) as 10, 20, 30, and 50 ms. In this table,

Table 1 The optimum results [19]

	TMD		ATMD		
st_max	2	3 and 4	2	3	4
m_d(t)	180	180	180	180	180
T_{atmd}(s)	0.9418	0.9434	0.8947	0.9941	1.0797
ξ_d(%)	5.64	4.69	28.77	11.86	10.15
K_p(Ns/m)	–	–	−10,000	−1157.6	8762.9
T_d(s)	–	–	67.39	599.42	−79.12
T_i(s)	–	–	8616	−7687.9	−2748.1
f(x) (m)	0.2820	0.2803	0.2469	0.1963	0.1898
g_1	1.9999	2.0755	1.9999	2.9989	3.2983

Table 2 FEMA P-695 far-field ground motion records [33]

Earthquake number	Date	Name	Component 1	Component 2
1	1994	Northridge	NORTHR/MUL009	NORTHR/MUL279
2	1994	Northridge	NORTHR/LOS000	NORTHR/LOS270
3	1999	Duzce, Turkey	DUZCE/BOL000	DUZCE/BOL090
4	1999	Hector Mine	HECTOR/HEC000	HECTOR/HEC090
5	1979	Imperial Valley	IMPVALL/H-DLT262	IMPVALL/H-DLT352
6	1979	Imperial Valley	IMPVALL/H-E11140	IMPVALL/H-E11230
7	1995	Kobe, Japan	KOBE/NIS000	KOBE/NIS090
8	1995	Kobe, Japan	KOBE/SHI000	KOBE/SHI090
9	1999	Kocaeli, Turkey	KOCAELI/DZC180	KOCAELI/DZC270
10	1999	Kocaeli, Turkey	KOCAELI/ARC000	KOCAELI/ARC090
11	1992	Landers	LANDERS/YER270	LANDERS/YER360
12	1992	Landers	LANDERS/CLW-LN	LANDERS/CLW-TR
13	1989	Loma Prieta	LOMAP/CAP000	LOMAP/CAP090
14	1989	Loma Prieta	LOMAP/G03000	LOMAP/G03090
15	1990	Manjil, Iran	MANJIL/ABBAR–L	MANJIL/ABBAR–T
16	1987	Superstition Hills	SUPERST/B-ICC000	SUPERST/B-ICC090
17	1987	Superstition Hills	SUPERST/B-POE270	SUPERST/B-POE360
18	1992	Cape Mendocino	CAPEMEND/RIO270	CAPEMEND/RIO360
19	1999	Chi-Chi, Taiwan	CHICHI/CHY101-E	CHICHI/CHY101-N
20	1999	Chi-Chi, Taiwan	CHICHI/TCU045-E	CHICHI/TCU045-N
21	1971	San Fernando	SFERN/PEL090	SFERN/PEL180
22	1976	Friuli, Italy	FRIULI/A-TMZ000	FRIULI/A-TMZ270

Table 3 The response of ATMD against time delay analysis results (st_max = 2)

td(ms)	10	20	30	50
f(x) (m)	0.2467	0.2489	0.2509	0.2549
g_1	2.0162	2.0263	2.0417	2.0878
Critical Earth. No	3 (Com2)	3 (Com2)	3 (Com2)	3 (Com2)
Fmax	6.2806e + 06	5.8445e + 06	4.9629e + 06	5.5083e + 06
F_{max} Earth. No	14 (1)	14 (1)	14 (1)	3 (1)

the critical excitation is also given with the component (com) number. The maximum control force (Fmax) and the corresponding earthquake excitation are also given.

Table 4 The response of ATMD against time delay analysis results (st_max = 3)

td(ms)	10	20	30	50
f(x) (m)	0.2000	0.2004	0.2053	0.2238
g_1	3.082	3.1785	3.260	3.4478
Critical Earth. No	3 (Com2)	3 (Com2)	3 (Com2)	3 (Com2)
Fmax	7.7155e + 06	6.3080e + 06	5.1633e + 06	5.9199e + 06
F_{max} Earth. No	14 (1)	14 (1)	14 (1)	12 (2)

Table 5 The response of ATMD against time delay analysis results (st_max = 4)

td(ms)	10	20	30	50
f(x) (m)	0.1986	0.2089	0.2205	0.2528
g_1	3.3457	3.4395	3.5571	3.8042
Critical Earth. No	3 (Com2)	3 (Com2)	3 (Com2)	3 (Com2)
Fmax	7.7694e + 06	6.3184e + 06	5.6201e + 06	6.6067e + 06
F_{max} Earth. No	14 (1)	14 (1)	3 (1)	3 (1)

4 Conclusions and Future Works

As expected, the performance of the optimum ATMD is reducing with the increase of the time delay. Despite which ATMD is also effective in the reduction of maximum displacement which is 0.41 m for uncontrolled structure. Also, the displacement can only reduce to 0.28 m by using a TMD with the same mass.

It can be observed that the time delay has more negative effects if st_max value is big. Also, by the effect of the time delay, the stroke limitation demand cannot be provided for time-delayed responses. For case st_max = 4, it has been provided since the value of g_1 is not at the limit, although a significant increase on g_1 value can be observed. It must be noted that the maximum force is needed for a different excitation which is not critical for the objective function value. This situation proved the importance of using different excitations in the optimization process.

By the increase of the time delay, a reduction of control force can be seen due to reduction of the performance of ATMD, but the control force shows a rapid increase for 50 ms time delay, and also the objective function is mostly affected one for 50 ms time delay.

According to the results, the optimized ATMD using PID controller is robust against time delay which may occur because of communication and production of control force. The optimization method which is the modified version of music-inspired metaheuristic algorithm harmony search developed by Geem et al. [22] is a feasible method. Despite that the maximum control force values may not be economical for practice. For that reason, the limitation of control force and consideration of time delay will be investigated in the optimization process.

References

1. A.Y.T. Leung, H. Zhang, Particle swarm optimization of tuned mass dampers. Eng. Struct. **31**(3), 715–728 (2009)
2. G. Bekdaş, S.M. Nigdeli, Metaheuristic based optimization of tuned mass dampers under earthquake excitation by considering soil-structure interaction. Soil Dyn. Earthquake Eng. **92**, 443–461 (2017)
3. G. Bekdaş, S.M. Nigdeli, Estimating optimum parameters of tuned mass dampers using harmony search. Eng. Struct. **33**, 2716–2723 (2011)
4. A. Farshidianfar, S. Soheili, ABC optimization of tmd parameters for tall buildings with soil structure interaction. Int. Multiscale Mech. **6**(4), 339–356 (2013)
5. G. Bekdaş, S.M. Nigdeli, X.S. Yang, A novel bat algorithm based optimum tuning of mass dampers for improving the seismic safety of structures. Eng. Struct. **159**, 89–98 (2018)
6. G. Bekdaş, S.M. Nigdeli, X.-S. Yang, Metaheuristic based optimization for tuned mass dampers using frequency domain responses, in *Harmony Search Algorithm. Advances in Intelligent Systems and Computing*, vol. 514, ed. J. Del Ser. (Springer, 2017), pp. 271–279
7. R. Adhikari, H. Yamaguchi, Sliding mode control of buildings with ATMD. Earthquake Eng. Struct. Dynam. **26**(4), 409–422 (1997)
8. Ozer, H. O., Sayin, A., Korkmaz, N. and Yagız, N., 2013, Sliding mode control optimized by genetic algorithm for building model, In Proceedings of International Conference on Vibration Problems, 1–10.
9. I. Nagashima, Optimal displacement feedback control law for active tuned mass damper. Earthquake Eng. Struct. Dynam. **30**(8), 1221–1242 (2001)
10. B. Fitzgerald, B. Basu, S.R. Nielsen, Active tuned mass dampers for control of in-plane vibrations of wind turbine blades. Structural Control and Health Monitoring **20**(12), 1377–1396 (2013)
11. A.S. Ahlawat, A. Ramaswamy, Multiobjective optimal fuzzy logic control system for response control of wind-excited tall buildings. Journal of Engineering Mechanics **130**(4), 524–530 (2004)
12. H.R. Owji, A.H.N. Shirazi, H.H. Sarvestani, A comparison between a new semi-active tuned mass damper and an active tuned mass damper. Procedia Engineering **14**, 2779–2787 (2011)
13. M. Aldawod, B. Samali, F. Naghdy, K.C. Kwok, Active control of along wind response of tall building using a fuzzy controller. Eng. Struct. **23**(11), 1512–1522 (2001)
14. B. Samali, M. Al-Dawod, K.C. Kwok, F. Naghdy, Active control of cross wind response of 76-story tall building using a fuzzy controller. Journal of Engineering Mechanics **130**(4), 492–498 (2004)
15. A.M. Aly, Vibration control of high-rise buildings for wind: a robust passive and active tuned mass damper. Smart Structures and Systems **13**(3), 473–500 (2014)
16. R. Guclu, H. Yazici, Vibration control of a structure with ATMD against earthquake using fuzzy logic controllers. J. Sound Vib. **318**(1–2), 36–49 (2008)
17. R. Guclu, H. Yazici, Seismic-vibration mitigation of a nonlinear structural system with an ATMD through a fuzzy PID controller. Nonlinear Dyn. **58**(3), 553 (2009)
18. R. Sugumar, S. Kajjam, T.K. Datta, Implementation of active seismic control of a building frame using ANSYS, *In International Conference on Engineering, Technology and Management* (ICETM Sep'12, 2012), 1,1.
19. A.E. Kayabekir, G. Bekdaş, S.M. Nigdeli, Z.W. Geem, Optimum design of PID controlled active tuned mass damper via modified harmony search. Applied Sciences **10**(8), 2976 (2020)
20. M.P. Singh, S. Singh, L.M. Moreschi, Tuned mass dampers for response control of torsional buildings. Earthq. Eng. Struct. D. **31**, 749–769 (2002)
21. FEMA P-695, Quantification of Building Seismic Performance Factors, Federal Emergency Management Agency. (Washington DC, 2009)
22. Z.W. Geem, J.H. Kim, G.V. Loganathan, A new heuristic optimization algorithm: harmony search. Simul **76**(2), 60–68 (2001)

Optimum Active Controlled SDOF Structures Under Pulse-Type Ground Motions via Harmony Search

Serdar Ulusoy, Sinan Melih Nigdeli, and Gebrail Bekdaş

Abstract In this study, the active controlled single degree of freedom (SDOF) structures with 0.5, 1.0 and 2.0 s period are investigated under various pulse-type ground motions (1.5, 2.0, and 2.5 s) to optimally tune the parameters of proportional–integral–derivative (PID)-type controllers. The damping ratio value is assumed as 5%. The paper proposes a methodology to determine the proportional gain (K_p), integral gain (T_i), and derivative gain (T_d) using Harmony Search algorithm (HS) which is one of the metaheuristic-based methods. The performance index of active controlled structures is the displacement feedback. Also, the control force limit as 50% of the total mass of the structure and the time delay effect 20 ms were considered in term of feasibility and stability of active controlled structures. According to the results, the optimum tuned PID controller parameters has major effects on the structural responses of SDOF structures.

Keywords Active structural control · PID controllers · Harmony search algorithm · Pulse-type ground motions

1 Introduction

Various types of control systems are used in the structure to reduce structural responses caused by dynamic loads. These control systems are classified as passive, semi-active, and active control systems. Active damping systems have a higher energy absorption capacity than other systems (passive and semi-active) due to the generation of a control force with the power source in the structures to respond to dynamic load.

S. Ulusoy (✉)
Department of Civil Engineering, Turkish-German University, 34820 Beykoz, Istanbul, Turkey
e-mail: serdar.ulusoy@tau.edu.tr

S. M. Nigdeli · G. Bekdaş
Department of Civil Engineering, Istanbul University-Cerrahpaşa, 34320 Avcılar, Istanbul, Turkey
e-mail: melihnig@istanbul.edu.tr

S. M. Nigdeli et al. (eds.), *Proceedings of 6th International Conference on Harmony Search, Soft Computing and Applications*, Advances in Intelligent Systems and Computing 1275, https://doi.org/10.1007/978-981-15-8603-3_7

Although theoretically active control systems have important advantages for structural reactions, these systems also have disadvantages such as time delay, stability problems, and applicability of the control force in the practical field. Therefore, active control systems should be investigated taking these factors into account.

In recent years, many studies have been conducted in which the structural reactions are significantly reduced by providing active control with different control techniques such as H_{inf} [1], direct adaptive control [2], Proportional–Integral–Derivative (PID) control [3], linear quadratic regulator (LQR) [4], wavelet-based adaptive pole assignment control [5], neural network control [6], equivalent-input-disturbance [7], fuzzy logic control [8], decentralized network control [9], and sliding mode control (SMC) [10].

The use of numerical algorithms is important for the solution to some complex problems encountered in engineering. The metaheuristic algorithms, one of these numerical algorithms, are the mathematical expressions created by inspiring from events that exist in nature. With metaheuristic algorithms (evolutionary algorithms, swarm intelligence, and the other metaheuristic algorithms), it is possible to obtain fast and optimum results in many fields. The most commonly used metaheuristic algorithms in the literature are Genetic Algorithm (GA) [11], Teaching Learning-Based Optimization (TLBO) [12], Jaya Algorithm (JA) [13], Artificial Bee Colony Algorithm (ABC) [14], Harmony Search Algorithm (HS) [15], Flower Pollination Algorithm (FPA) [16], Particle Swarm Optimization (PSO) [17], and Bat Algorithm (BA) [18].

In this study, different SDOF structures with tendons are actively controlled. PID controller is chosen as the control technique. HS algorithm is used to determine the controller parameters of PID controller such as proportional gain (K_p), integral time (T_i), and derivative time (T_d) optimally. Thus, the change in the maximum structural reactions of the controlled and uncontrolled SDOF structures with different periods such as 0.5 s, 1 s, 1.5 s, and 2.0 s are significant under pulse-type ground motions.

2 Methodology

The equation of motion of uncontrolled SDOF structures is presented in Eq. 1. m_1, k_1, and c_1 are the mass, stiffness, and damping coefficient of SDOF structures, respectively. $a_g(t)$ is the ground acceleration, $a_1(t)$ is the acceleration, $v_1(t)$ is the velocity, and $x_1(t)$ is the displacement of SDOF. The equation of motion of active controlled structures is given in Eq. 2. k_c is the stiffness of tendon, α is the angle of tendon with respect to ground, and $u_1(t)$ is the control signal. Also, the force of tendon in the static and dynamic states is shown in Fig. 1. In the dynamic state, the force in the tendon is calculated $4k_c u_1(t)\cos\alpha$ and is considered in the equation of motion of active controlled structure.

$$m_1 a_1(t) + c_1 v_1(t) + k_1 x_1(t) = -m_1 a_g(t) \tag{1}$$

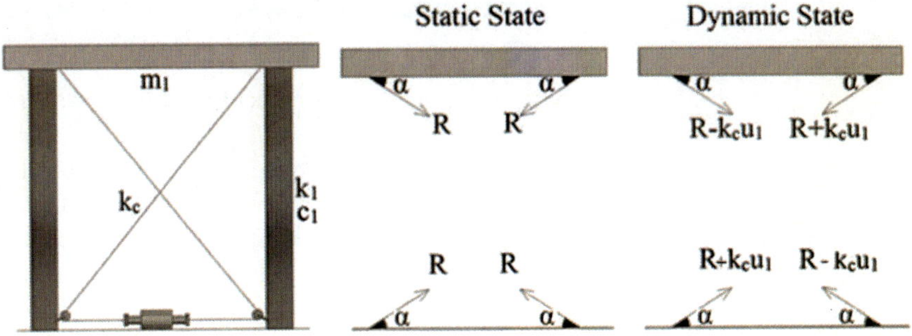

Fig. 1 The static and dynamic State of active controlled structures

$$m_1a_1(t) + c_1v_1(t) + k_1x_1(t) = -m_1a_g(t) - 4k_cu_1(t)cos\alpha \tag{2}$$

The formulation of PID controller given in Eq. 3 is used to determine the control signal $u_1(t)$. The sum of the error signal, division of the integral gain with the integrated error signal, and multiplication of the derivative gain with derived error signal are equal to the control signal. The error signal e(t) is the displacement of SDOF structures due to choosing displacement as feedback.

$$u_1(t) = K(e(t) + \frac{1}{T_i}\int e(t)dt + T_d\frac{de(t)}{d(t)} \tag{3}$$

The near-fault ground motion has impulsive features such as distinct velocity pulse in the time history and has two components called the directivity effect (the fault-normal component) and the flint step (fault-parallel component) [19]. It is known that the directivity effect causes more damage to the structures than the flint step [20]. The equations of Makris (1997) are used for the expression of components of near-fault ground motion [21]. These equations are described as follows:

$$a_g(t) = \omega_pV_pcos(\omega_pt)0 \le t \le T_p \tag{4}$$

$$a_g(t) = \omega_p\frac{V_p}{2}sin(\omega_pt)0 \le t \le T_p\text{fortheflintstep} \tag{5}$$

T_p, V_p, and ω_p are the pulse period, peak ground velocity, and the pulse frequency of near-fault ground motion, respectively. These values are calculated with Eqs. 6–8 depending on Moment Magnitude (M_w) and directivity angle (ϕ) and soil type from the studies of Cox and Ashford (2002) and Sommerville (2003) [22, 23].

$$LogT_p = -3.17 + 0.5M_w \text{ for rock site} \tag{6}$$

$$LogT_p = -2.02 + 0.346M_w \text{ for soft site} \tag{7}$$

$$LogV_p = 6.444 - 0.0187\phi - 5.022Log(M_w) \tag{8}$$

Harmony search algorithm is a metaheuristic algorithm developed by Geem [15] and inspired by a musician who wants to please his audience. In this algorithm, it is aimed to reveal the best music by playing popular notes in musician memory, new notes added to known music or different notes. With this approach (attainment of the best music), fast solutions are obtained in a short time in some complex engineering problems. The mathematical expression of the harmony search algorithm is as follows:

$$x_i^{t+1} = x_{min} + rand(1)(x_{max} - x_{min}) \text{ if HMCR} > r_1 \tag{9}$$

$$x_i^{t+1} = x_j^t \text{ if HMCR} \le r_1 \text{andPAR} > r_2 \tag{10}$$

$$x_i^{t+1} = x_{j+ni}^t \text{ if HMCR} \le r_1 \text{ and PAR} \le r_2 \text{ (for discrete variables)} \tag{11}$$

$$x_i^{t+1} = x_j^t + b_w(r_3) \quad \text{if HMCR} \le r_1 \text{ and PAR} \le r_2 \text{ (for continuous variables)} \tag{12}$$

Harmony memory considering rate (HMCR) and pitch adjusting rate (PAR) which vary between 0 and 1 in the optimization process are two parameters of harmony search algorithm to avoid the local optimization. x_i^{t+1} is the new solution, x_j is the existing solution, x_{max} and x_{min} are the max and min values of design variables, respectively, rand(1), r_1, and r_2 are the random numbers between 0 and 1, r_3 is the random number between 1 and –1, b_w is an arbitrary distance bandwidth, and ni is the neighborhood index. The flowchart diagram of harmony search algorithm is given in Fig. 2.

3 The Numerical Example

The active controlled SDOF structures with different periods are investigated. For each SDOF structures, the stiffness and angle of tendon are $\alpha = 36^0$ and $k_c = 372.1$ KN/m, respectively [24]. Also, the mass and the damping coefficient of SDOF structures are 1000 kg and %5. Three different directivity pulses with 1.5, 2.0, and 2.5 s periods and 230 cm/s peak ground velocity (PGV) presented in Fig. 3 are used in the optimization process. The optimum results for SDOF structures under different directivity pulses are given in Table 1. Also, the time history of the maximum displacement

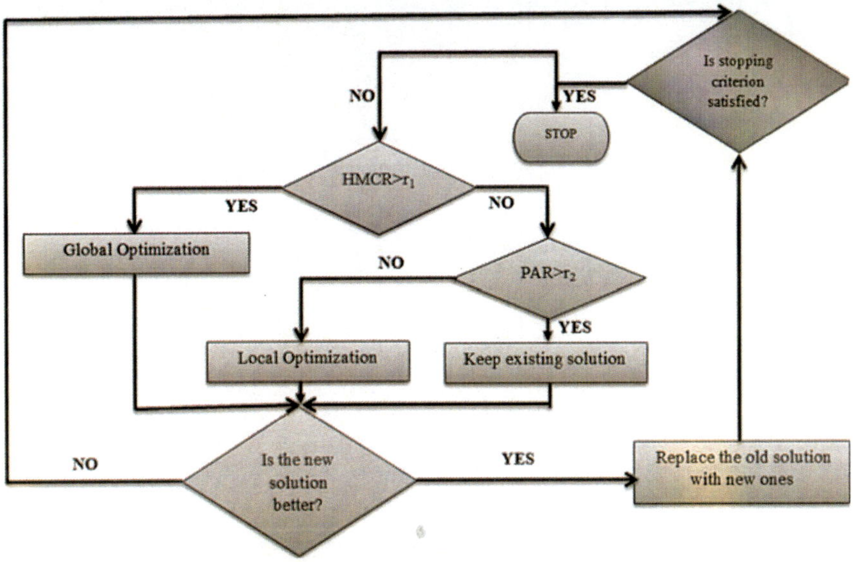

Fig. 2 The Flowchart diagram of harmony search algorithm

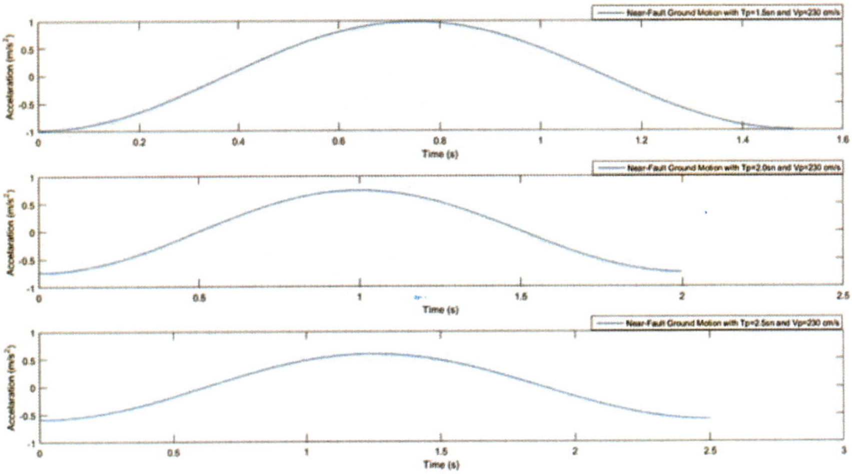

Fig. 3 Three different directivity Pulses with 1.5 s, 2.0 s, and 2.5 s periods and 230 cm/s PGV

and total acceleration of three different uncontrolled and active controlled structures are presented in Figs. 4–9 under directivity pulses.

Table 1 The optimum results for SDOF structures under different Pulse types

Structures	SDOF with 0.5 s period			SDOF with 1.0 s period			SDOF with 2.0 s period		
Pulse-Type	1.5 s	2.0 s	2.5 s	1.5 s	2.0 s	2.5 s	1.5 s	2.0 s	2.5 s
K	−0.0158	−0.0362	−0.0730	−0.0043	−0.0219	−0.0351	−0.0008	0.0025	−0.0073
T_d	0.4897	0.3001	0.1983	0.9228	0.1900	0.2172	1.4125	−1.9870	1.1401
T_i	3.0613	3.4664	4.4804	0.1113	−4.5476	−2.9465	−0.3383	−0.0999	0.4293
X (cm)	5.88	3.75	2.39	32.06	12.04	7.09	88.35	66.19	24.59

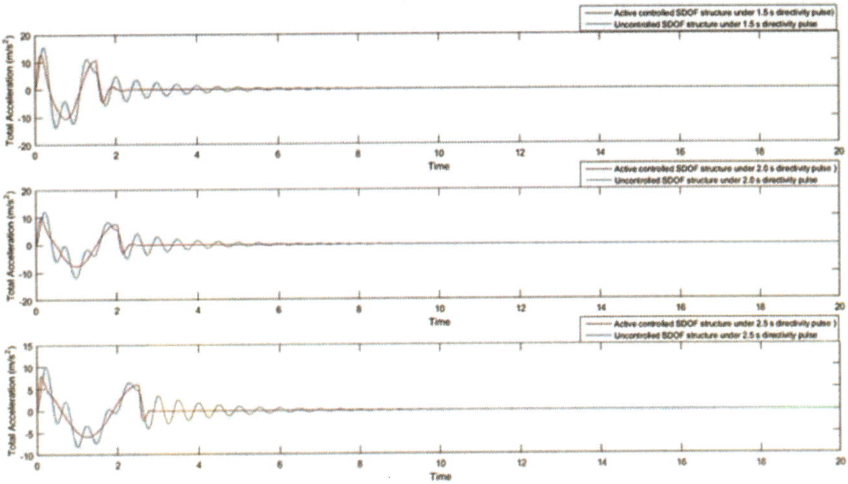

Fig. 4 The time history of total acceleration of SDOF structure with 0.5 s period under pulse-type ground motions

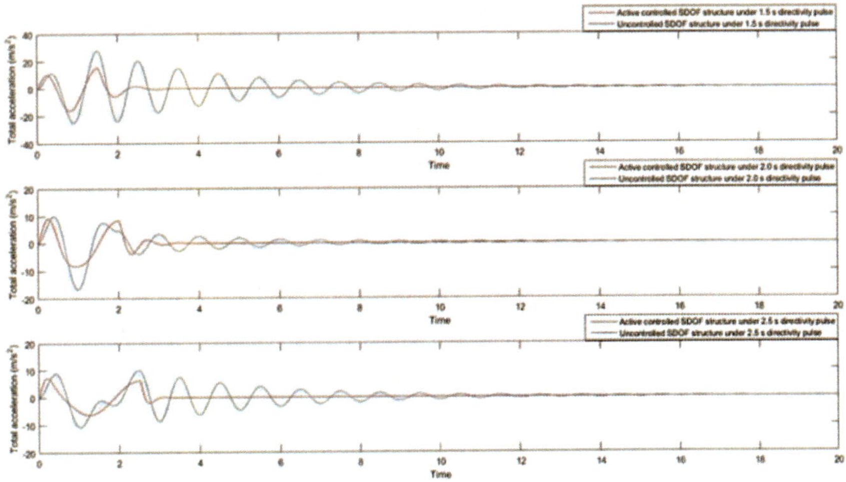

Fig. 5 The time history of total acceleration of SDOF structure with 1.0 s period under pulse-type ground motions

4 Conclusions

The maximum responses of SDOF structures for different periods under pulse-type ground motions with 230 cm/s peak ground velocity are given in Table 2. The maximum displacements for uncontrolled structure with 0.5 s period are 9.77, 7.78, and 6.41 cm under 1.5, 2.0, and 2.5 s pulse-type ground motions, respectively. These

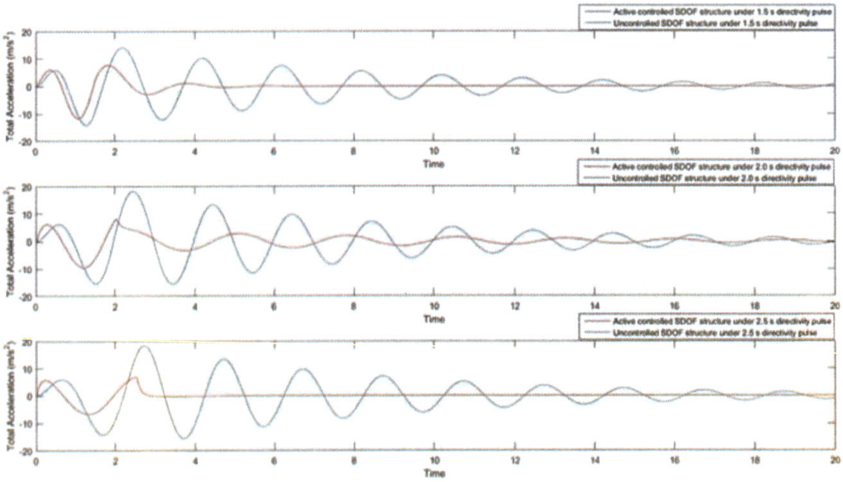

Fig. 6 The time history of total acceleration of SDOF structure with 2.0 s period under pulse-type ground motions

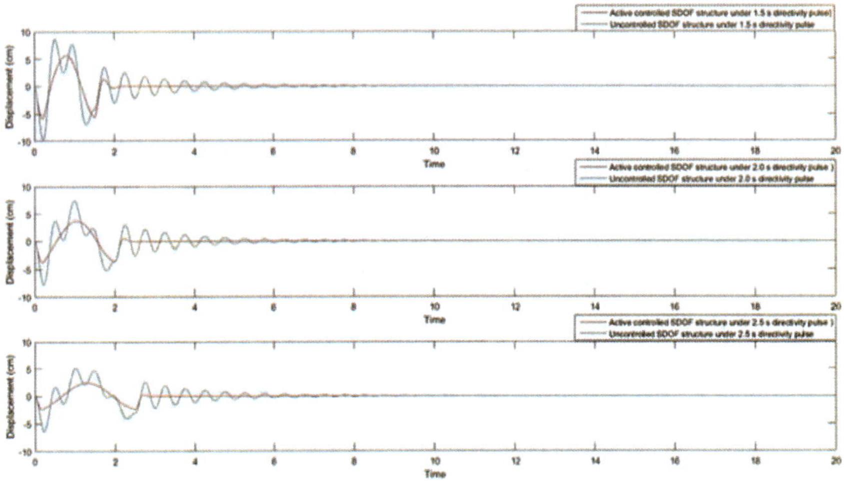

Fig. 7 The time history of displacement of SDOF structure with 0.5 s period under pulse-type ground motions

displacements are 70.59, 42.08, and 27.03 cm for uncontrolled structure with 1.0 s period, while they are 142.35, 184.23 (in the case of resonance), and 184.15 cm for uncontrolled structure with 2.0 s period. The displacement reductions of three different structures under pulse-type ground motions ranged from 38 to 86%. Also, the maximum total accelerations of all structures under all pulse-type ground motions

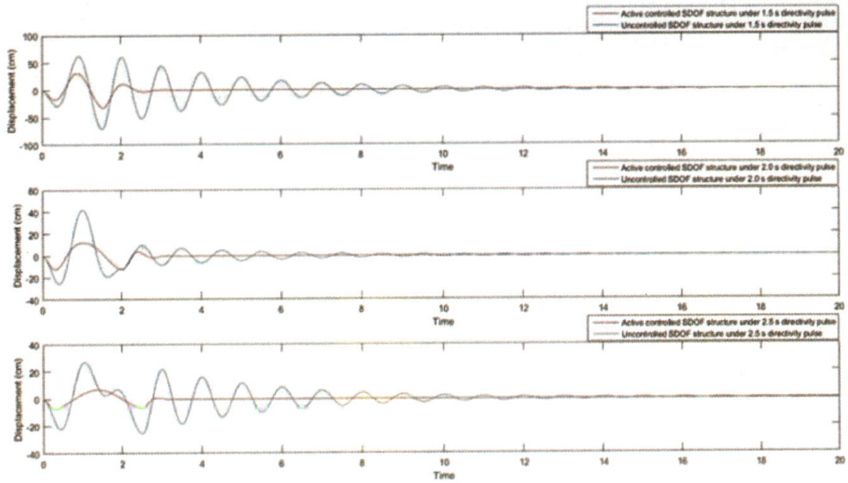

Fig. 8 The time history of displacement of SDOF structure with 1.0 s period under pulse-type ground motions

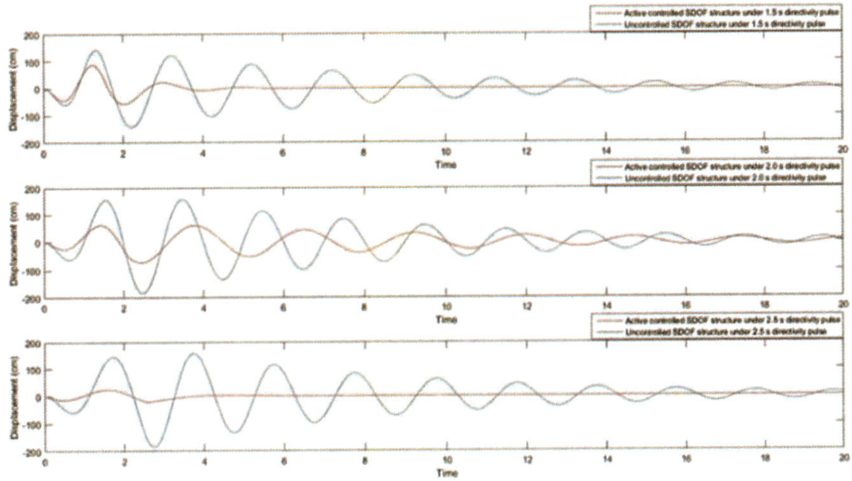

Fig. 9 The time history of displacement of SDOF structure with 2.0 s period under pulse-type ground motions

were reduced significantly considering 50% control force limit. As a result, it is determined that the use of the harmony search algorithm is very effective for the optimum tuned PID parameters of active controlled structures.

Table 2 The Maximum Responses of SDOF structures with different periods under pulse-type ground motions

Impulsive Motion	Structures	Displacement (cm)	Total Acceleration (m/s^2)	Control Signal (cm)	Control Force (KN)
1.5 s	Uncontrolled with 0.5 s	9.77	15.48	–	–
	Controlled with 0.5 s	5.88	12.80	0.41	4.90
2.0 s	Uncontrolled with 0.5 s	7.78	12.32	–	–
	Controlled with 0.5 s	3.75	9.76	0.41	4.90
2.5 s	Uncontrolled with 0.5 s	6.41	10.15	–	–
	Controlled with 0.5 s	2.39	7.96	0.41	4.90
1.5 s	Uncontrolled with 1.0 s	70.59	28.00	–	–
	Controlled with 1.0 s	32.08	15.97	0.40	4.83
2.0 s	Uncontrolled with 1.0 s	42.08	16.66	–	–
	Controlled with 1.0 s	12.04	9.12	0.40	4.76
2.5 s	Uncontrolled with 1.0 s	27.03	10.70	–	–
	Controlled with 1.0 s	7.09	7.23	0.41	4.88
1.5 s	Uncontrolled with 2.0 s	142.35	14.12	–	–
	Controlled with 2.0 s	88.35	11.71	0.40	4.86
2.0 s	Uncontrolled with 2.0 s	184.23	18.28	–	–
	Controlled with 2.0 s	66.19	9.25	0.40	4.87
2.5 s	Uncontrolled with 2.0 s	184.15	18.27	–	–
	Controlled with 2.0 s	24.59	6.83	0.41	4.91

References

1. C.C. Chang, C.C. Lin, H∞ drift control of time-delayed seismic structures. Earthquake Engineering and Engineering Vibration **8**(4), 617–626 (2009)
2. M. Bitaraf, S. Hurlebaus, L.R. Barroso, Activeand semi-active adaptive control for undamaged and damaged building structures under seismic load. ComputerAided Civil and Infrastructure Engineering **27**(1), 4864 (2012)
3. S. S. Melih Nigdeli, M. Hasan Boduroğlu, Active tendon control of torsionally irregular structures under near-fault ground motion excitation. *Computer-Aided Civil and Infrastructure Engineering* 28, no. 9 (August 20, 2013): 718–736. doi:https://doi.org/10.1111/mice.12046
4. U. Aldemir, M. Bakioglu, Active structural control based on the prediction and degree of stability. J. Sound Vib. **247**(4), 561–576 (2001)
5. F. Amini, M.Z. Samani, A wavelet based adaptive pole assignment method for structural control. Comput. Aided Civil and Infrastruct. Eng. **29**(6), 464–477 (2014)
6. N. Wang, H. Adeli, Self-constructing wavelet neural network algorithm for rnonlinear control of large structures. Eng. Appl. Artif. Intell. **41**, 249–258 (2015)
7. K. Miyamoto, J. She, J. Imani, X. Xin, D. Sato, Equivalent-input-disturbance approach to active structural control for seismically excited buildings. Eng. Struct. **125**, 392–399 (2016)
8. E. Nazarimofrad, S.M. Zahrai, Fuzzy control of asymmetric plan buildings with active tuned mass damper considering soil-structure interaction. *Soil Dynamics and Earthquake Engineering* (2017)
9. L. Bakule, B. Rehk, M. Papk, Decentralized networked control of building structures. Comput. Aided Civil Infrastruct. Eng. **31**(11), 871–886 (2016)
10. R. Guclu, Sliding mode and PID control of a structural system against earthquake. Mathemat. Comput. Modell. **44**, 210–217 (2006)
11. J.H. Holland, *Adaptation in Natural and Artificial Systems* (University of Michigan Press, Ann Arbor, Michigan, 1975)
12. R.V. Rao, V.J. Savsani, D.P. Vakharia, Teaching–learning-based optimization: a novel method for constrained mechanical design optimization problems. Comput. Aided Des. **43**(3), 303–315 (2011)
13. R. Rao, Jaya: a simple and new optimization algorithm for solving constrained and unconstrained optimization problems. Int. J. Ind. Eng. Computat. **7**(1), 19–34 (2016)
14. D. Karaboga, B. Basturk, A powerful and efficient algorithm for numerical function optimization: artificial bee colony (ABC) algorithm. J. Global Optim. **39**(3), 459–471 (2007)
15. Z.W. Geem, J.H. Kim, G.V. Loganathan, A new heuristic optimization algorithm: *harmony search*. Simulation **76**, 60–68 (2001)
16. X.S. Yang, Flower pollination algorithm for global optimization. *In International conference on unconventional computing and natural computation* (pp. 240–249). Springer, Berlin, Heidelberg.
17. J. Kennedy, R.C. Eberhart, Particle swarm optimization. *Proceedings of IEEE International Conference on Neural Networks* No. IV, November 27-December 1, pp. 1942–1948, (Perth Australia, 1995).
18. X.S. Yang, A new metaheuristic bat-inspired algorithm. *In Nature inspired cooperative strategies for optimization (NICSO 2010)* (pp. 65–74), (Springer, Berlin, Heidelberg, 2010).
19. A. Ghobarah, Response of structures to near-fault ground motion, *13th World conference on Earthquake Engineering,* (Vancouver, B.C, Canada, 2004)
20. A.K. Chopra, C. Chintanapakdee, Comparing response of SDF systems to near-fault and far-fault earthquake motions in the context of spectral regions. *Earthquake Engineering and Structural dynamics* 30 (2001).
21. N. Makris, Rigidity-plasticity-viscosity: can electrorheological dampers protect base-isolated structures from near-source ground motions? Earthquake Eng. Struct. Dynam. **26**, 571–591 (1997)

22. K.E. Cox, S.A. Ashford, Characterization of large velocity pulses for laboratory testing. Report 2002/22, *Pacific Earthquake Engineering Research Center*, University of California, Berkeley, California, USA (2002).
23. P.G. Sommerville, Magnitude scaling of the near fault rupture directivity pulse. Phys. Earth Planet. Inter. **137**, 201–212 (2003)
24. L.L. Chung, R.C. Lin, A.M. Reinhorn, T.T. Soong, Experimental study of active control for MDOF seismic structures. Journal of Structural Engineering (ASCE) **115**, 1609–1627 (1989)

Hybrid Harmony Search Algorithm for Optimum Design of Vibration Absorber System for Adjacent Buildings

Sinan Melih Nigdeli and Gebrail Bekdaş

Abstract The pounding of building blocks is a dangerous happening during an earthquake. The collision forces occurring between two buildings may result with particular damage at the collision point, and this damage will lead to the collapse of buildings. For that reason, a required seismic gap must be provided, but the gap may not be provided in the required amounts. Also, if the structures have a flexible base like seismic isolated structures, base displacement must be limited in order to protect isolators from yielding and collision of blocks. For that reason, several control methods can be used, but the essential factor is a good optimization of parameters of the design. In this paper, the optimum parameters of an absorber system connecting both buildings were proposed. In the optimization method, a hybrid Harmony Search (HS) algorithm was proposed. The combined algorithm is Jaya algorithm (JA). The proposed method is more effective than classical HS approach for reduction of the amount of the required seismic gap.

Keywords Vibration absorber system · Adjacent buildings · Harmony search · Jaya algorithm · Hybrid algorithms

1 Introduction

If the seismic gap between structure blocks is insufficient, these blocks may collide. According to Jeng an Tzeng [1], pounding of structures are defined in five types such as mid-column pounding, heavier adjacent building pounding, taller adjacent building pounding, eccentric pounding and end building pounding. The most dangerous type is mid-column pounding, and columns of a building may be critically damaged because of the impact (hammering) of the story level of the other structure. This may lead to the collapse of a structure with a column damage.

S. M. Nigdeli (✉) · G. Bekdaş
Department of Civil Engineering, Istanbul University-Cerrahpaşa, 34320 Avcılar, Istanbul, Turkey
e-mail: melihnig@istanbul.edu.tr

S. M. Nigdeli et al. (eds.), *Proceedings of 6th International Conference on Harmony Search, Soft Computing and Applications*, Advances in Intelligent Systems and Computing 1275, https://doi.org/10.1007/978-981-15-8603-3_8

By using structural control systems, it is possible to protect the structures from pounding. The recently proposed methods are listed as follows:

- Interconnected adjacent structures with nonlinear hysteretic dampers [2]
- Stochastic optimal coupling-control of adjacent buildings [3]
- Magnetorheological (MR) dampers connecting adjacent structures [4–6]
- Connecting of adjacent buildings with hydraulics actuators [7]
- Bumper-type collision shear walls [8]
- Viscoelastic dampers connecting adjacent structures [9]
- Active controlled adjacent structures [10]
- Rubber shock absorber [11]
- Optimization of passive dissipation devices connecting adjacent structures [12]
- Variable damping semi-active control system [13]
- Optimum combination passive dampers connecting adjacent structures [14]
- Direct and indirect implementations of Mass Proportional Damping System [15]
- Optimum tuned mass dampers [16–19]
- A vibration absorber device connecting structure with a mass, stiffness element and dampers [16, 20–21]

In this study, the absorber device proposed by Nigdeli and Bekdaş [16] was optimized via a hybrid metaheuristic method. The proposed hybrid method combines the music-inspired Harmony Search (HS) [22] and a single-phase metaheuristic algorithm called Jaya Algorithm (JA) [23] to minimize the required seismic gap between controlled adjacent structures.

2 The Methodology

2.1 Equations of Motion

In Fig. 1, adjacent structures implemented with the proposed absorber system are shown. The single degree of freedom shear buildings are subjected to ground acceleration ($\ddot{x}_g$) resulting from earthquakes. The response of the first structure is shown as x_1, displacement. The properties of the first structure are mass (m_1), stiffness (k_1) and damping (c_1) coefficients. The optimized design variables are mass (m_d), stiffness (k_d, k_d) and damping values (c_d, c_d) of the absorber system. The displacement of the absorber system is x_d. The response, properties and design variables connected to the second structure are shown in italics.

The equation of the coupled adjacent structures is shown as in Eq. (1) in matrix form.

$$M\ddot{x} + C\dot{x} + Kx = -M\{1\}\ddot{x}_g \quad (1)$$

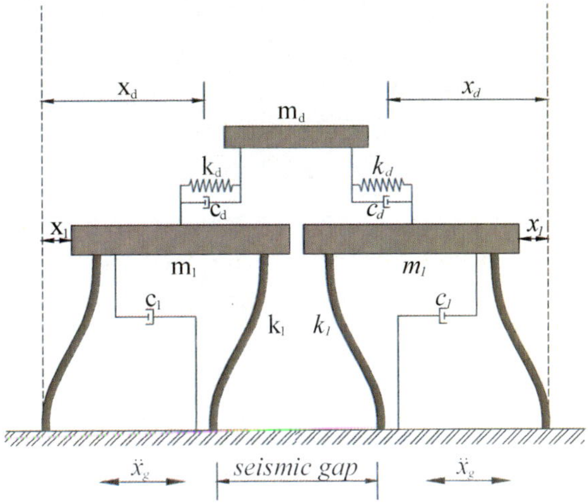

Fig. 1 The absorber system with adjacent SDOF structures

In Eq. (1), the matrices M, C and K shown as in Eqs. (2)–(4) represent mass, stiffness and damping matrices, respectively. The displacement vector (x) is shown as in Eq. (5).

$$\mathrm{M} = \mathrm{diag}\left[m_1 \; m_d \; \mathbf{m_1}\right] \tag{2}$$

$$K = \begin{bmatrix} k_1 + k_d & -k_d & 0 \\ k_d & k_d + \mathbf{k_d} & -\mathbf{k_d} \\ 0 & -\mathbf{k_d} & \mathbf{k_1} + \mathbf{k_d} \end{bmatrix} \tag{3}$$

$$C = \begin{bmatrix} c_1 + c_d & -c_d & 0 \\ c_d & c_d + \mathbf{c_d} & -\mathbf{c_d} \\ 0 & -\mathbf{c_d} & \mathbf{c_1} + \mathbf{c_d} \end{bmatrix} \tag{4}$$

$$x = \left[\mathrm{x}_1 \; x_d \; x_1\right]^T \tag{5}$$

The optimization methodology.

The aim of the optimization is to minimize the required gap between two structures. For that reason, the minimization of the relative displacement of a structure with respect to the other one is considered as the objective function (OF) as shown in Eq. (6). The design variables are the mechanical properties of the absorber system such as $\mathrm{m_d}$, $\mathrm{k_d}$, k_d, $\mathrm{c_d}$ and c_d. The general process of the optimization is summarized in the flow chart given as in Fig. 2.

In this study, four different algorithms were investigated. The first one is the modified HS for the absorber system developed by Nigdeli and Bekdaş [16]. In the algorithm, a new design variable (X_{new}) is modified according to Eqs. (7)–(8) for global and local searches of HS, respectively. The search type is chosen by a

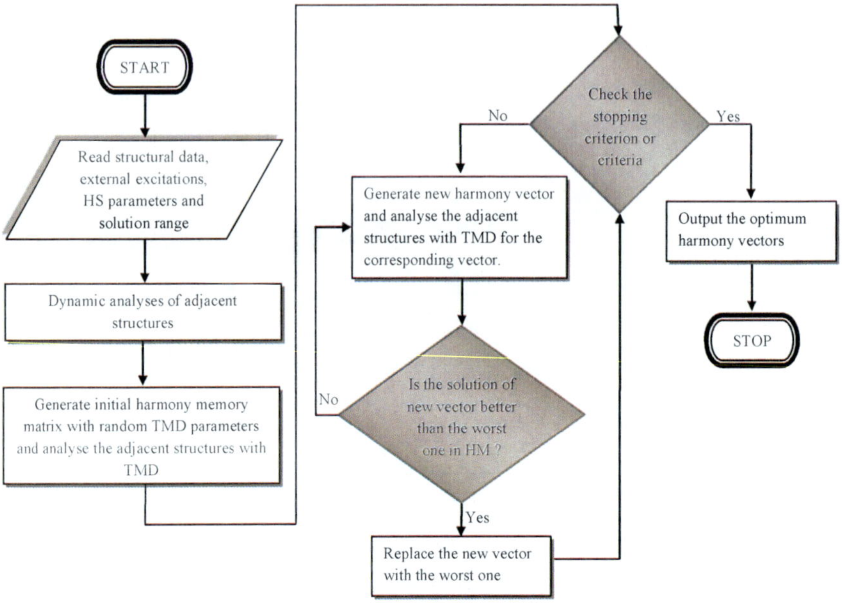

Fig. 2 The flow chart of the optimization process

parameter called Harmony Memory Considering Rate (HMCR), which is taken as 0.5 in the numerical case of this study. X_{min} and X_{max} are the minimum and maximum ranges of the design variables, respectively. PAR is a parameter called Pitch Adjusting Rate, which is taken as 0.5 in the current study. A random number between 0 and 1 is shown as rand. X_r is a randomly chosen existing value of a design variable.

As seen in the equation of HS, two user-defined parameters exist. In the first modification of HS called HS_{pf}, HMCR and PAR values are assigned with a random number between 0 and 1 for each iteration. As the second modified HS, the algorithm is hybridized with JA (HSJA). JA is a single-phase algorithm, and the global search phase of HS was swapped with the equation of JA as seen in Eq. (9). X_{best} and X_{worst} define the existing best and worst solutions according to the OF value, respectively.

$$OF = \max(|x_1 - \mathbf{x_1}|) \quad (6)$$

$$X_{new} = X_{\min} + rand(X_{\max} - X_{\min}) \quad (7)$$

$$X_{new} = X_r + rand(PAR)(X_{\max} - X_{\min}) \quad (8)$$

$$X_{new} = X_r + rand(X_{best} - |X_r|) - rand(X_{worst} - |X_r|) \quad (9)$$

The third one is the parameter-free version of the second modification ($HSJA_{pf}$). Same as the first modification, HMCR and PAR are assigned with a random number in each iteration.

3 The Numerical Example

As the numerical example, two single degree of freedom (SDOF) adjacent structures were investigated for an optimum absorber device. The first structure has 2 s period, while the second one has 4 s period. Both structures have the same mass as 100t and 5% damping ratio. The optimization process was performed by considering 6 different earthquake records as given in Table 1.

The peak ground acceleration (PGA), peak velocity (PGV) and peak ground displacement (PGD) of these records are given, and it is clearly seen from PGV values that these excitations are near-fault records.

The optimum results of different methods are given in Table 2. The employed ranges are 0.01–200 kN/m for stiffness and 0.01–1 kNs/m for damping values. The mass ranges are between 0.02–10 t.

Table 1 The optimization ground motions

Earthquake	Date	Station	Component	PGA (g)	PGV (cm/s)	PGD (cm)
Cape Mendocino	1992	Petrolia	PET090	0.662	89.7	29.55
Kobe	1995	0 KJMA	KJM000	0.821	81.3	17.68
Erzincan	1992	95 Erzincan	ERZ-NS	0.515	83.9	27.35
Northridge	1994	Rinaldi	RRS228	0.838	166.1	28.78
Northridge	1994	24,514 Sylmar	SYL360	0.843	129.6	32.68
Loma Prieta	1989	16 LGPC	LGP000	0.563	94.8	41.18

Table 2 The optimization ground motion

	HS	HS_{pf}	HSJA	$HSJA_{pf}$
m_d (t)	8.0729	7.5036	9.7743	9.8558
k_d (kN/m)	191.2000	144.9463	200.0000	183.4361
k_d (kN/m)	108.9460	103.5329	95.6404	92.1170
c_d (kNs/m)	0.3511	0.1305	1.0000	0.8352
c_d (kNs/m)	0.3202	0.9506	0.8135	1.0000
OF	0.3395	0.3349	0.3281	0.3202

4 Conclusions

The OF value without any absorber device is 1.2566 m and, it is occurring under the LGP000 record. This value is reduced to very low values by using the optimally tuned absorber device. The objective function values are close to each other for different approaches, but the best one is $HSJA_{pf}$. It seems that the user-defined parameter-free technique is a good option without calibrating the best parameter for the problem. As a conclusion, HS hybridized with JA is an effective approach for the optimal tuning of parameters of the absorber device proposed for adjacent structures.

References

1. V. Jeng, W.L. Tzeng, Assessment of seismic pounding hazard for Taipei City. Eng. Struct. **22**, 459–471 (2000)
2. Y.Q. Ni, J.M. Ko, Z.G. Ying, Random seismic response analysis of adjacent buildings coupled with non-linear hysteretic dampers. J. Sound Vib. **246**(3), 403–417 (2001)
3. Z.G. Ying, Y.Q. Ni, J.M. Ko, Stochastic optimal coupling-control of adjacent building structures. Comput. Struct. **81**(30-31), 2775–2787 (2003)
4. S.D. Bharti, S.M. Dumne, M.K. Shrimali, Seismic response analysis of adjacent buildings connected with MR dampers. Eng. Struct. **32**, 2122–2133 (2010)
5. M.N. Sheikh, J. Xiong, W.H. Li, Reduction of seismic pounding effects of base-isolated RC highway bridges using MR damper. Structural Engineering and Mechanics **41**(6), 791–803 (2012)
6. G.C. Kim, J.W. Kang, Seismic response control of adjacent building by using hybrid control algorithm of MR. Procedia Engineering **14**, 1013–1020 (2011)
7. Y.L. Xu, W.S. Zhang, Closed-form solution for seismic response of adjacent buildings with linear quadratic Gaussian controllers. Earthquake Engng. Struct. Dyn. **31**, 235–259 (2002)
8. S.A. Anagnostopoulos, C.E. Karamaneas, Use of collision shear walls to minimize seismic separation and to protect adjacent buildings from collapse due to earthquake-induced pounding. Earthquake Engng. Struct. Dyn. **37**, 1371–1388 (2008)
9. V.A. Matsagar, R.S. Jangid, Viscoelastic damper connected to adjacent structures involving seismic isolation. J. Civil Eng. Manag. **11**(4), 309–322 (2005)
10. K.-S. Park, S.-Y. Ok, Optimal design of actively controlled adjacent structures for balancing the mutually conflicting objectives in design preference aspects. Eng. Struct. **45**, 213–222 (2012)
11. P.C. Polycarpou, P. Komodromos, A.C. Polycarpou, "A nonlinear impact model for simulating the use of rubber shock absorbers for mitigating the effects of structural pounding during earthquakes", Earthquake Engng Struct. Dyn. **42**, 81–100 (2013)
12. G.P. Cimellaro, D. Lopez-Garcia, Algorithm for design of controlled motion of adjacent structures. Struct. Control Health Monit. **18**, 140–148 (2011)
13. O. Cundumi, L.E. Suarez, Numerical investigation of a variable damping semiactive device for the mitigation of the seismic response of adjacent structures. Computer-Aided Civil and Infrastructure Engineering **23**, 291–308 (2008)
14. K. Bigdeli, W. Hare, Tesfamariam S. Configuration optimization of dampers for adjacent buildings under seismic excitations. Eng Optim **44**, 1491–1509 (2012)
15. T. Trombetti, S. Silvestri, Novel schemes for inserting seismic dampers in shear-type systems based upon the mass proportional component of the Rayleigh damping matrix. J. Sound Vib. **302**, 486–526 (2007)
16. S.M. Nigdeli, G. Bekdas, Optimum tuned mass damper approaches for adjacent structures. Earthquakes and Structures **7**(6), 1071–1091 (2014)

17. G. Bekdaş, S.M. Nigdeli, Preventing the Pounding of Adjacent Buildings with Harmony Search Optimized Tuned Mass Damper. 3rd European Conference of Civil Engineering, 2–4 December 2012, (Paris, France, 2012)
18. G. Bekdas, S.M. Nigdeli, Optimization of tuned mass damper for adjacent buildings with equal properties. 11th International Conference of Numerical Analysis and Applied Mathematics, 21–27 September 2013, (Rhodes, Greece, 2013)
19. G. Bekdas, S.M. Nigdeli, Optimum tuning of mass damper for adjacent structures subjected near fault ground motions. 2nd International Conference on Civil and Environmental Engineering (ICOCEE 2017), 8–10 May 2017, (Cappadocia, Turkey, 2017)
20. G. Bekdaş, S.M. Nigdeli, Vibration absorber system for adjacent structures subjected to near-fault excitations. 15th international conference of numerical analysis and applied mathematics, 25–30 September 2017, (Thessaloniki, Greece, 2017)
21. S.M. Nigdeli, G. Bekdaş, The effect of mass ratio on optimum design of a vibration absorber system for adjacent structures. 12th International Congress on Mechanics (HSTAM2019), 22–25 September 2019, (Thessaloniki Greece, 2019)
22. Z.W. Geem, J.H. Kim, G.V. Loganathan, A new heuristic optimization algorithm: harmony search. *simulation*, *76*(2), 60–68 (2001)
23. R. Rao, Jaya: a simple and new optimization algorithm for solving constrained and unconstrained optimization problems. Int. J. Ind. Eng. Comput. **7**(1), 19–34 (2016)

The Effect of Initial Values on Metaheuristic-Based Optimum Design of Tuned Mass Dampers

Aylin Ece Kayabekir, Gebrail Bekdaş, and Sinan Melih Nigdeli

Abstract In metaheuristic-based optimum design methodologies, algorithm parameters may play an important role in three factors. These factors are finding optimum solutions with the lowest objective function, reaching to final results in a short time and robustness of the method for different runs of the same methodology. Although parameter-free techniques or parameter-free algorithms exist, the population number (the number of evaluated design variables) and the choice of initial values are the two problems of all optimization methodologies. Due to these reasons, an alternative approach for the development of initial candidate values of design variables is proposed for optimum design of tuned mass dampers for structures under earthquake excitations. In the proposed strategy, the initial values are assigned with constant values by dividing the range of design variables according to population number. This strategy is compared with the process using randomly defined variables within the selected range. The results of the two approaches were compared according to standard deviation, mean value and minimum value of different runs. The alternative approach is effective in finding robust optimum solutions and fast computational effort.

Keywords Tuned mass dampers · Structural control · Optimization · Initial value · Metaheuristics

1 Introduction

Metaheuristic algorithms are the methods used in finding design variables of a problem with an objective function, which are minimized or maximized. By the development of new algorithms, it is possible to solve all engineering problems

A. E. Kayabekir (✉) · G. Bekdaş · S. M. Nigdeli
Department of Civil Engineering, Istanbul University-Cerrahpaşa, 34320 Avcılar, Istanbul, Turkey
e-mail: ecekayabekir@gmail.com

S. M. Nigdeli
e-mail: melihnig@istanbul.edu.tr

S. M. Nigdeli et al. (eds.), *Proceedings of 6th International Conference on Harmony Search, Soft Computing and Applications*, Advances in Intelligent Systems and Computing 1275, https://doi.org/10.1007/978-981-15-8603-3_9

with a nonlinear formulation on finding optimum values for constrained problems. In addition to finding the best optimum results, robustness and computational time needed to find the final results are the most important factor since new algorithms are perfectly successful on a high variety of optimization problems.

Metaheuristic methods use the inspiration of a metaphor, which is a process related to making that process excellent. Up to now, there are a lot of metaheuristics and all of them include a metaphor, but the consideration of different strategies on metaheuristic-based methods are more important than a metaphor.

The metaphor of early metaheuristics developed in the method-centric period (1980–2000) [1] includes human memory for Tabu Search [2], a process of material production in Simulated Annealing [3], the behavior of swarms in Particle Swarm Optimization [4], the behavior of ants in Ant Colony Optimization [5] and classical features of Genetic Algorithm [6–7] such as mutation, crossover and selection in differential evaluation [8]. After that period, specific processes are used in the inspiration to solve problems robustly and faster. After 2000, one of the best-known algorithms is the music-inspired Harmony Search [9]. The other examples include Big Bang–Big Crunch [10], Artificial Bee Colony Algorithm [11], Cuckoo Search Algorithm [12], Firefly Algorithm [13], Bat Algorithm [14], Teaching Learning-Based Optimization [15], Flower Pollination Algorithm [16], Grey Wolf Optimizer [17] and Jaya Algorithm [18].

In the present study, an alternative approach for the selection of initial values was proposed. The investigation was done on the tuned mass damper optimization problem which has been previously optimized via several metaheuristic methods [19–26].

2 Methodology

2.1 Optimum Design of Tuned Mass Dampers via Jaya Algorithm

Jaya algorithm (JA) [18] developed by Rao is a parameter-free metaheuristic algorithm using a single phase formulated as in Eq. (1).

$$X_{new,i} = X_{old,i} + r_1(X_{best} - |X_{old,i}|) - r_2(X_{worst} - |X_{old,i}|) \tag{1}$$

As the given Sanskrit name "Jaya" means victory, the method converges to the best solution by using the best (X_{best}) and the worst (X_{worst}) existing set of design variables with the best objective function. In Eq. (1), two randomly selected numbers (r_1 and r_2) between 0 and 1 are used to converge the best solution while diverge to the worst one. As the final process is done according to Eq. (1), a new set of solutions ($X_{new,i}$) is generated with respect to the existing set ($X_{old,i}$) of ith individual of the population. A set of design variables contains two design variables. These design variables are the period (T_d) and damping ratio (ξ_d) of the TMD, which are formulated as in Eqs. (2) and (3), respectively.

$$T_d = 2\pi\sqrt{\frac{m_d}{k_d}} \tag{2}$$

$$\xi_d = \frac{c_d}{2m_d\sqrt{\frac{k_d}{m_d}}} \tag{3}$$

The objective function of the problem is given as in Eq. (4).

$$f(X) = \max|x_N \ of\ set\ of\ earthquakes| \tag{4}$$

The objective function is related to the minimization of the top story displacement (X_N) of an N-story structure. In the optimization process, the analysis is done for several earthquake excitations, since the exact behavior and data of the ground acceleration of the earthquake cannot be known.

The displacement of the structure is found by solving equations of motion of a structure given as in Eq. (5) in matrix form.

$$\mathbf{M}\ddot{x}(t) + \mathbf{C}\dot{x}(t) + \mathbf{K}x(t) = -\mathbf{M}\{\mathbf{1}\}\ddot{x}_g(t) \tag{5}$$

The coupled equations of motions are solved by using numerical integrations with respect to the 4th order Runge–Kutta method withthe help of Simulink [27]. The earthquake ground acceleration ($\ddot{x}_g$ (t)) are taken as the near-fault ground motion set given FEMA: Quantification of Building Seismic Performance Factors [28].

M, C and K are the mass, damping and stiffness matrices given as in Eqs. (6)–(8), respectively. m_i, c_i and k_i represent the mass, damping coefficient and stiffness of the ith story of the structure. m_d, c_d and k_d are the properties of TMD such as mass, damping coefficient and stiffness, respectively.

M = diag[$m_1 m_2 \ldots m_N m_d$](6)

$$\mathbf{C} = \begin{bmatrix} (c_1+c_2) & -c_2 & & & & \\ -c_2 & (c_2+c_3) & -c_3 & & & \\ & \cdot & \cdot & & & \\ & \cdot & \cdot & \cdot & & \\ & & \cdot & \cdot & \cdot & \\ & & & -c_N & (c_N+c_d) & -c_d \\ & & & & -c_d & c_d \end{bmatrix} \tag{7}$$

$$K = \begin{bmatrix} (k_1 + k_2) & -k_2 & & & & \\ -k_2 & (k_2 + k_3) & -k_3 & & & \\ & \cdot & \cdot & & & \\ & & \cdot & \cdot & \cdot & \\ & & & \cdot & \cdot & \cdot \\ & & & -k_N & (k_N + k_d) & -k_d \\ & & & & -k_d & k_d \end{bmatrix} \quad (8)$$

The displacement vector (x(t)) contains story displacement (x_1, x_2,…x_N) and TMD displacement (x_d) as given in Eq. (9).

The optimization problem is constrained by the limitation of the stroke capacity of TMD. A normalized limit (g_1) was used in the optimization process as seen in Eq. (10). st_max is chosen according to the preference of the designer.

$$x(t) = \begin{Bmatrix} x_1(t) \\ x_2(t) \\ \vdots \\ x_N(t) \\ x_d(t) \end{Bmatrix} \quad (9)$$

$$g_1 = \frac{\max(|x_d - x_N|)_{with\,TMD}}{\max(|x_N|)_{without\,TMD}} < st\_\max \quad (10)$$

2.2 *The Alternative Approach for Initial Values*

The definition of initial values before the iteration process may play an important role in the optimum design. Generally, the initial values are randomly defined within the selected range of design variables with (X_{min}) and maximum (X_{max}). In that case, the initial set of design variables (X_{in}) is defined as in Eq. (11). rand is a random number between 0 and 1.

$$X_{in} = \frac{(X_{\max} - X_{\min})}{2} + (rand - 0.5)(X_{\max} - X_{\min}) \quad (11)$$

In the proposed alternative approach, initial values of the ith initial value ($X_{in,i}$) are defined according to Eq. (12) in order to define constant values by dividing the range to equal parts according to the population number (pn) of the optimization process.

$$X_{in,i} = X_{\min} + \frac{(X_{\max} - X_{\min})}{(pn - 1)}(i - 1) \text{ for } i = 1 - pn \quad (12)$$

3 The Numerical Example

The investigation was done by using a ten-story structure for a TMD application. For each story, the properties such as mass, stiffness and damping coefficients are 360t, 650MN/m and 6.2MNs/m, respectively [29]. The mass of TMD was taken as 180t.

Since pn may be affected by the responses, the investigation was done for different pn values. Also, statistical investigation was done by applying the same optimization process for 10 runs. The st_max values are tested for two values such as 1 and 2.

All of the optimization process is repeated for 10 runs. In Table 1, the optimum results for cases of st_max = 1 are given. The first three cases are for the random selection of initial values, and cases 4–6 are for the proposed alternative approach.

Figure 1 shows the comparison of Case 1 (classical approach) and Case 4 (proposed approach). Looking at the figure, it can be seen that the proposed approach starts with a lower initial value and converges to the optimum result more quickly than Case 1. It must be also noted that the proposed approach has a better standard deviation value (see Table 1).

In the comparison made in Fig. 2 for Cases 2 and 5 where the number of population is ten, it is understood that the rate of convergence to the optimum result of Case 5 is better in the classical approach, although the initial value of the proposed method is higher than Case 2.

Referring to Fig. 3, when the population number is 20, the proposed method has a relatively large initial value like Case 2. When examined the detailed convergence graphs, although the classical method (Case 3) started with a closer value to the optimum result, it could not achieve the best result as the proposed approach even in 100 iterations. However, the proposed method (Case 6) finds the optimum result (in the 95th iteration) more quickly than the classical method. When Case 2,5 and Case 3,6 are evaluated in terms of standard deviation, it can be seen from Table 1 that the proposed approach is slightly better than the classical approach.

In Table 2, the optimum results for cases of st_max = 2 are given. The Cases 7–9 are for random selection of initial values, and Cases 10–12 are for the proposed alternative approach.

If Cases 7 and 10 are evaluated from Fig. 4, optimum results and convergence to

Table 1 The optimum results for st_max = 1

Case No.	pn	T_d (s)	ξ_d	f(X) (m)	Standard deviation
1	5	0.902289927526370	0.282137049670333	0.320235107710295	0.0106
2	10	0.902284959758051	0.282135529748869	0.320235373167212	1.8019e-04
3	20	0.902162008917412	0.282066031075431	0.320237781816171	2.3848e-05
4	5	0.902263262774798	0.282115678263766	0.320234592199187	3.1899e-04
5	10	0.902222505799621	0.282088439287840	0.320234776289344	2.7066e-05
6	20	0.902065597761129	0.281993427106426	0.320237130933011	9.8007e-06

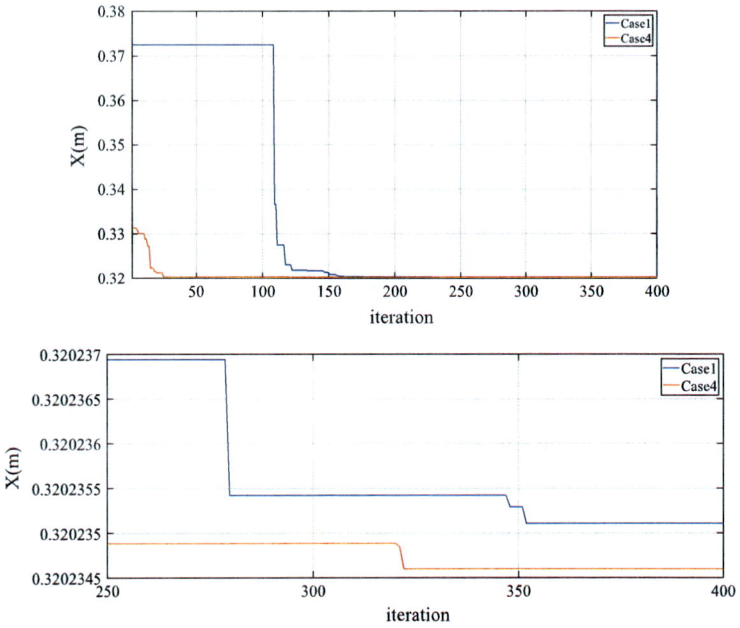

Fig. 1 Comparison of Cases 1 and 4 (upper), detailed graph for convergence (bottom)

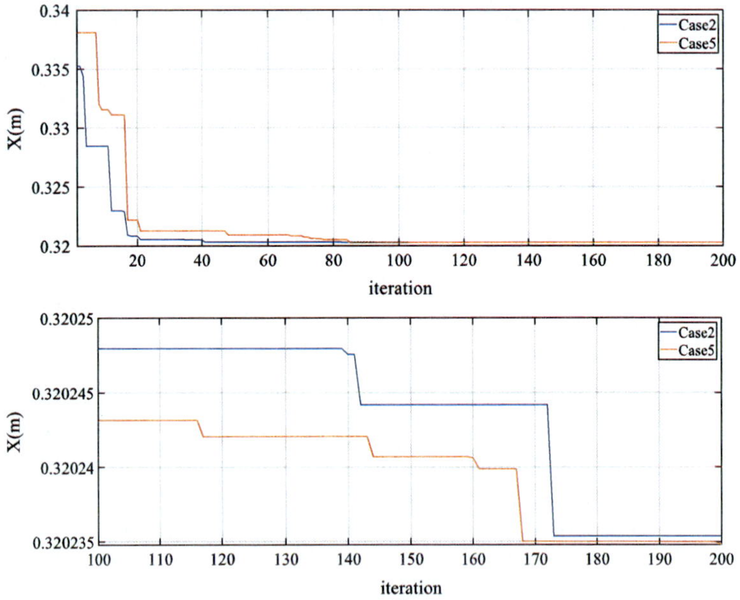

Fig. 2 Comparison of Cases 2 and 5 (upper), detailed graph for convergence (bottom)

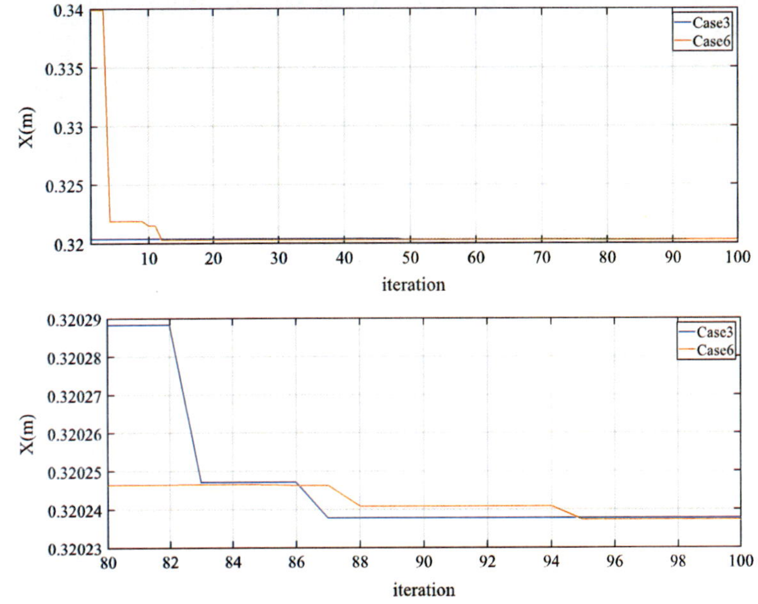

Fig. 3 Comparison of Cases 3 and 6 (upper), detailed graph for convergence (bottom)

Table 2 The optimum results for st_max = 2

Case No.	pn	T_d (s)	ξ_d	f(X) (m)	Standard deviation
7	5	0.941807619345672	0.056338387496699	0.282033459957915	0.0049
8	10	0.941807619345672	0.056338387496703	0.282033459957916	1.2717e-14
9	20	0.941807618598387	0.056338393869750	0.282033461189836	2.5498e-08
10	5	0.941807619345672	0.056338387496699	0.282033459957915	5.8514e-17
11	10	0.941807619345671	0.056338387496700	0.282033459957915	1.0099e-14
12	20	0.941807624254981	0.056338403805017	0.282033464579141	3.2473e-08

the optimum value for both classical and proposed approaches appear approximately with the same characteristics. In contrast, standard deviation value of the proposed approach is much better than that of the classical one.

In Cases 8–11 (Fig. 5) and Cases 9–12 (Fig. 6), standard deviation values are close to each other. Looking at the graphs, it is understood that both methods achieve optimum results. When the approach speeds are compared, it is seen that the methods reached the optimum result in the nearly same number of iterations or the proposed method achieved a slightly faster result.

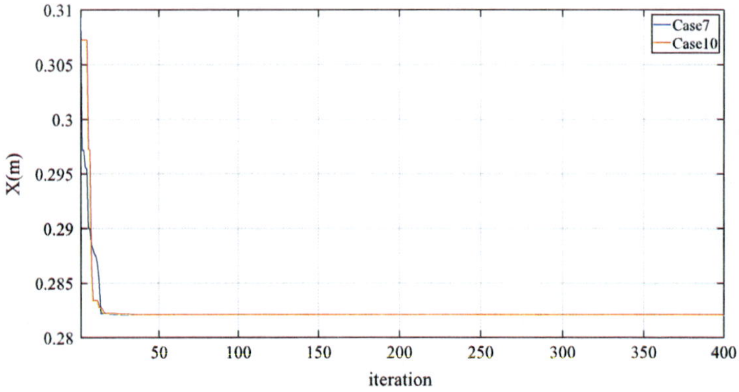

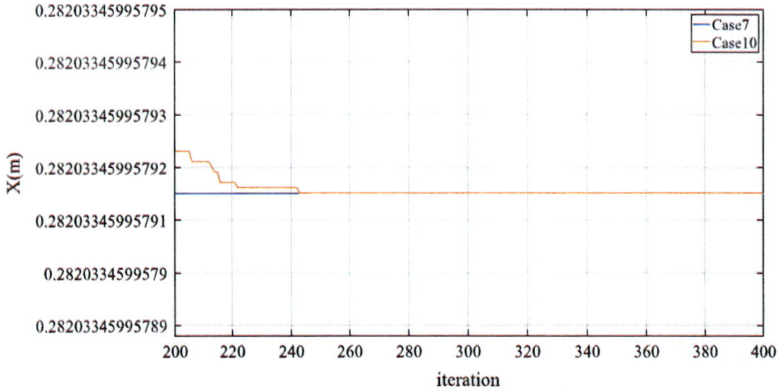

Fig. 4 Comparison of Cases 7 and 10 (upper), detailed graph for convergence (bottom)

4 Conclusions

In the optimization methods based on metaheuristic algorithms, finding the best result in accordance with the objective function, reaching the optimum result in a short time and giving stable results in the repeated analysis with the same method are the three important factors. In this study, the effect of population number and initial candidate values on the optimization process are examined. Therefore, as an alternative to the classical method where the initial values are randomly selected within the selected range, a method in which the initial values are assigned by constant values by dividing the range of design variables by the number of populations is proposed. This proposed approach has been run for finding optimum TMD parameters using different population numbers.

The results obtained from the analysis can be summarized as follows.

- Both approaches are successful in finding optimum results.
- In a general sense, the proposed approach is achieved more quickly to the optimum result than the classical one.

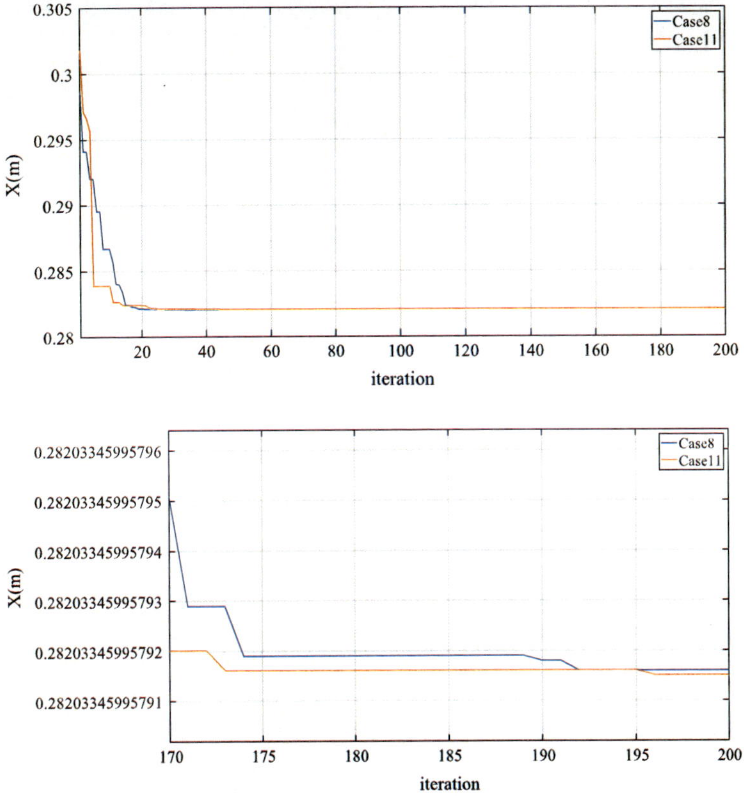

Fig. 5 Comparison of Cases 8 and 11 (upper), detailed graph for convergence (bottom)

- For all population numbers 10 and 20, although standard deviation values for the proposed approach are smaller than the classical one, there are no significant differences between the methods. But, for the analysis of population number 5, a remarkable difference is seen between the proposed and classical approaches.

Considering these results, it can be concluded that the proposed method seems relatively better. It is thought that it will be beneficial to test the method in other optimization problems especially in problems with a high number of design variables.

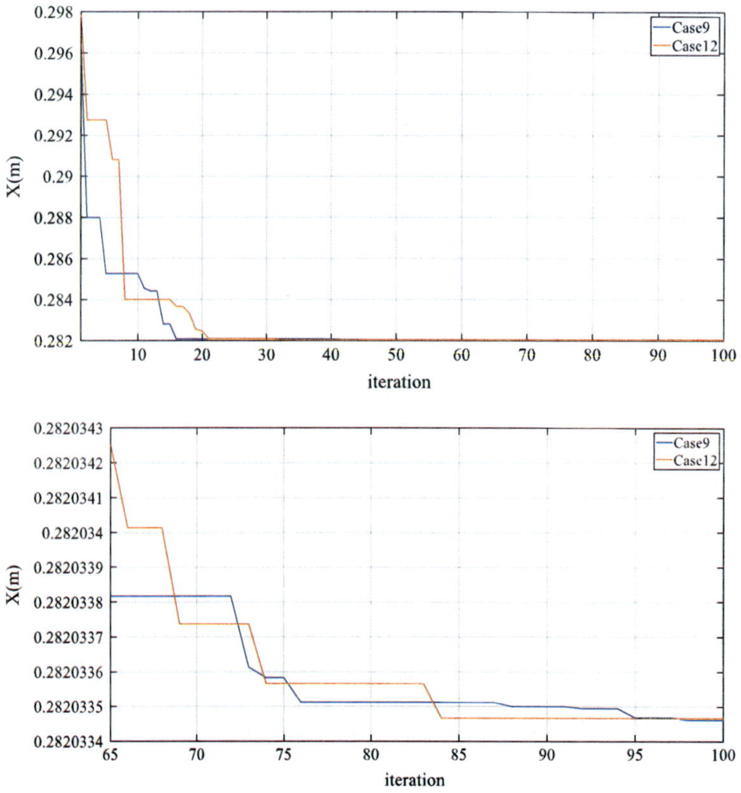

Fig. 6 Comparison of Cases 9 and 12 (upper), detailed graph for convergence (bottom)

References

1. K. Sörensen, M. Sevaux, F. Glover, A history of metaheuristics. *Handbook of heuristics* (2018), 1–18
2. F. Glover, Future paths for integer programming and links to artificial intelligence. Comput. Oper. Res. **13**(5), 533–549 (1986)
3. S. Kirkpatrick, C.D. Gelatt, M.P. Vecchi, Optimization by simulated annealing. Science **220**(4598), 671–680 (1983)
4. J. Kennedy, R.C. Eberhart, Particle swarm optimization. *Proceedings of IEEE International Conference on Neural Networks No. IV*, November 27-December 1, pp. 1942–1948, (Perth Australia, 1995)
5. M. Dorigo, V. Maniezzo, A. Colorni, The ant system: optimization by a colony of cooperating agents. IEEE Transactions on Systems Man and Cybernet B **26**, 29–41 (1996)
6. D.E. Goldberg, M.P. Samtani, Engineering optimization via genetic algorithm. *Proceedings of Ninth Conference on Electronic Computation* (ASCE, New York, NY, 1986), pp. 471–482
7. J.H. Holland, *Adaptation in Natural and Artificial Systems* (University of Michigan Press, Ann Arbor, Michigan, 1975)
8. R. Storn, K. Price, Differential evolution–a simple and efficient heuristic for global optimization over continuous spaces. J. Global Optim. **11**(4), 341–359 (1997)

9. Z.W. Geem, J.H. Kim, G.V. Loganathan, A new heuristic optimization algorithm: harmony search. Simulation **76**, 60–68 (2001)
10. O.K. Erol, I. Eksin, A new optimization method: big bang–big crunch. Adv. Eng. Softw. **37**(2), 106–111 (2006)
11. D. Karaboga, B. Basturk, A powerful and efficient algorithm for numerical function optimization: artificial bee colony (ABC) algorithm. J. Global Optim. **39**(3), 459–471 (2007)
12. X.S. Yang, S. Deb, Engineering optimisation by cuckoo search. Int. J. Mathemat. Model. Numerical Optim. **1**(4), 330–343 (2010)
13. X.S. Yang, Firefly algorithm, stochastic test functions and design optimisation. Int. J. Bio-Inspired Comput. **2**(2), 78–84 (2010)
14. X.S. Yang, A new metaheuristic bat-inspired algorithm. In *Nature inspired cooperative strategies for optimization (NICSO 2010)* (Springer, Berlin, Heidelberg, 2010), pp. 65–74
15. R.V. Rao, V.J. Savsani, D.P. Vakharia, Teaching–learning-based optimization: a novel method for constrained mechanical design optimization problems. Comput. Aided Des. **43**(3), 303–315 (2011)
16. X.S. Yang, Flower pollination algorithm for global optimization. In *International conference on unconventional computing and natural computation* (Springer, Berlin, Heidelberg, 2012), pp. 240–249
17. S. Mirjalili, S.M. Mirjalili, A. Lewis, Grey wolf optimizer. Adv. Eng. Softw. **69**, 46–61 (2014)
18. R. Rao, Jaya: A simple and new optimization algorithm for solving constrained and unconstrained optimization problems. Int. J. Ind. Eng. Comput. **7**(1), 19–34 (2016)
19. A.Y.T. Leung, H. Zhang, Particle swarm optimization of tuned mass dampers. Eng. Struct. **31**(3), 715–728 (2009)
20. G. Bekdaş, S.M. Nigdeli, Estimating optimum parameters of tuned mass dampers using harmony search. Eng. Struct. **33**, 2716–2723 (2011)
21. S. Pourzeynali, S. Salimi, H.E. Kalesar, Robust multi-objective optimization design of tmd control device to reduce tall building responses against earthquake excitations using genetic algorithms. Scientia Iranica **20**(2), 207–221 (2013)
22. Y. Arfiadi, Reducing response of structures by using optimum composite tuned mass dampers. Proce. Eng. **161**, 67–72 (2016)
23. A. Farshidianfar, S. Soheili, ABC optimization of tmd parameters for tall buildings with soil structure interaction. Int. and Multiscale Mech. **6**(4), 339–356 (2013)
24. G. Bekdaş, S.M. Nigdeli, X.S. Yang, A novel bat algorithm based optimum tuning of mass dampers for improving the seismic safety of structures. Eng. Struct. **159**, 89–98 (2018)
25. G. Bekdaş, S.M. Nigdeli, X.-S. Yang, Metaheuristic based optimization for tuned mass dampers using frequency domain responses, in *Harmony Search Algorithm. Advances in Intelligent Systems and Computing*, vol 514, ed. J Del Ser, (Springer, 2017), pp. 271–279
26. S.M. Nigdeli, G. Bekdaş, E. Ayyıldız, Optimization of parameters of tuned mass dampers by using hybrid metaheuristic algorithms. *3rd International Conference on Engineering Technology and Innovation (ICETI),* 17–21 April, (Belgrade, Serbia, 2019)
27. T. MathWorks, *Matlab R2018a* (Natick, MA, 2018)
28. FEMA P-695, Quantification of Building Seismic Performance Factors, Federal Emergency Management Agency, (Washington DC, 2009)
29. M.P. Singh, S. Singh, L.M. Moreschi, Tuned mass dampers for response control of torsional buildings. Earthq. Eng. Struct. D. **31**, 749–769 (2002)

The Comparison of Classical and Artificial Neural Network-Based Formulations for Tuned Mass Damper Optimization

Melda Yücel, Sinan Melih Nigdeli, and Gebrail Bekdaş

Abstract Tuned mass dampers (TMDs) were used to damp vibration of mechanical systems. TMDs were also used on structures to reduce the effects of strong winds and earthquakes. For the efficiency of TMD, the optimization of TMD parameters is needed. Several classical formulations were proposed, but metaheuristic methods are generally used to find the best result. Also, the metaheuristic-based optimum results are used in machine learning of artificial intelligence models like artificial neural networks (ANN). These ANN models are also used in the development of tuning equation via curve fitting. The classical and ANN-based formulations were found according to frequency-domain responses. In the present study, the classical and ANN-based formulations were compared on time-history responses of seismic structures. In comparison, near-fault ground motion records including directivity pulses are used. The ANN-based methods have advantages by providing smaller stroke requirement and damping for TMDs.

Keywords Artificial neural networks · Tuned mass dampers · Structural control · Optimization · Near-fault · Metaheuristic

1 Introduction

In order to reduce structural vibrations, several control methods have been proposed. These methods may be passive, active, semi-active or hybrid control methods. All systems of different control strategies need to be optimized to find a perfect tuning

M. Yücel (✉) · S. M. Nigdeli · G. Bekdaş
Department of Civil Engineering, Istanbul University-Cerrahpaşa, 34320 Avcılar, Istanbul, Turkey
e-mail: melda.yucel@yahoo.com.tr

S. M. Nigdeli
e-mail: melihnig@istanbul.edu.tr

G. Bekdaş
e-mail: bekdas@istanbul.edu.tr

S. M. Nigdeli et al. (eds.), *Proceedings of 6th International Conference on Harmony Search, Soft Computing and Applications*, Advances in Intelligent Systems and Computing 1275, https://doi.org/10.1007/978-981-15-8603-3_10

of parameters according to controlled structure subjected to vibration sources like earthquakes, winds and traffics.

Tuned mass dampers are passive vibration control devices which are used in general mechanical systems. For this purpose, several simplified equations are proposed [1–5]. These equations were derived according to mathematical equations, curve fitting of numerical iteration and optimum results as detailed in Sect. 2.1.

In recent years, the most chosen methods for optimum tuning of mass dampers is the usage of metaheuristic methods. As examples, the usage of harmony search (HS) [6, 7], particle swarm optimization [8], genetic algorithm [9–11], artificial bee colony optimization [12], simulated annealing [13], cuckoo search [14], gravitational algorithm [15], bat algorithm [16] and flower pollination (FPA) [17] were proposed to optimize different types of TMDs for different objectives about responses of structures.

The process of numerical optimization method may need a high amount of computation time. For that reason, Yucel et al. [5] used the optimum TMD parameters on machine learning of an artificial neural network (ANN) model, and several equations were developed for the easy use of engineers.

The classical and ANN-based formulations were found according to the frequency-domain response of structures. The exact effect of TMDs can be only seen on time-history responses. For that reason, several formulations are investigated on time-history responses of seismic structures by using near-fault ground motions. As seen in Fig. 1, near-fault ground motions have two significant impulsive motions.

Directivity pulses, which are critically effective on structures, occur in the direction perpendicular to a fault [18]. These pulses have a long period and high peak ground velocity. In FEMA P-695, Quantification of Building Seismic Performance Factors [19], several near-fault ground motion records with pulses are grouped, and these records given in Table 1 were used in the present study.

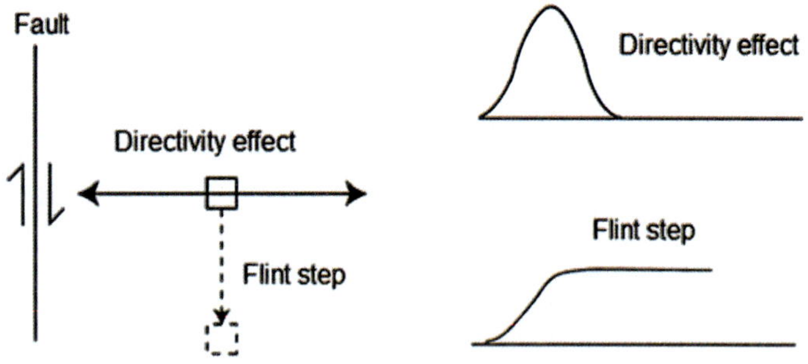

Fig. 1 Impulsive motions

Table 1 Earthquake set for near-field excitations with pulses [19]

Earthquake no.	Earthquake name	Recording station	Year	Magnitude
1	Imperial Valley-06	El Centro Array #6	1979	6.5
2	Imperial Valley-06	El Centro Array #7	1979	6.5
3	Irpinia, Italy-01	Sturno	1980	6.9
4	Superstition Hills-02	Parachute Test Site	1987	6.5
5	Loma Prieta	Saratoga—Aloha	1989	6.9
6	Erzican, Turkey	Erzican	1992	6.7
7	Cape Mendocino	Petrolia	1992	7.0
8	Landers	Lucerne	1992	7.3
9	Northridge-01	01 Rinaldi Receiving Sta	1994	6.7
10	Northridge-01	01 Sylmar—Olive View	1994	6.7
11	Kocaeli, Turkey	Izmit	1999	7.5
12	Chi-Chi, Taiwan	TCU065	1999	7.6
13	Chi-Chi, Taiwan	TCU102	1999	7.6
14	Duzce, Turkey	Duzce	1999	7.1

2 The Design of Tuned Mass Dampers

2.1 The Equations of Motion

In Fig. 2, a TMD attached to a single degree of freedom (SDOF) structure is shown. $\ddot{x}_g$ is the recorded acceleration of ground motions. The response of the structure is the displacement, x. The structural parameters taken as the design constants are the mass (m), stiffness (k) and the damping coefficient (c), which are found according to the inherent damping (ξ) of the structure as given in Eq. (1).

$$c = 2m\xi\sqrt{\frac{k}{m}} \tag{1}$$

The parameters of TMD are mass (m_d), stiffness (k_d) and damping coefficient (c_d) of the structure. Generally, the mass of TMD is optimum at the highest allowed value according to the axial loading capacity of the structure. In tuning formulations, mass ratio (μ) is defined as given in Eq. (2).

$$\mu = \frac{m_d}{m} \tag{2}$$

In matrix form, the equation of motion of the structure coupled with a TMD is given as in Eq. (3). The mass, damping and stiffness matrices are shown as M, C and K, respectively. These matrices and the displacement vector (x(t)) are shown in Eqs. (4)–(7).

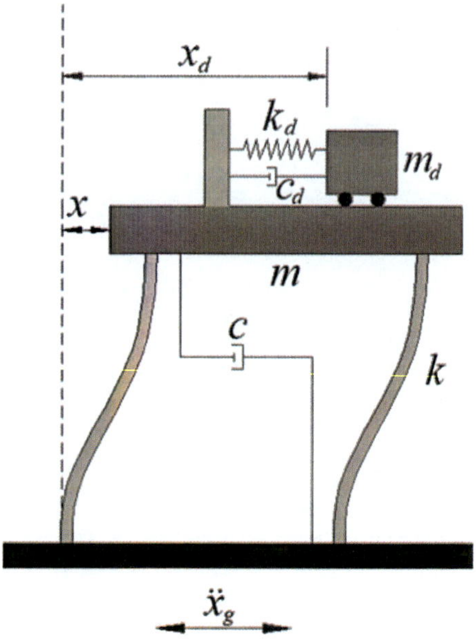

Fig. 2 SDOF structure with TMD

$$\mathbf{M}\ddot{x}(t) + \mathbf{C}\dot{x}(t) + \mathbf{K}x(t) = -\mathbf{M}\{\mathbf{1}\}\ddot{x}_g(t) \tag{3}$$

$$\mathrm{M} = \begin{bmatrix} m & 0 \\ 0 & m_d \end{bmatrix} \tag{4}$$

$$\mathrm{C} = \begin{bmatrix} c + c_d & -c_d \\ -c_d & c_d \end{bmatrix} \tag{5}$$

$$\mathrm{K} = \begin{bmatrix} k + k_d & -k_d \\ -k_d & k_d \end{bmatrix} \tag{6}$$

$$x(t) = \begin{Bmatrix} x \\ x_d \end{Bmatrix} \tag{7}$$

The dots on x(t) represent a derivative of x(t) with respect to time. The solving of x needs long calculation time, and the results are specific for the excited ground acceleration. Due to that, the response can be investigated in the frequency domain (ω) by calculating the amplitude (f) of the transfer function (TF(ω)). TF(ω) is the ratio of Laplace transformations of acceleration of the system and ground. TF(ω) and f are formulated as in Eqs. (8) and (9), respectively.

$$TF(\omega) = \begin{bmatrix} TF \\ TF_d \end{bmatrix} = \left[-\mathrm{M}\omega^2 + \mathrm{C}\omega j + \mathrm{K}\right]^{-1}\mathrm{M}\omega^2\{\mathbf{1}\} \tag{8}$$

$$f = 20Log_{10}|\max(TF)| \tag{9}$$

2.2 The Tuning Equations for TMDs

In Table 2, the compared tuning equations with ANN-based equations are given. The equations are for optimum frequency ratio (f_{opt}) and optimum damping ratio TMD (ξ_{dopt}). f_{opt} and ξ_{dopt} are formulated as in Eqs. (10) and (11), respectively.

$$f_{opt} = \frac{\omega_d}{\omega_s} \tag{10}$$

$$\xi_{dopt} = \frac{c_d}{2m_d\omega_d} \tag{11}$$

ω_{dopt} is the frequency of the TMD as given in Eq. (12), and ω_s is the frequency of a single degree of freedom (SDOF) structure as shown in Eq. (13).

$$\omega_d = \sqrt{\frac{k_d}{m_d}} \tag{12}$$

$$\omega_s = \sqrt{\frac{k}{m}} \tag{13}$$

The equation of Den Hartog was found for harmonic excitations [1]. Warburton derived the formulation for white-noise excitation [2]. Sadek et al. [3] used curve fitting of numerical optimization results to find a formulation including the effect of inherent damping of the structure [3]. Leung and Zhang [4] employed PSO to find the TMD tuning equations.

Table 2 The frequency and damping ratio expressions of the compared methods

Method	f_{opt}	ξ_{dopt}
Den Hartog 1947 [1]	$\frac{1}{1+\mu}$	$\sqrt{\frac{3\mu}{8(1+\mu)}}$
Warburton 1982 [2]	$\frac{\sqrt{1-(\mu/2)}}{1+\mu}$	$\sqrt{\frac{\mu(1-\mu/4)}{4(1+\mu)(1-\mu/2)}}$
Sadek et al 1997 [3]	$\frac{1}{1+\mu}\left[1-\xi\sqrt{\frac{\mu}{1+\mu}}\right]$	$\frac{\xi}{1+\mu}+\sqrt{\frac{\mu}{1+\mu}}$
Leung and Zhang 2009 [4]	$\frac{\sqrt{1-(\mu/2)}}{1+\mu}$ $+(-4.9453+20.2319\sqrt{\mu}-37.9419\mu)\sqrt{\mu}\xi$ $+(-4.8287+25.0000\sqrt{\mu})\sqrt{\mu}\xi^2$	$\sqrt{\frac{\mu(1-\mu/4)}{4(1+\mu)(1-\mu/2)}}$ $-5.3024\xi^2\mu$

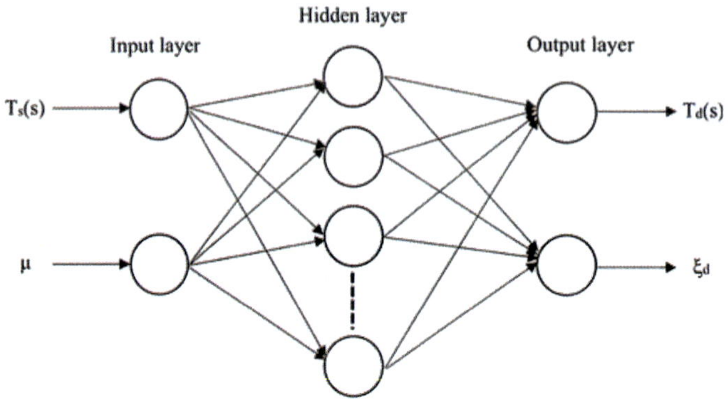

Fig. 3 ANN structure

2.3 *ANN-Based Formulation for TMD Optimization*

Artificial neural network (ANN) is an artificial intelligence modeling tool, and it based on simulating human brain behavior on different systems. Via ANN, it is possible to estimate final results without processing the solving of problems. Yucel et al. [5] developed an ANN model for the optimization of TMDs providing their optimum period (T_d) and damping ratio (ξ_{dopt}). The relationship between T_d and stiffness of TMD (k_d) is given as in Eq. (14).

$$T_d = 2\pi\sqrt{\frac{m_d}{k_d}} \quad (14)$$

The structure of the developed ANN model is shown in Fig. 3. The inputs are the period (T_s) of SDOF structure given as in Eq. (15) and mass ratio (μ).

$$T_s = 2\pi\sqrt{\frac{m}{k}} \quad (15)$$

In machine learning of the ANN model, the FPA algorithm developed by Yang [20] was used. The methodology given in Bekdaş et al. [17] was used to find the optimum TMD values for different cases of SDOF structures. The developed ANN model was used to find the simplified equations given in Table 3.

3 Comparison of the Methods

The investigations were done for four cases of SDOF structure period. All structures were also tested for 3 mass ratio values. The cases are shown in Tables 4 and 5.

Table 3 ANN-based formulations [5]

Method	f_{opt}	ξ_{dopt}
Linear	$f_{opt} = -0.6438\,\mu + 0.9966$	$\xi_{d,opt} = 0.5673\,\mu + 0.1235$
Polynomial	$f_{opt} = -249.91\,\mu^5 + 400.09\,\mu^4$ $-208.03\,\mu^3 + 43.801\,\mu^2$ $-4.1453\,\mu + 1.0675$	$\xi_{d,opt} = -54.673\,\mu^4 + 54.639\,\mu^3$ $-19.274\,\mu^2 + 3.2302\,\mu$ $+0.0237$
Exponential	$f_{opt} = 1.0038\ e^{-0.747\,\mu}$	$\xi_{d,opt} = 1.1258\ e^{2.8573\,\mu}$

Table 4 The numerical cases for periods

Case	T_s(s)
1	0.5
2	1
3	2
4	4

Table 5 The numerical cases for mass ratio

Case	μ
a	0.05
b	0.10
c	0.20

The results of different methods are presented in Tables 6–9 including the sub-cases of mass ratio. The results include the optimum TMD values, maximum displacement (x) and acceleration (a) of structure. Also, stroke values are presented for the TMD. The presented stroke value is a normalized value as given in Eq. (16).

$$stoke = \frac{\max(|x_d|) - \max(|x|)}{\max(|x|)_{structure\ without\ TMD}} \tag{16}$$

The results also contain the response of the SDOF structure without TMD. The maximum responses are given for the critical earthquake for the corresponding case where the maximum displacement is found. Also, the mean values of responses under ground motion records are also given.

4 Conclusions

In general, the ANN-based formulations are more effective than all the formulations other than Sadek et al. [3]. When the optimum damping ratios of TMDs are investigated, Sadek et al. [3] has a very big optimum value compared to the ANN-based method. Thus, the optimum TMD designed according to Sadek et al. [3] has

Table 6 The numerical values Case 1

		SDOF without TMD	Den Hartog [1]	Warburton [2]	Sadek et al. [3]	Leung and Zhang [4]	Linear [5]	Polynomial [5]	Exponential [5]
$\mu = 0.05$									
	T_d (s)		0.525	0.532	0.531	0.547	0.518	0.528	0.517
	ξ_d (%)		10.911	10.981	26.584	10.914	15.187	14.351	14.512
Critical	x (m)	0.124	0.116	0.116	0.114	0.116	0.115	0.115	0.115
	a (m/s^2)	19.761	17.391	17.426	17.077	17.492	17.224	17.295	17.233
	stroke		3.076	3.185	1.972	3.457	2.499	2.702	2.544
Mean	x (m)	0.051	0.048	0.048	0.049	0.048	0.048	0.048	0.048
	a (m/s^2)	8.119	7.280	7.284	7.314	7.308	7.268	7.272	7.267
	stroke		2.380	2.428	1.540	2.543	1.999	2.124	2.033
$\mu = 0.10$									
	T_d (s)		0.550	0.564	0.558	0.587	0.536	0.543	0.537
	ξ_d (%)		15.076	15.273	34.697	15.140	18.023	20.315	16.741
Critical	x (m)	0.124	0.120	0.120	0.118	0.119	0.120	0.119	0.120
	a (m/s^2)	19.761	17.169	17.158	16.844	17.133	17.073	17.026	17.109
	stroke		2.806	2.924	1.821	3.130	2.446	2.345	2.542
Mean	x (m)	0.051	0.048	0.048	0.049	0.047	0.049	0.048	0.049
	a (m/s^2)	8.119	6.891	6.896	7.010	6.911	6.886	6.889	6.885
	stroke		2.025	2.094	1.357	2.216	1.800	1.733	1.863
$\mu = 0.20$									
	T_d (s)		0.600	0.632	0.613	0.695	0.576	0.564	0.578
	ξ_d (%)		20.412	20.972	44.991	20.707	23.696	24.842	22.277

(continued)

Table 6 (continued)

		SDOF without TMD	Den Hartog [1]	Warburton [2]	Sadek et al. [3]	Leung and Zhang [4]	Linear [5]	Polynomial [5]	Exponential [5]
Critical	x (m)	0.124	0.124	0.123	0.126	0.120	0.126	0.127	0.126
	a (m/s^2)	19.761	16.234	16.211	16.378	16.212	16.284	16.308	16.273
	stroke		2.636	2.787	1.710	3.054	2.321	2.198	2.402
Mean	x (m)	0.051	0.049	0.047	0.050	0.046	0.050	0.050	0.049
	a (m/s^2)	8.119	6.326	6.317	6.585	6.382	6.369	6.405	6.360
	stroke		1.841	1.945	1.253	2.170	1.627	1.545	1.682

Table 7 The numerical values Case 2

		SDOF without TMD	Den Hartog [1]	Warburton [2]	Sadek et al. [3]	Leung and Zhang [4]	Linear [5]	Polynomial [5]	Exponential [5]
$\mu = 0.05$									
	T_d (s)		1.050	1.063	1.062	1.093	1.037	1.057	1.034
	ξ_d (%)		10.911	10.981	26.584	10.914	15.187	14.351	14.512
Critical	x (m)	0.454	0.404	0.404	0.420	0.403	0.410	0.408	0.409
	a (m/s^2)	17.987	15.466	15.510	15.915	15.604	15.560	15.582	15.532
	stroke		2.878	2.959	1.821	3.149	2.368	2.543	2.414
Mean	x (m)	0.165	0.142	0.142	0.150	0.141	0.145	0.144	0.145
	a (m/s^2)	6.553	5.448	5.457	5.684	5.490	5.512	5.506	5.500
	stroke		2.297	2.326	1.468	2.402	1.940	2.044	1.976
$\mu = 0.10$									
	T_d (s)		1.100	1.129	1.117	1.174	1.073	1.086	1.073
	ξ_d (%)		15.076	15.273	34.697	15.140	18.023	20.315	16.741
Critical	x (m)	0.454	0.393	0.393	0.400	0.392	0.395	0.396	0.394
	a (m/s^2)	17.987	14.630	14.675	14.687	14.744	14.590	14.614	14.589
	stroke		2.524	2.618	1.573	2.770	2.217	2.132	2.299
Mean	x (m)	0.165	0.139	0.138	0.144	0.137	0.141	0.141	0.140
	a (m/s^2)	6.553	5.138	5.142	5.283	5.182	5.146	5.154	5.142
	stroke		1.814	1.855	1.201	1.932	1.622	1.555	1.680
$\mu = 0.20$									
	T_d (s)		1.200	1.265	1.225	1.389	1.152	1.128	1.157
	ξ_d (%)		20.412	20.972	44.991	20.707	23.696	24.842	22.277

(continued)

Table 7 (continued)

		SDOF without TMD	Den Hartog [1]	Warburton [2]	Sadek et al. [3]	Leung and Zhang [4]	Linear [5]	Polynomial [5]	Exponential [5]
Critical	x (m)	0.454	0.382	0.380	0.398	0.377	0.387	0.389	0.385
	a (m/s^2)	17.987	13.520	13.654	13.620	13.899	13.425	13.384	13.430
	stroke		2.054	2.188	1.567	2.502	1.882	1.812	1.948
Mean	x (m)	0.165	0.136	0.134	0.142	0.133	0.138	0.139	0.138
	a (m/s^2)	6.553	4.756	4.786	4.849	4.882	4.731	4.731	4.731
	stroke		1.519	1.587	1.041	1.746	1.354	1.290	1.400

Table 8 The numerical values Case 3

		SDOF without TMD	Den Hartog [1]	Warburton [2]	Sadek et al. [3]	Leung and Zhang [4]	Linear [5]	Polynomial [5]	Exponential [5]
$\mu = 0.05$									
	T_d (s)		2.100	2.127	2.123	2.186	2.074	2.114	2.068
	ξ_d (%)		10.911	10.981	26.584	10.914	15.187	14.351	14.512
Critical	x (m)	0.830	0.745	0.748	0.754	0.753	0.744	0.747	0.743
	a (m/s^2)	8.237	7.316	7.364	7.258	7.458	7.223	7.291	7.218
	stroke		3.291	3.362	1.872	3.524	2.678	2.861	2.737
Mean	x (m)	0.382	0.344	0.344	0.356	0.343	0.348	0.347	0.348
	a (m/s^2)	3.788	3.303	3.311	3.392	3.329	3.325	3.327	3.320
	stroke		2.337	2.374	1.485	2.465	1.975	2.086	2.011
$\mu = 0.10$									
	T_d (s)		2.200	2.257	2.234	2.349	2.145	2.173	2.147
	ξ_d (%)		15.076	15.273	34.697	15.140	18.023	20.315	16.741
Critical	x (m)	0.830	0.665	0.666	0.704	0.667	0.671	0.676	0.667
	a (m/s^2)	8.237	6.442	6.533	6.571	6.677	6.376	6.427	6.369
	stroke		2.608	2.655	1.586	2.776	2.281	2.145	2.385
Mean	x (m)	0.382	0.328	0.327	0.346	0.325	0.333	0.335	0.331
	a (m/s^2)	3.788	3.053	3.072	3.178	3.108	3.065	3.085	3.055
	stroke		1.902	1.950	1.253	2.045	1.692	1.627	1.755
$\mu = 0.20$									
	T_d (s)		2.400	2.530	2.450	2.779	2.305	2.256	2.313
	ξ_d (%)		20.412	20.972	44.991	20.707	23.696	24.842	22.277

(continued)

Table 8 (continued)

		SDOF without TMD	Den Hartog [1]	Warburton [2]	Sadek et al. [3]	Leung and Zhang [4]	Linear [5]	Polynomial [5]	Exponential [5]
Critical	x (m)	0.830	0.613	0.616	0.641	0.624	0.620	0.624	0.618
	a (m/s^2)	8.237	5.700	5.808	5.611	6.016	5.609	5.574	5.621
	stroke		2.301	2.442	1.508	2.685	2.026	1.917	2.094
Mean	x (m)	0.382	0.310	0.307	0.335	0.305	0.318	0.321	0.316
	a (m/s^2)	3.788	2.751	2.798	2.877	2.893	2.741	2.742	2.735
	stroke		1.600	1.670	1.064	1.811	1.424	1.356	1.472

Table 9 The numerical values Case 4

		SDOF without TMD	Den Hartog [1]	Warburton [2]	Sadek et al. [3]	Leung and Zhang [4]	Linear [5]	Polynomial [5]	Exponential [5]
$\mu = 0.05$									
	T_d (s)		4.200	4.254	4.246	4.372	4.148	4.228	4.137
	ξ_d (%)		10.911	10.981	26.584	10.914	15.187	14.351	14.512
Critical	x (m)	1.936	1.585	1.580	1.655	1.569	1.612	1.600	1.610
	a (m/s^2)	4.797	3.707	3.707	3.918	3.707	3.768	3.753	3.760
	stroke		2.928	2.999	1.770	3.165	2.415	2.589	2.461
Mean	x (m)	0.639	0.591	0.591	0.599	0.590	0.593	0.592	0.592
	a (m/s^2)	1.591	1.448	1.450	1.450	1.454	1.444	1.447	1.443
	stroke		2.209	2.226	1.364	2.271	1.861	1.951	1.901
$\mu = 0.10$									
	T_d (s)		4.400	4.514	4.467	4.698	4.291	4.346	4.294
	ξ_d (%)		15.076	15.273	34.697	15.140	18.023	20.315	16.741
Critical	x (m)	1.936	1.517	1.503	1.628	1.483	1.557	1.565	1.547
	a (m/s^2)	4.797	3.399	3.403	3.667	3.412	3.462	3.496	3.437
	stroke		2.378	2.466	1.451	2.619	2.033	1.950	2.130
Mean	x (m)	0.639	0.571	0.570	0.582	0.570	0.573	0.574	0.573
	a (m/s^2)	1.591	1.372	1.378	1.367	1.388	1.364	1.365	1.365
	stroke		1.710	1.732	1.080	1.781	1.535	1.461	1.593
$\mu = 0.20$									
	T_d (s)		4.800	5.060	4.900	5.558	4.609	4.513	4.627
	ξ_d (%)		20.412	20.972	44.991	20.707	23.696	24.842	22.277

(continued)

Table 9 (continued)

		SDOF without TMD	Den Hartog [1]	Warburton [2]	Sadek et al. [3]	Leung and Zhang [4]	Linear [5]	Polynomial [5]	Exponential [5]
Critical	x (m)	1.936	1.408	1.382	1.597	1.345	1.474	1.502	1.456
	a (m/s^2)	4.797	2.989	3.023	3.366	3.104	3.065	3.097	3.033
	stroke		2.077	2.148	1.185	2.218	1.775	1.662	1.863
Mean	x (m)	0.639	0.542	0.540	0.563	0.537	0.548	0.551	0.546
	a (m/s^2)	1.591	1.264	1.276	1.252	1.302	1.251	1.246	1.253
	stroke		1.329	1.358	0.859	1.423	1.194	1.142	1.236

a high cost. By the increase of the damping, control can be provided with a small stroke value. The ANN-based methods also have small stroke values compared to the others. By the increase of mass ratio, the damping values are increasing, and the values of Sadek et al. [3] (4.5%) may need a high-cost design. The effectiveness of the ANN-based formulations can be best seen in the average results of response under different excitations.

While ANN-based methods [5] and Sadek et al. [3] are the best in the reduction of maximum responses, the classical methods like Den Hartog and Warburton may be more effective for long period structures. For the same cases of the 4 s period SDOF structure, the best method is the PSO-based formulation [4].

In total, the ANN-based method provides comparative solutions with a feasible TMD design. The success of optimum values is both seen for the displacement and the acceleration values. By the increase of the mass ratio, a small advantage of the exponential equations can be seen, while all three types have similar performances.

References

1. J.P. Den Hartog, *Mechanical Vibrations*, 3rd edn. (McGraw-Hill, New York, 1947)
2. G.B. Warburton, Optimum absorber parameters for various combinations of response and excitation parameters. Earthq. Eng. Struct. D. **10**, 381–401 (1982)
3. F. Sadek, B. Mohraz, A.W. Taylor, R.M. Chung, A method of estimating the parameters of tuned mass dampers for seismic applications. Earthq. Eng. Struct. D. **26**, 617–635 (1997)
4. A.Y.T. Leung, H. Zhang, Particle swarm optimization of tuned mass dampers. Eng. Struct. **31**(3), 715–728 (2009)
5. M. Yucel, G. Bekdaş, S.M. Nigdeli, S. Sevgen, Estimation of optimum tuned mass damper parameters via machine learning. J. Build. Eng. **26**, 100847 (2019)
6. G. Bekdaş, S.M. Nigdeli, Estimating optimum parameters of tuned mass dampers using harmony search. Eng. Struct. **33**, 2716–2723 (2011)
7. S.M. Nigdeli, G. Bekdaş, Optimum tuned mass damper design in frequency domain for structures. KSCE J. Civil Eng. **21**(3), 912–922 (2017)
8. M. Khatibinia, H. Gholami, S.F. Labbafi, Multi–objective optimization of tuned mass dampers considering soil–structure interaction. Iran Uni. Sci. Technol. **6**(4), 595–610 (2016)
9. S. Pourzeynali, S. Salimi, H.E. Kalesar, Robust multi-objective optimization design of tmd control device to reduce tall building responses against earthquake excitations using genetic algorithms. Sci Iranica **20**(2), 207–221 (2013)
10. M. Mohebbi, K. Shakeri, Y. Ghanbarpour, H. Majzoub, Designing optimal multiple tuned mass dampers using genetic algorithms (gas) for mitigating the seismic response of structures. J. Vib. Control **19**(4), 605–625 (2013)
11. Y. Arfiadi, Reducing response of structures by using optimum composite tuned mass dampers. Procedia Eng **161**, 67–72 (2016)
12. A. Farshidianfar, S. Soheili, ABC optimization of tmd parameters for tall buildings with soil structure interaction. Int Multiscale Mechan **6**(4), 339–356 (2013)
13. Y. Yang, C. Li, Performance of tuned tandem mass dampers for structures under the ground acceleration. Struct Cont Health Monitor **24**(10), e1974 (2016)
14. S. Etedali, N. Mollayi, Cuckoo search-based least squares support vector machine models for optimum tuning of tuned mass dampers. Int. J. Struct. Stab. Dyn. **18**(02), 1850028 (2018)
15. M. Khatibinia, H. Gholami, R. Kamgar, Optimal design of tuned mass dampers subjected to continuous stationary critical excitation. Int. J. Dyn Control **6**(3), 1094–1104 (2018)

16. G. Bekdaş, S.M. Nigdeli, X.S. Yang, A novel bat algorithm based optimum tuning of mass dampers for improving the seismic safety of structures. Eng. Struct. **159**, 89–98 (2018)
17. G. Bekdaş, S.M. Nigdeli, X.-S. Yang, Metaheuristic based optimization for tuned mass dampers using frequency domain responses, *in* Harmony Search Algorithm. Advances in Intelligent Systems and Computing, vol 514, ed. J. Del Ser, (Springer, 2017), pp. 271–279
18. J.P. Steward, S. Chiou, J.D. Bray, R.W. Graves, P.G. Somerville, N.A. Abrahamson, *Ground Motion Evalution Procedures for Performance-Based Design*. Pacific Earthquake Engineering Research Center. Report 2001/09; September (2001)
19. FEMA P-695, Quantification of Building Seismic Performance Factors, Federal Emergency Management Agency, (Washington DC, 2009)
20. X.-S. Yang, Flower pollination algorithm for global optimization. In unconventional computation and natural computation, (2012), pp. 240–249

Passive Control of Frame Structures by Optimum Tuned Mass Dampers

Apaer Mubuli, Sinan Melih Nigdeli, and Gebrail Bekdaş

Abstract Tuned Mass Dampers (TMDs) were used in control of mechanical systems including structures. The civil structure may expose to high vibrations due to earthquakes. In the present study, TMDs positioned on the top of the structure were optimized via a metaheuristic method called Jaya Algorithm. The structure models are modelled in a finite element analysis software to obtain mass and stiffness properties. Then, the structure is modelled as a shear building in the optimization process. The optimum TMD was applied to the example structures in the finite element method analysis software to find structural response including base shear force. According to the results, the optimum TMDs are excellently effective in reduction of the total base shear force of frame structures.

Keywords Tuned mass dampers · Frame structures · Earthquake · Finite element method · Jaya algorithm

1 Introduction

The structures subjected to seismic effects can be damaged or collapse. Although structure is strong enough to resist all forces resulting from the ground acceleration, non-constructive members such as shelfs, cabinets, electronics and valuable items like artefacts in museums can be damaged via strong ground motion. For that reason, passive and active control methods are used for civil structures. Tuned Mass Dampers (TMDs) are an option, which have been applied to various civil structures. The most

A. Mubuli · S. M. Nigdeli (✉) · G. Bekdaş
Department of Civil Engineering, Istanbul University-Cerrahpaşa, 34320 Avcılar, Istanbul, Turkey
e-mail: melihnig@istanbul.edu.tr

A. Mubuli
e-mail: apaer.mubuli@ogr.iu.edu.tr

G. Bekdaş
e-mail: bekdas@istanbul.edu.tr

S. M. Nigdeli et al. (eds.), *Proceedings of 6th International Conference on Harmony Search, Soft Computing and Applications*, Advances in Intelligent Systems and Computing 1275, https://doi.org/10.1007/978-981-15-8603-3_11

Fig. 1 The Berlin TV Tower

popular TMD is to sphere-shape TMD of Taipei 101 building. Generally, TMDs have been added to the structure to retrofit it. For example, a 1.5t TMD was added to the Berlin TV Tower as shown in Fig. 1 to eliminate wind-indicated vibrations. For earthquake, the Lax Theme Building in the Los Angeles International Airport was retrofitted by adding a TMD with 20% mass ratio [1]. For efficiency of TMD, an additional vibration mode close to the critical vibration mode of structure is added. For that reason, TMDs must be optimally tuned according to the characteristics of the structure.

In optimum design of the TMDs, two main ways can be chosen. In several studies, simplified formulations were proposed. These formulations were found for single degree of freedom main system, but the formulation can be used by considering the critical vibration mode only. In these formulations, a frequency ratio of TMD and structure (f_{opt}) and optimum TMD damping ratio ($\xi_{d,opt}$) are proposed. These formulations were derived by theoretical formulations by making some assumptions [2–4] or by using metaheuristic algorithms such as Particle Swarm Optimization [5] and Artificial Neural Network [6].

The second way is to couple a dynamic analysis code or software within the optimization process code. In optimization of TMDs, the most popular approaches are based on metaheuristic algorithms, and the algorithms such as Genetic Algorithm

[7, 8], Particle Swarm Optimization [5–9], Harmony Search [10, 11], Ba Algorithm [12, 13], Ant Colony Optimization [14], Flower Pollination Algorithm [15], Teaching Learning Based Optimization and Jaya Algorithm (JA) [16] have been employed.

In the present study, two different frame structure models were modelled in a finite element method based-computer software called SAP2000 [17]. The mass and period properties of the structure were used to generate a shear building model to find the optimum TMD properties by using an optimization approach using Matlab with Simulink [18] and JA [16] firstly developed by Rao [19]. Then, the structural responses on SAP2000 were compared for the frame structures with and without TMD.

2 The Optimum Design of TMDs

The equation of motion of a N-storey shear structure including a TMD on the top given in Fig. 2 can be written as Eq. (1).

$$\mathbf{M}\ddot{x}(t) + \mathbf{C}\dot{x}(t) + \mathbf{K}x(t) = -\mathbf{M}\{\mathbf{1}\}\ddot{x}_g(t) \tag{1}$$

Equation (1) is solved via a numerical integration method. In the numerical analyses, the equations of motions were solved with the help of Simulink [18] by using 4th order Runge–Kutta method. The solution of Eq. (1), which is shown as $x(t)$, was

Fig. 2 N-storey shear with TMD

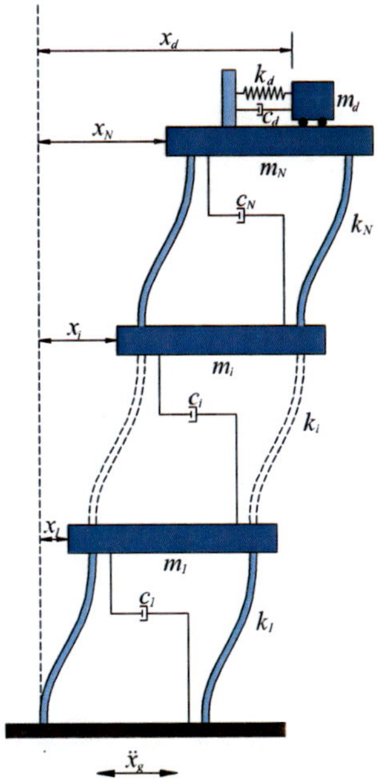

used as the objective by considering the maximum absolute value of $x(t)$ for the top storey (x_N). The responses were found for a set of earthquake motions given in FEMA P:695: Quantification of Building Seismic Performance Factors [20] as far-fault ground motions. The most critical excitation with the highest top storey response was taken into consideration. The response vector is given as Eq. (2), matrices mass (M), damping (C) and stiffness (K) matrices are found according to the structural properties such as mass (m_i), stiffness (k_i) and damping coefficient (c_i) of i-th storey.

$$x(t) = \begin{Bmatrix} x_1(t) \\ x_2(t) \\ M \\ x_N(t) \\ x_d(t) \end{Bmatrix} \tag{2}$$

x_i represents the displacement of ith storey of N-storey structure, while the displacement of TMD was shown as x_d. The optimum results were found according to the following constraint given as Eq. (3).

$$g_1 = \frac{\max(|x_d - x_N|)_{\text{with TMD}}}{\max(|x_N|)_{\text{without TMD}}} < 2 \tag{3}$$

By using the constraint (g_1), stroke of optimized TMD is limited.

The structure of the optimization process is as follows:

i. Define structural properties of the structure, earthquake ground accelerations, the design variable ranges and population number (taken as 10).
ii. Analyze the structure without TMD to use the results of maximum x_N in constraints g_1.
ii. Generate an initial set of design variables according to the population number by using random values within the selected range. The design variables are period (T_d) and damping ratio (ξ_d) of TMD. These variables are used to find the stiffness (k_d) and damping (c_d) of TMD as given in Eqs. (4) and (5), respectively.

$$k_d = 4\pi^2 \frac{m_d}{T_d^2} \tag{4}$$

$$c_d = 2m_d\xi_d\sqrt{\frac{k_d}{m_d}} \tag{5}$$

Also, the objective function and g_1 are calculated for all sets.

iv. Update set of solutions via JA by using Eq. (6)

$$X_{new,i} = X_{old,i} + r_1(X_{best} - |X_{old,i}|) - r_2(X_{worst} - |X_{old,i}|) \quad (6)$$

In Eq. (6), $X_{new,i}$ is the updated solution of i-th population, while $X_{old,i}$ is the existing solution. r_1 and r_2 are two random numbers between 0 and 1. X_{best} and X_{worst} are the best and worst existing solutions, respectively. The best and the worst solutions are chosen according to the minimization of the objective function. If the objective function values have violated g_l values, these are not considered as best vectors, and the most violated one is chosen as the worst one.

v. Repeat the process of iv for several iterations (200 iterations for examples).

3 Numerical Examples

Two 10 storey structures were investigated in the study. The structural models are firstly modelled at SAP2000 [17] to find the period of the critical mode and mass of the structure. The damping ratio of Reinforced Concrete (RC) structures was assumed as 5%. The beams and slabs are taken as rigid to represent a shear structure. According to this finding, the structural parameters were defined in the optimization code to find the optimum parameters of TMD via JA-based method. The Time History Analyses (THA) were investigated in SAP2000 [17] for the critical excitation which is RIO360 record of the 1992 Cape Mendocino earthquake.

3.1 Structure Model 1

In this study, the analytical model of the 10 storey (32 m) regular and symmetrical reinforced concrete frame building was created by the SAP2000 [17] with and without Tuned Mass Damper (TMD) to study the behaviour of the building and reactions under the effect of earthquake load were investigated. In this model, the dampers are placed on the top floor of the building. The 10 storey RC building with 5 bays in X- and Y-direction and each bay with 7 m in the horizontal directions, also this model is with uniform height of storey 3.2 m in vertical direction. The plan and elevation of model are shown in Fig. 3. The columns, beams, slabs and other features of frame structure are summarized in Table 1.

Tuned Mass Damper (TMD) is placed on the top storey defined as damper link member with one end fixed with structure and other end is freely attached by mass of damper and the free link ends are connected by infinite rigid beams with zero weights shown in Fig. 4.

The optimum parameter of TMD is shown in Table 2 for the model with 0.4297 s critical period at the first mode. The mass of the damper was taken as 5% of the total mass of the structure.

With and without tuned mass damper, Time History Analysis (THA) was done under the effect of the critical earthquake acceleration, and plots for displacement,

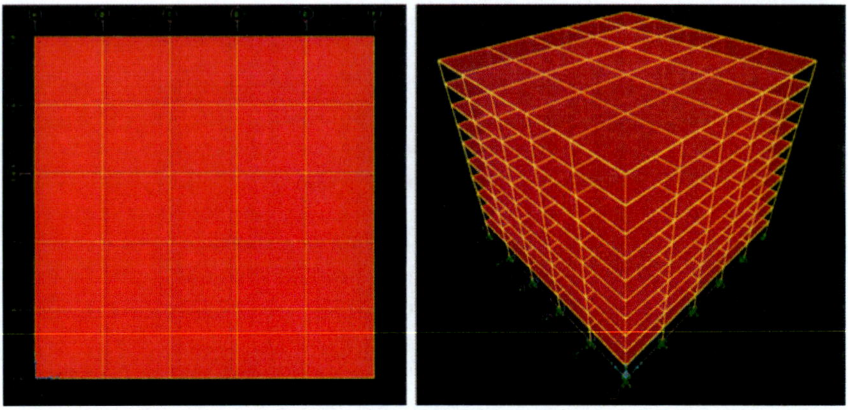

Fig. 3 Plan and three-dimensional view of model 1 without TMD

Table 1 Structural properties of model 1

Materials and section properties		
Variable	Data	
Type of the building	RC	
Number of stories	10	
Floor height	3.2 m	
Materials	Concrete C30	
Density of concrete	25 kN/m^3	
Modulus of Elasticity, E	32,000 Mpa	
Poisson ratio	0.2	
Rebar	S420	
	Fy = 420 Mpa	Fu = 500 Mpa
Rebar Modulus of Elasticity, E	200,000 Mpa	
Size of the column	700 × 700 mm	
Size of the beams	300 × 700 mm	
Depth of the slab	120 mm	

acceleration values of the top floor and displacement and acceleration values of the middle floor are shown as Figs. 5 and 6.

As given in Fig. 7, the stroke displacement of TMD is at feasible levels to apply. The essential effect of TMD-based control can be seen on the total base shear force plots given as Fig. 8.

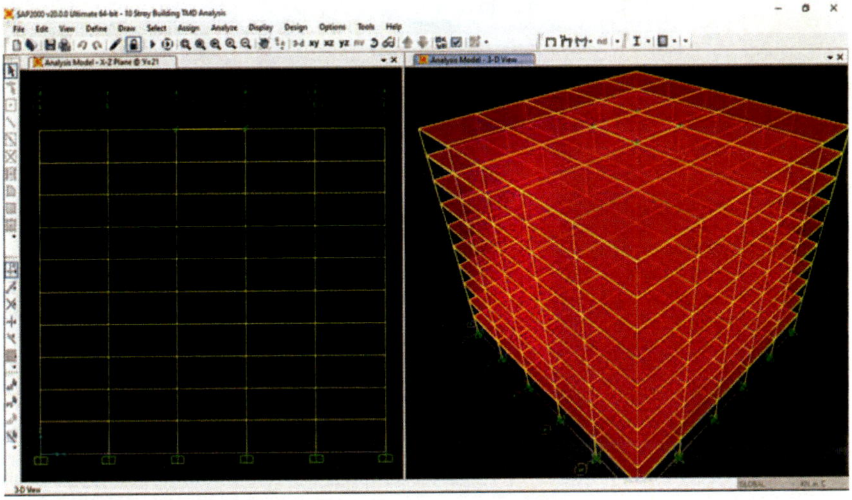

Fig. 4 Plan and three-dimensional view of model 1 with TMD

Table 2 Properties of Tuned Mass Damper (TMD) for model 1

Properties of Tuned Mass Damper (TMD)	
Parameter	Value
Mass of Damper m_d	356.908 ton
Optimum Frequency Ratio T_d	0.5289 s
Optimum Damping Ratio ξ_d	0.1432
Spring Stiffness of Damper k_d	50,363 kN/m
Damping of Damper c_d	1215 kNs^2/m

3.2 *Structure Model 2*

The 10 storey RC building with 4 bays in X- and Y-direction and each bay with 6 m in the horizontal directions, also this model has 3.2 m uniform height in vertical direction. The plan and elevation of model 2 are shown in Fig. 9. The structural parameters are presented in Table 3.

Tuned Mass Damper (TMD) is placed on the top storey defined as damper link member with one end fixed with structure and other end is freely attached by mass of damper shown in Fig. 10. The optimum parameter of TMD is shown in Table 4 for the model with 0.4794 s critical period at the first model.

Time History Analysis (THA) of model 2 were done under the effect of the critical earthquake. The plots are shown as Figs. 11, 12, 13 and 14. The optimum TMD is also effective for the second RC frame model.

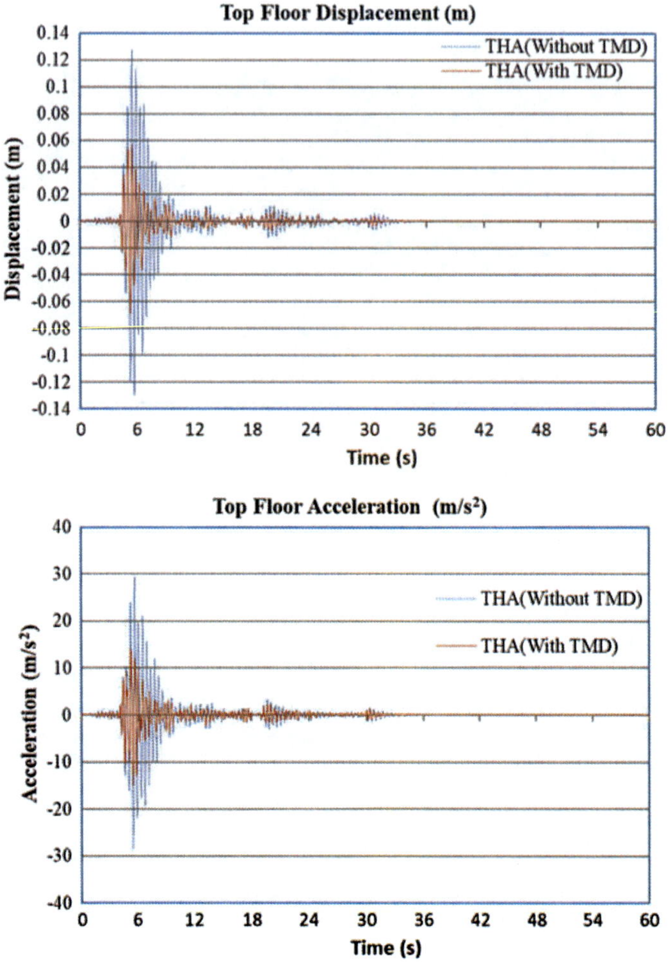

Fig. 5 Top storey responses for structure model 1

3.2.1 Conclusions

The results in this study suggest that the application of TMD is appropriate to mitigate the dynamic response of the structures subjected to seismic ground motion. From reviewing the results obtained in this study, the following conclusions can be drawn.

Lateral deflection, acceleration and base shear after the use of TMD are as follows for the first structural model:

- On the Top Floor the minimum displacement reduction is 46.91% and the maximum displacement reduction is 55.71%
- On the 5th floor the minimum displacement reduction is 44.93% and the maximum displacement reduction is 52.17%

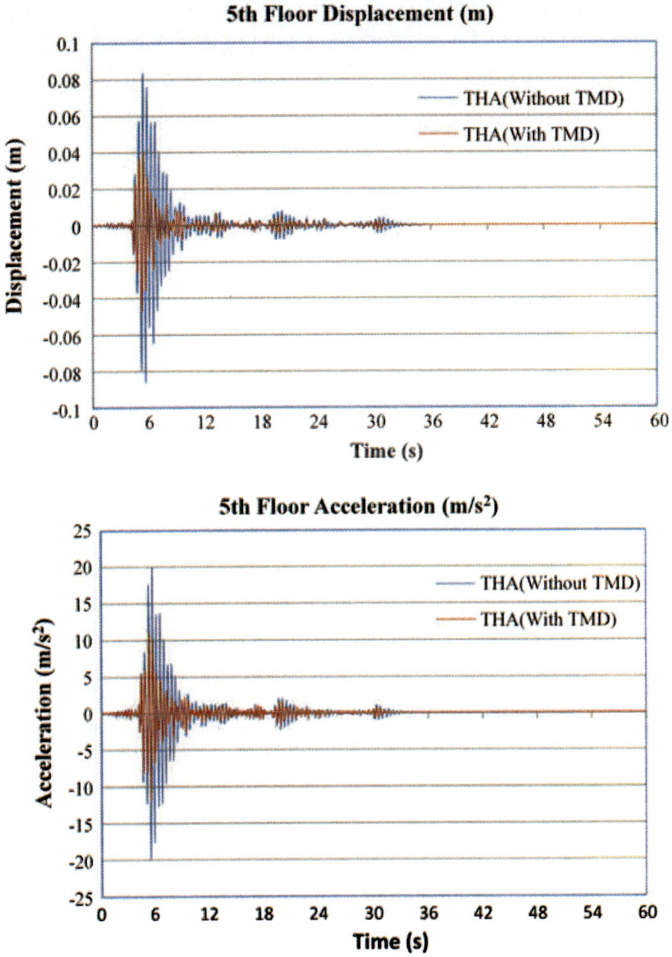

Fig. 6 5th storey responses for structure model 1

- On the Top Floor the minimum acceleration reduction is 47.3% and the maximum acceleration reduction is 52.05%
- On the 5th floor the minimum acceleration reduction is 40.54% and the maximum acceleration reduction is 44.30%
- The minimum base shear reduction is 50.76% and the maximum base shear reduction is 42.56%.

The same maximum response reductions for the second model are as follows:

- On the Top Floor the minimum displacement reduction is 26.96% and the maximum displacement reduction is 43.27%

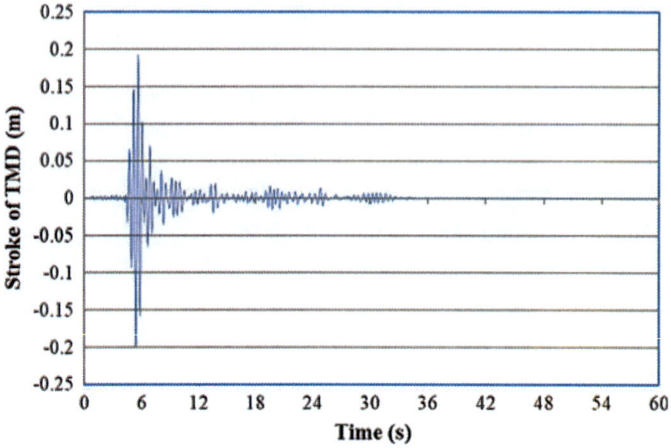

Fig. 7 Stroke of TMD

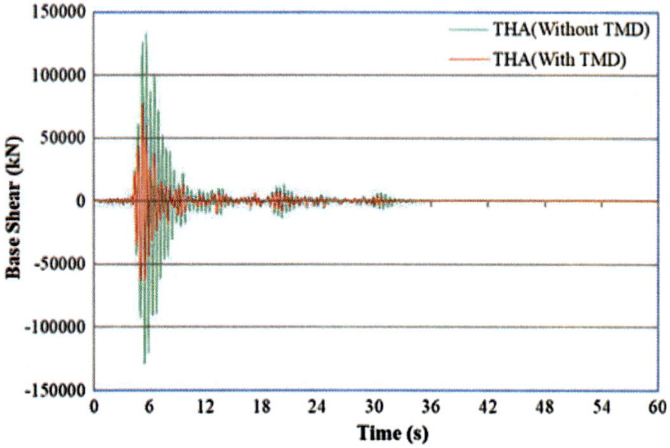

Fig. 8 Base shear force for model 1

- On the 5th floor the minimum displacement reduction is 27.54% and the maximum displacement reduction is 42.04%
- On the Top Floor the minimum acceleration reduction is 36.86% and the maximum acceleration reduction is 35.70%
- On the 5th floor the minimum acceleration reduction is 36.29% and the maximum acceleration reduction is 28.03%
- The minimum base shear reduction is 39.72% and the maximum base shear reduction is 28%.

As the conclusion, the optimized TMDs for frame structures idealized as a shear structure are effective to reduce structural vibrations. The SAP2000 analysis results

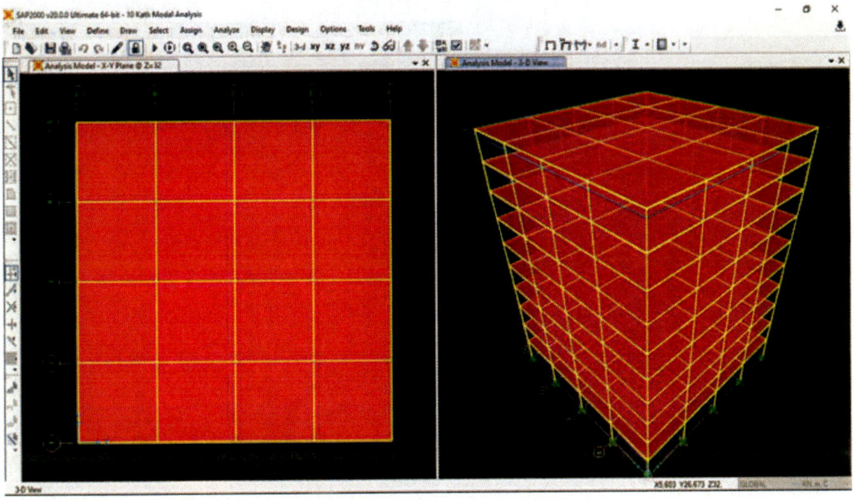

Fig. 9 Plan and three-dimensional view of model 2 without TMD

Table 3 Structural properties of model 2

Materials and aection properties		
Variable	Data	
Type of the building	RC	
Number of stories	10	
Floor height	3.2 m	
Materials	Concrete C30	
Density of concrete	25 kN/m^3	
Modulus of Elasticity, E	32,000 Mpa	
Poisson ratio	0.2	
Rebar	S420	
	Fy = 420 Mpa	Fy = 420 Mpa
Rebar Modulus of Elasticity, E	200,000 Mpa	
Size of the column	700 × 700 mm	
Size of the beams	300 × 700 mm	
Depth of the slab	120 mm	

support the effectiveness of the optimum values of shear structures applied for frame structures.

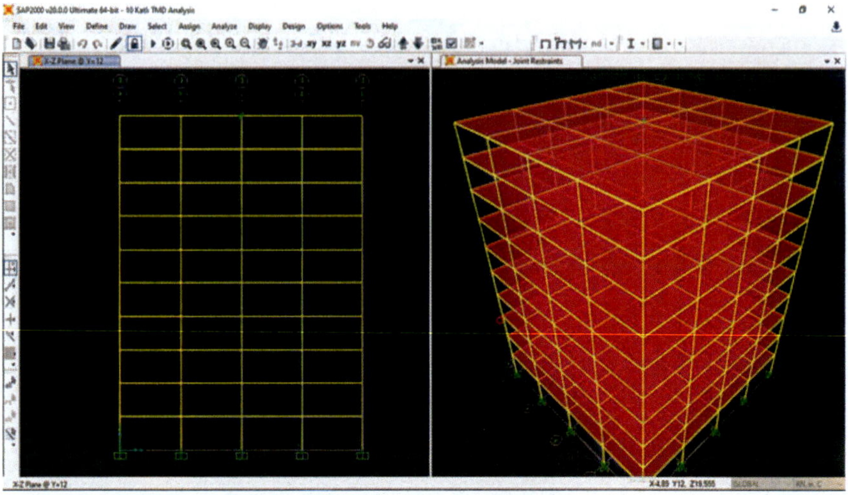

Fig. 10 Plan and three-dimensional view of model 2 with TMD

Table 4 Properties of Tuned Mass Damper (TMD) for model 2

Properties of Tuned Mass Damper (TMD)	
Parameter	Value
Mass of Damper m_d	179.825 ton
Optimum Frequency Ratio T_d	0.5085 s
Optimum Damping Ratio ξ_d	0.12
Spring Stiffness of Damper k_d	27431.401 kN/m
Damping of Damper c_d	533.377 kNs^2/m

Fig. 11 Top storey responses for structure model 2

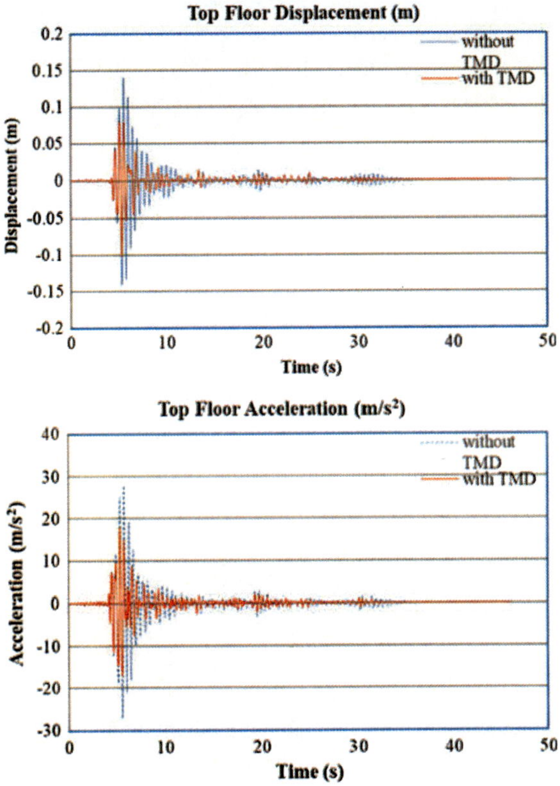

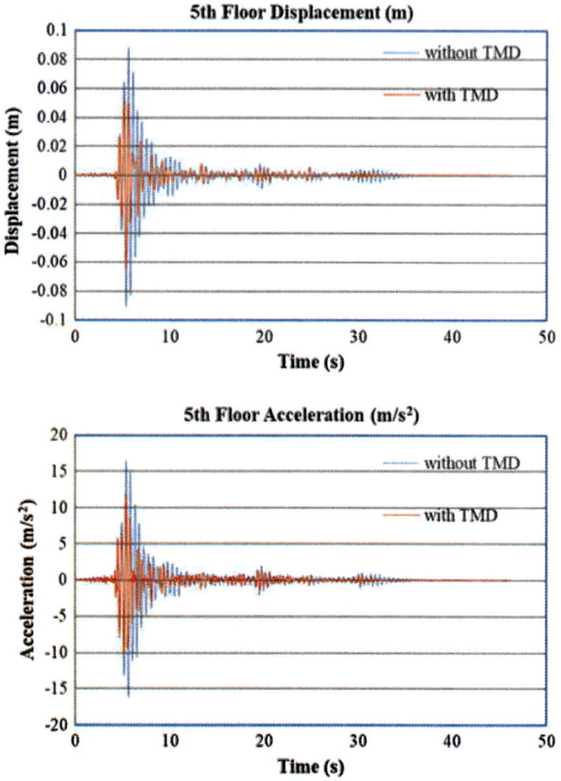

Fig. 12 5th storey responses for structure model 2

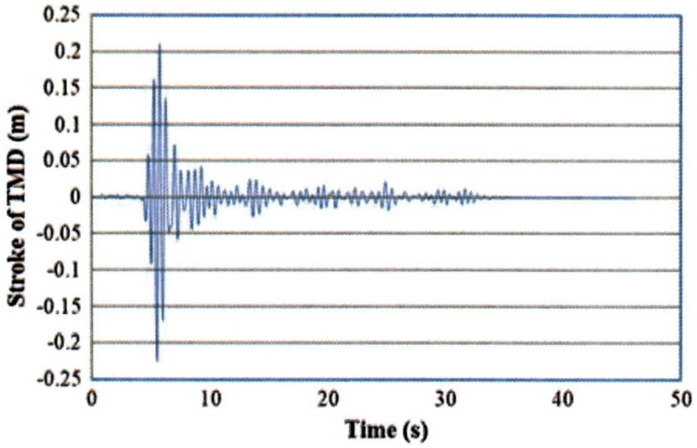

Fig. 13 Stroke of TMD

Fig. 14 Base shear force for model 2

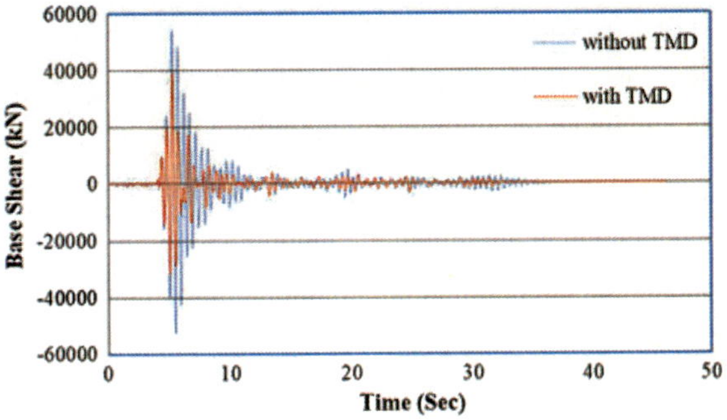

References

1. H.K. Miyamoto, A.S.J. Gilani, Y.Z. Gndodu, Innovative seismic retroit of an iconic building, in *Seventh National Conference on Earthquake Engineering*, 30 May–3 June, Istanbul, Turkey (2011)
2. J.P. Den Hartog, *Mechanical Vibrations*, 3rd edn. (McGraw-Hill, New York, 1947)
3. G.B. Warburton, Optimum absorber parameters for various combinations of response and excitation parameters. Earthq. Eng. Struct. D. **10**, 381–401 (1982)
4. F. Sadek, B. Mohraz, A.W. Taylor, R.M. Chung, A method of estimating the parameters of tuned mass dampers for seismic applications. Earthq. Eng. Struct. D. **26**, 617–635 (1997)
5. A.Y.T. Leung, H. Zhang, Particle swarm optimization of tuned mass dampers. Eng. Struct. **31**(3), 715–728 (2009)
6. M. Yucel, G. Bekdaş, S.M. Nigdeli, S. Sevgen, Estimation of optimum tuned mass damper parameters via machine learning. J. Build. Eng. **26**, 100847 (2019)
7. M.N.S. Hadi, Y. Arfiadi, Optimum design of absorber for MDOF structures. J. Struct. Eng. ASCE **124**, 1272–1280 (1998)
8. G.C. Marano, R. Greco, B. Chiaia, A comparison between di erent optimization criteria for tuned mass dampers design. J. Sound Vib. **329**, 4880–4890 (2010)
9. A.Y.T. Leung, H. Zhang, C.C. Cheng, Y.Y. Lee, Particle swarm optimization of TMD by non-stationary base excitation during earthquake. Earthquake Eng. Struct. Dynam. **37**, 1223–1246 (2008)
10. G. Bekdaş, S.M. Nigdeli, Estimating optimum parameters of tuned mass dampers using harmony search. Eng. Struct. **33**, 2716–2723 (2011)
11. S.M. Nigdeli, G. Bekdaş, Optimum tuned mass damper design in frequency domain for structures. KSCE J. Civil Eng. **21**(3), 912–922 (2017)
12. G. Bekdaş, S.M. Nigdeli, Metaheuristic based optimization of tuned mass dampers under earthquake excitation by considering soil-structure interaction. Soil Dyn. Earthq. Eng. **92**, 443–461 (2017)
13. G. Bekdaş, S.M. Nigdeli, X.S. Yang, A novel bat algorithm based optimum tuning of mass dampers for improving the seismic safety of structures. Eng. Struct. **159**, 89–98 (2018)
14. A. Farshidianfar, S. Soheili, Ant colony optimization of tuned mass dampers for earthquake oscillations of high-rise structures including soil-structure interaction. Soil Dyn. Earthquake Eng. **51**, 14–22 (2013)
15. G. Bekdaş, S.M. Nigdeli, X.-S. Yang, Metaheuristic based optimization for tuned mass dampers using frequency domain responses, in *Harmony Search Algorithm*, ed. by J. Del Ser. Advances in Intelligent Systems and Computing, vol. 514 (Springer, 2017) , pp. 271–279
16. S.M. Nigdeli, G. Bekdaş, E. Ayyıldız, Optimization of parameters of tuned mass dampers by using hybrid metaheuristic algorithms. in *3rd International Conference on Engineering Technology and Innovation (ICETI)*, 17–21 April 2019, Belgrade, Serbia (2019)

17. SAP2000, C. S. I., *Integrated Software for Structural Analysis & Design* (Computer and Structures. Inc., Berkeley, 2007)
18. T. MathWorks, *Matlab R2018a* (Natick, MA, 2018)
19. R. Rao, Jaya: a simple and new optimization algorithm for solving constrained and unconstrained optimization problems. Int. J. Ind. Eng. Comput. **7**(1), 19–34 (2016)
20. FEMA P-695, Quantification of seismic performance factors. FEMA P-695 report, the Applied Technology Council for the Federal Emergency Management Agency, Washington, DC (2009)

Total Potential Optimization Using Metaheuristics: Analysis of Cantilever Beam via Plane-Stress Members

Yusuf Cengiz Toklu, Gebrail Bekdaş, Aylin Ece Kayabekir, Sinan Melih Nigdeli, and Melda Yücel

Abstract In addition to classical methods in structural analysis, the exact solution of the deformed shape of the structure is by using a metaheuristic method. According to the theory of total potential energy minimization, the static condition of the deformed shape of structures can be found directly by assigning the coordinates of the deformed shape of the structure and finding the case with the minimum potential energy. In that case, the analysis process is an optimization process, and metaheuristics are effective in this process. The total potential optimization using metaheuristic algorithms (TPO/MA) is an effective approach for several including plane-stress members. A cantilever beam is presented in the study, and it is solved via plane-stress members using TPO/MA. The problem is presented via two meshing options. For the frequent meshing, a hybrid algorithm of Jaya Algorithm (JA) using student phase of Teaching–Learning-Based Optimization (TLBO) is presented. TPO/MA is effective to find similar results with the finite element method.

Keywords TPO/MA · Metaheuristics · Hybrid algorithms · Structural analysis · Optimization

---

Y. C. Toklu
Department of Civil Engineering, Istanbul Beykent University, 34398 Sariyer, Istanbul, Turkey
e-mail: cengiztoklu@gmail.com

G. Bekdaş · A. E. Kayabekir · S. M. Nigdeli (✉) · M. Yücel
Department of Civil Engineering, Istanbul University-Cerrahpaşa, 34320 Avcılar, Istanbul, Turkey
e-mail: melihnig@istanbul.edu.tr

G. Bekdaş
e-mail: bekdas@istanbul.edu.tr

A. E. Kayabekir
e-mail: ecekayabekir@gmail.com

M. Yücel
e-mail: melda.yucel@yahoo.com.tr

S. M. Nigdeli et al. (eds.), *Proceedings of 6th International Conference on Harmony Search, Soft Computing and Applications*, Advances in Intelligent Systems and Computing 1275, https://doi.org/10.1007/978-981-15-8603-3_12

1 Introduction

The classical analysis of structures involves the generation of a matrix equation with stiffness matrix, loads, and deflections. These matrices are generated according to the properties of structural members, and the member values are merged to generate the system matrices. In the merging of values, transformation to the global coordinates is needed. By solving the matrix equation, the deflections are found. In this process, it is needed to take the inverse of the stiffness matrix which may have a big size. To take this process, computers help engineers. The described process is for only linear systems. For non-linear systems, it is needed to add iterative methods [1–4]. Whereas, the deformation can be directly found by assigning candidate values. Also, as we know that the systems are stable at the level of minimum total potential energy. So, minimization of total potential energy gives the exact deformed shape of the system, and it is not needed to add a different method for non-linear analysis. By using metaheuristic algorithms, the design variables such as coordinates of deformed shape of structures can be iteratively generated. By using total potential optimization using metaheuristic algorithms (TPO/MA), several types of structures such as trusses, cable systems, and plane-stress members [5–8] have been solved.

In the present study, a cantilever beam is presented for analysis via TPO/MA for plane-stress members. Two meshing options are given, and a hybrid metaheuristic algorithm is proposed for the frequent meshing case to find the same values with Finite Element Method (FEM).

Fig. 1 The triangular element

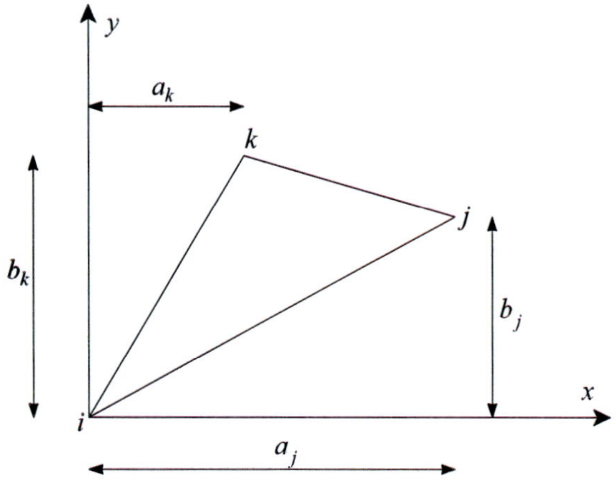

2 The Total Potential Energy of Plane-Stress Members

In the formulations, the plates are considered as several plane-stress member in the form of the triangular element given in Fig. 1. The nodal displacements are u_i and v_i in x- and y-directions, respectively.

The displacement fields such as $u(x, y)$ and $v(x, y)$ are given as Eqs. (1) and (2), respectively.

$$u(x, y) = u_i + C_1 x + C_2 y \tag{1}$$

$$v(x, y) = v_i + C_3 x + C_4 y \tag{2}$$

The particular derivatives, which are the stain in x-direction (ε_x), y-direction (ε_y), and shear strain (γ), are constants such as C_1, C_2, C_3, and C_4 as follows:

$$\varepsilon_x = \frac{\partial u}{\partial x} = C_1 \tag{3}$$

$$\varepsilon_y = \frac{\partial v}{\partial y} = C_4 \tag{4}$$

$$\gamma = \frac{\partial u}{\partial y} + \frac{\partial v}{\partial x} = C_2 + C_3 \tag{5}$$

If the node i is the origin (0, 0), the other nodal coordinates are as in Eqs. (6)–(8).

$$u(0, 0) = u_i,\ \ v(0, 0) = v_i \tag{6}$$

$$u(a_j, b_j) = u_j,\ v(a_j, b_j) = v_j \tag{7}$$

$$u(a_k, b_k) = u_k,\ v(a_k, b_k) = v_k \tag{8}$$

As a matrix equation, the relationship of nodal displacements is written as Eq. (9).

$$\begin{bmatrix} u_i \\ v_j \\ u_k \\ v_k \end{bmatrix} = \begin{bmatrix} a_j & b_j & 0 & 0 \\ 0 & 0 & a_j & b_j \\ a_k & b_k & 0 & 0 \\ 0 & 0 & a_k & b_k \end{bmatrix} \begin{bmatrix} C_1 \\ C_2 \\ C_3 \\ C_4 \end{bmatrix} + \begin{bmatrix} u_i \\ v_i \\ u_i \\ v_i \end{bmatrix} \tag{9}$$

By solving Eq. (9),

$$C_1 = \frac{b_k(u_j - u_i)}{a_j b_k - a_k b_j} + \frac{b_j(u_k - u_i)}{a_k b_j - a_j b_k} \tag{10}$$

$$C_2 = \frac{a_k(u_j - u_i)}{a_k b_j - a_j b_k} + \frac{a_j(u_k - u_i)}{a_j b_k - a_k b_j} \tag{11}$$

$$C_3 = \frac{b_k(v_j - v_i)}{a_j b_k - a_k b_j} + \frac{b_j(v_k - v_i)}{a_k b_j - a_j b_k} \tag{12}$$

$$C_4 = \frac{a_k(v_j - v_i)}{a_k b_j - a_j b_k} + \frac{a_j(v_k - v_i)}{a_j b_k - a_k b_j} \tag{13}$$

As known, the strain energy density resulted due to deformation is given as Eq. (14). The formulation is for a two-dimensional elastic body.

$$e = \int_{\varepsilon=0}^{\varepsilon} \sigma d\varepsilon = \frac{1}{2}(\sigma_x \varepsilon_x + \sigma_y \varepsilon_y + \tau\gamma) \tag{14}$$

When the solution of constants is used in Eq. (3)–(5),

$$\varepsilon_x = \frac{b_k(u_j - u_i)}{a_j b_k - a_k b_j} + \frac{b_j(u_k - u_i)}{a_k b_j - a_j b_k} \tag{15}$$

$$\varepsilon_y = \frac{a_k(v_j - v_i)}{a_k b_j - a_j b_k} + \frac{a_j(v_k - v_i)}{a_j b_k - a_k b_j} \tag{16}$$

$$\gamma = \frac{1}{a_k b_j - a_j b_k} + \big(a_k(u_j - u_i) + a_j(u_i - u_k) + b_k(v_i - v_j) + b_j(v_k - v_i)\big) \tag{17}$$

For linear stress–strain relationship, the normal stress in x (σ_x) and y (σ_y) and shear stress (τ) are written as Eqs. (18)–(20), where E and ν are elasticity modulus and Poisson's ratio, respectively.

$$\sigma_x = \frac{E}{1-\nu^2}(\varepsilon_x + \nu\varepsilon_y) \tag{18}$$

$$\sigma_y = \frac{E}{1-\nu^2}(\varepsilon_y + \nu\varepsilon_x) \tag{19}$$

$$\tau = \frac{E}{1-\nu^2}(\frac{1-\nu}{2}\gamma) \tag{20}$$

According to nodal coordinates, the stress is as follows:

$$\begin{aligned} \sigma_x =& \frac{E}{(\nu^2-1)(a_k b_j - a_j b_k)} \nu\big(a_k(v_i - v_j) + a_j(v_k - v_i)\big) \\ &+ (\nu-1)\big(b_k(u_i - u_j) - b_j(u_i - u_k)\big) \end{aligned} \tag{21}$$

$$\sigma_y = \frac{E}{(\nu^2 - 1)(a_k b_j - a_j b_k)} a_k(v_i - v_j) + a_j(v_k - v_i) + \nu b_k(u_j - u_i) + \nu b_j(u_i - u_k) \quad (22)$$

$$\tau = \frac{E}{(\nu^2 - 1)(a_k b_j - a_j b_k)} (\frac{1-\nu}{2} a_k(u_j - u_i) + a_j(u_i - u_k) + b_k(v_i - v_j) + b_j(v_k - v_i) \quad (23)$$

Finally, the strain energy density of a linear system can be written as Eq. (24).

$$e = \frac{E}{4(1-\nu^2)(a_k b_j - a_j b_k)^2}(-(\nu-1)((a_k(u_j - u_i) + a_j(u_i - u_k) + b_k(v_i - v_j) + b_j(v_k - v_i))^2 +2(a_k(v_j - v_i) + a_j(v_i - v_k))(\nu(a_k(v_j - v_i) + a_j(v_i - v_k)) + b_k(u_i - u_j) + b_j(u_k - u_i)) +2(a_k(v_j - v_i) + a_j(v_i - v_k))(a_k(v_j - v_i) + a_j(v_i - v_k)) + \nu b_k(u_i - u_j) + \nu b_j(u_k - u_i))) \quad (24)$$

If the strain energy density of mth element (e_m) is multiplied with the volume of it (V_m), the strain energy is written as follows:

$$U_i = e_m V_m \quad (25)$$

The volume of mth element is given as

$$V_m = \frac{(a_j b_k - a_k b_j)t}{2} \quad (26)$$

The sum of all strain energies of members for a n member structure is as follows:

$$U = \sum_{m=1}^{n} U_m \quad (27)$$

The total potential energy (Π_P) is found as Eq. (28) if m_{xi} and P_{yi} are the nodal point forces of ith node for a system with p nodes.

$$\Pi_p = U - \sum_{i=1}^{p} (P_{xi} u_i + P_{yi} v_i) \quad (28)$$

3 The Methodology of the Minimization of Total Potential Energy

In TPO/MA, the design variables are the nodal displacements (u_i and v_i for $i =$ 1 to p). These values are assigned with candidate values according to the rules of metaheuristic algorithms. The flowchart of TPO/MA for a two-phase algorithm is summarized in Fig. 2. For the numerical example, harmony search (HS) [9], Teaching Learning-Based Optimization (TLBO) [10], Flower Pollination (FPA) [11], and Jaya Algorithm (JA) [12] are employed. Since the classical forms of these algorithms are not effective for the example case with high number of design variables, a hybrid algorithm is proposed.

JA is a single-phase algorithm, but it has a great convergence ability in optimization problems including analysis via TPO/MA. It may trap to local optimums. To avoid it, JA is hybridized with the student phase of TLBO. In that case, a hybrid algorithm called Jaya Algorithm with Probabilistic Student Phase (JA1SP) is developed, and the first phase is done according to the formulation of JA given as Eq. (29).

$$X_{new} = X_{old} + \text{rand}(0,1)(g^* - X_{old}) - \text{rand}(0,1)(g^w - X_{old}) \tag{29}$$

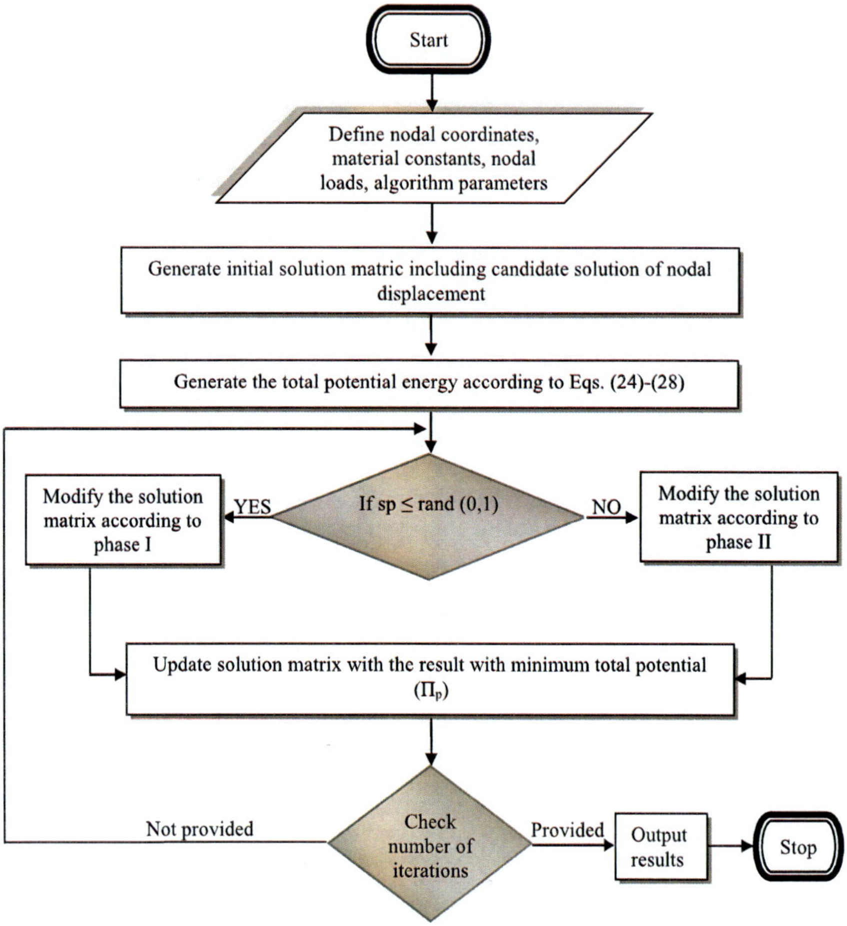

Fig. 2 The flowchart of the TPO/MA for two-phase metaheuristic algorithms

The new solution (X_{new}) is generated according to the existing solution (X_{old}), the solutions with minimum (g^*), and maximum (g^w) total potential energy values, and a random number between 0 and 1 is shown as rand(0,1). For the second phase, the student phase of TLBO using two randomly chosen solutions (X_{rand1} and X_{rand2}) from the existing ones are used (Eq. (30)). A switch probability is used to choose one of the phases, and it is taken as 0.5 in the numerical examples for equal probability.

$$X_{new} = \begin{cases} X_{old} + rand(0,1)(X_{rand1} - X_{rand2}); & f(X_{rand1}) > f(X_{rand2}) \\ X_{old} + rand(0,1)(X_{rand2} - X_{rand1}); & f(X_{rand1}) < f(X_{rand2}) \end{cases} \tag{30}$$

4 Numerical Examples

4.1 Cantilever Beam Under a Concentered Load

In Fig. 3, the cantilever beam which was previously considered by Cook [13, 14] is given. It is loaded at the end (point C) in +y-direction with load P taken as 20 N. The elasticity modulus, Poisson's ratio of the material, and thickness of the beam are 100 N/mm^2, 0.3, and 10 mm, respectively. The meshing case is shown in Fig. 4. In this case, the structure has 15 free nodes and 30 displacement values. According to the results presented in Table 1, TPOMA using FPA, JA, and TLBO has similar solutions with FEM. These three algorithms are effective for this case, but hybrid algorithms are needed for frequent meshing according to the presented results in Sect. 4.2.

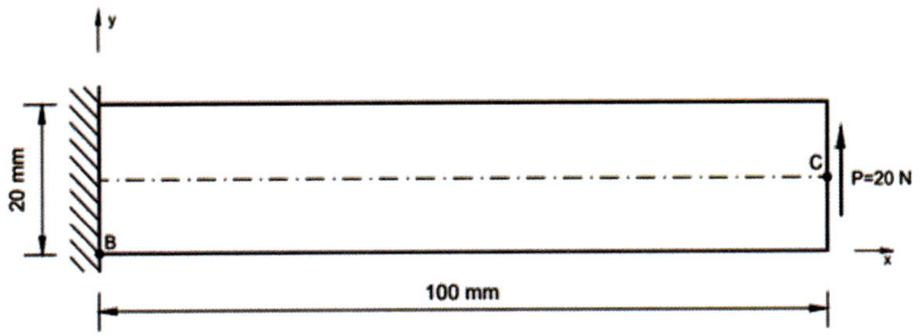

Fig. 3 Cantilever beam

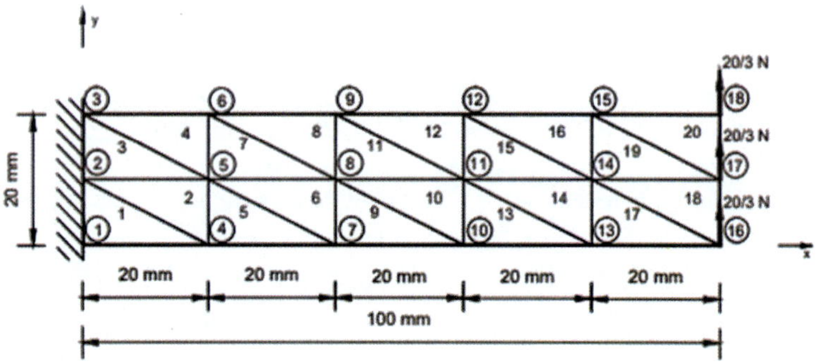

Fig. 4 Meshing of cantilever beam

4.2 Cantilever Beam Under a Concentered Load with More Frequent Meshing

The structure is also considered with more frequent meshing (160 members and 105 nodes) as shown in Fig. 5. The total potential energy of the structure with 200 design variables is given as Table 2. By using a single algorithm, only FPA is effective to find a close solution to FEM results. The hybrid algorithm JA1SP is effective in finding the minimum energy.

The match of displacements of structure by using JA1SP and FEM is clearly seen in Fig. 6. The displacement obtained by TLBO has no match with FEM results as given in Fig. 7. Especially, JA has displacement values quite different than those obtained by FEM as shown in Fig. 8.

5 Conclusions

This study is another proof of using TPOMA on general types of structures. For more detailed analysis of structural system with high number of plane-stress members, investigations are needed to carry out by orientating hybrid algorithms.

In the first meshing option, the exact results with FEM can be handled by classical form of employed algorithms. When frequent meshing of this example is investigated, the importance of using a hybrid method is seen. Only, FPA can converge to the minimum energy value for structure 4 with 105 nodes, but FPA is not also effective to find the exact solution of the system.

Table 1 Solutions for the cantilever beam with 20 plane members

	TPOMA (FPA)		TPOMA (JA)		TPOMA (HS)		TPOMA (TLBO)		FEM	
Node	Δx (mm)	Δy (mm)	Δx (mm)	Δy (mm)	Δx (mm)	Δy (mm)	Δx (mm)	Δy (mm)	Δx (mm)	Δy (mm)
4	0.19369	0.25151	0.19369	0.25151	0.20715	0.27089	0.19369	0.25151	0.1937	0.2515
5	0.00059	0.24176	0.00059	0.24176	−0.00003	0.26078	0.00059	0.24176	0.0006	0.2418
6	−0.19912	0.25858	−0.19912	0.25858	−0.21326	0.27713	−0.19912	0.25858	−0.1991	0.2586
7	0.34701	0.84814	0.34701	0.84814	0.36992	0.90709	0.34701	0.84814	0.347	0.8481
8	−0.00314	0.8425	−0.00314	0.8425	−0.0046	0.90067	−0.00314	0.8425	−0.0031	0.8425
9	−0.35408	0.85627	−0.35408	0.85627	−0.37917	0.91413	−0.35408	0.85627	−0.3541	0.8563
10	0.4547	1.70967	0.4547	1.70967	0.48581	1.83084	0.4547	1.70967	0.4547	1.7097
11	−0.00576	1.70704	−0.00576	1.70704	−0.00686	1.82991	−0.00576	1.70704	−0.0058	1.707
12	−0.46619	1.71762	−0.46619	1.71762	−0.50155	1.8422	−0.46619	1.71762	−0.4662	1.7176
13	0.51858	2.74696	0.51858	2.74696	0.55003	2.93925	0.51858	2.74696	0.5186	2.747
14	−0.0084	2.74754	−0.0084	2.74754	−0.01011	2.94063	−0.0084	2.74754	−0.0084	2.7475
15	−0.53422	2.75474	−0.53422	2.75473	−0.57002	2.94848	−0.53422	2.75473	−0.5342	2.7547
16	0.53861	3.87428	0.53861	3.87428	0.57192	4.13881	0.53861	3.87428	0.5386	3.8743
17	−0.01054	3.87495	−0.01054	3.87495	−0.01274	4.14002	−0.01054	3.87495	−0.0105	3.8749
18	−0.55939	3.88219	−0.55939	3.88219	−0.59412	4.14664	−0.55939	3.88219	−0.5594	3.8822
Π_p (Nmm)	−38.8		38.8		−38.6		−38.8		−38.8	
Analysis number	999,103		672,375		999,343		987,341		–	

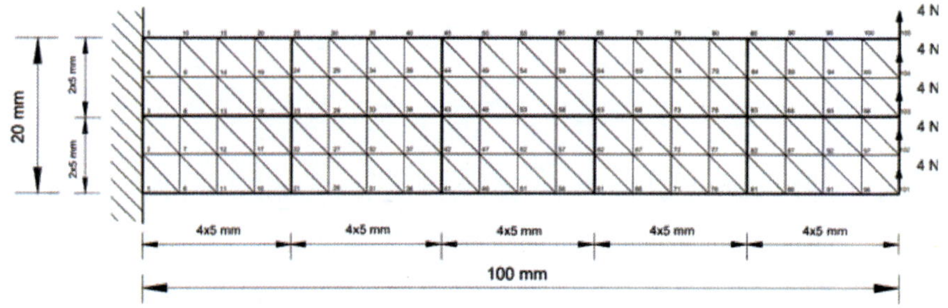

Fig. 5 Mathematical models prepared for the analyses of cantilever beam with more frequent meshing

Table 2 Minimum objective function values for frequently meshed structure

	TPOMA (FPA)	TPOMA (JA)	TPOMA (TLBO)	TPOMA (JA1SP)	FEM
Π_p (Nmm)	−84.3	−5.7	−51.4	−84.5	−84.5

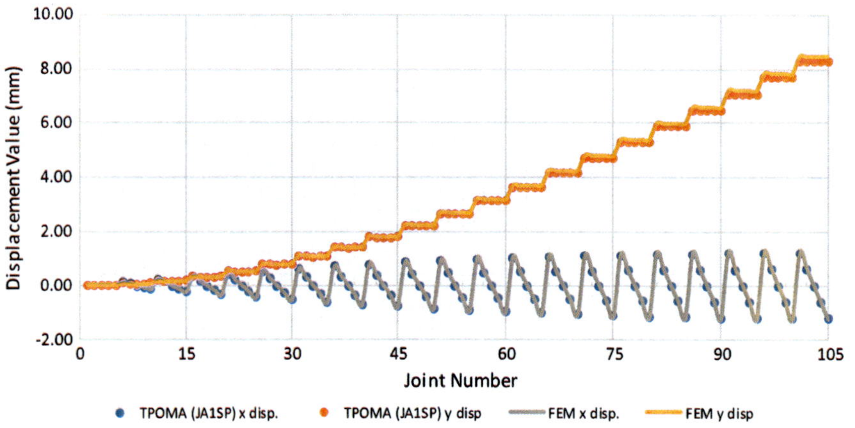

Fig. 6 Nodal displacement of frequently meshed structure for both directions (JA1SP)

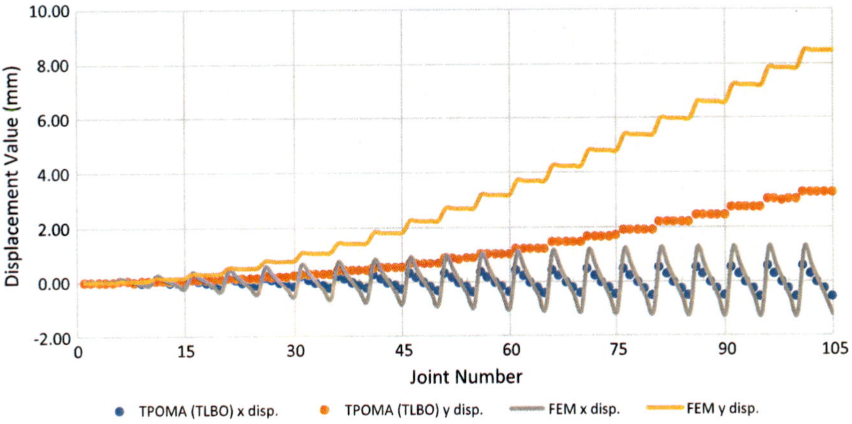

Fig. 7 Nodal displacement of frequently meshed structure for both directions (TLBO)

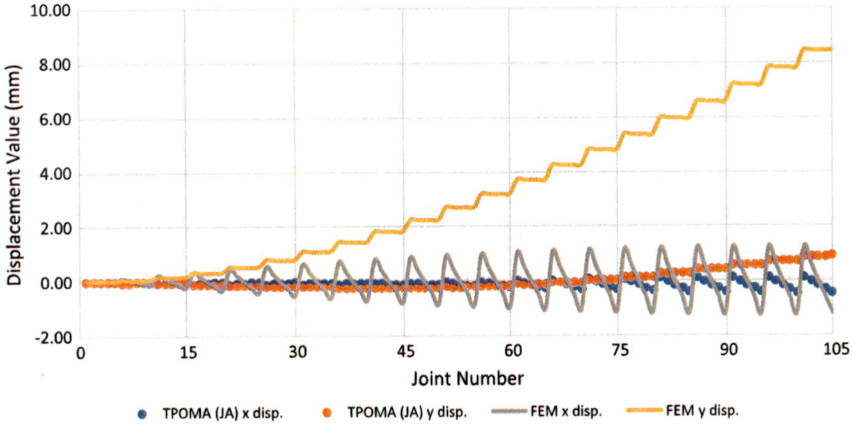

Fig. 8 Nodal displacement of frequently meshed structure for both directions (JA)

References

1. R. Levy, W.R. Spillers, *Analysis of Geometrically Nonlinear Structures*, 2nd edn. (Kluwer Academic Publishers, Dordrecht, 2003)
2. H. Saffari, M.J. Fadaee, R. Tabatabaei, Nonlinear analysis of space trusses using modified normal flow algorithm. J. Struct. Eng. **134**(6), 998–1005 (2008)
3. S. Krenk, *Non-Linear Modeling and Analysis of Solids and Structures* (Cambridge University Press, Cambridge, 2009)
4. M. Greco, R.C.G. Menin, I.P. Ferreira, F.B. Barros, Comparison between two geometrical nonlinear methods for truss analyses. Struct. Eng. Mech. **41**(6), 735–750 (2012)
5. Y.C. Toklu, Nonlinear analysis of trusses through energy minimization. Comput. Struct. **82**(20–21), 1581–1589 (2004)

6. G. Bekdas, A.E. Kayabekir, S.M. Niğdeli, Y.C. Toklu, Advanced Energy Based Analyses of Trusses employing Hybrid Metaheuristics. Struct. Des. Tall Spec. Build. **28**(9), e1609 (2019)
7. Y.C. Toklu, F. Uzun, Analysis of tensegric structures by total potential optimization using metaheuristic algorithms. J. Aerosp. Eng. **29**(5), 04016023 (2016)
8. A.E. Kayabekir, Y.C. Toklu, G. Bekdaş, S.M. Nigdeli, M. Yücel, Z.W. Geem, A novel hybrid harmony search approach for the analysis of plane stress systems via total potential optimization. Appl. Sci. **10**(7), 2301 (2020)
9. Z.W. Geem, J.H. Kim, G.V. Loganathan, A new heuristic optimization algorithm: harmony search. Simulation **76**, 60–68 (2001)
10. X.S. Yang, Flower pollination algorithm for global optimization, in *Unconventional computation and natural computation* (2012), pp. 240–249
11. R.V. Rao, V.J. Savsani, D.P. Vakharia, Teaching–learning-based optimization: a novel method for constrained mechanical design optimization problems. Comput. Aided Des. **43**(3), 303–315 (2011)
12. R. Rao, Jaya: A simple and new optimization algorithm for solving constrained and unconstrained optimization problems. Int. J. Ind. Eng. Comput. **7**(1), 19–34 (2016)
13. R.D. Cook, *Finite Element Modelling for Stress Analysis* (Wiley, USA, 1995) 0-471-10774-3
14. H. Gazi, Sonlu Elemanlar Yönemi ile Çalışan Bir Bilgisayar Programının Geliştirilmesi ve Düzlem Gerilme – Düzlem Deformasyon Problemlerinin Analizi, MSc thesis, Istanbul University, Turkey (2008)

Performance of the Whale Optimization Algorithm in Space Steel Frame Optimization Problems

Bahar Neslişah Kale, İbrahim Aydoğdu, and Emre Demir

Abstract Frame optimization that contains highly non-linear and irregular functions and discrete design variables is one of the most challenging optimization problems. Therefore, gradient-based optimization techniques cannot be successful in such problems. Metaheuristic techniques, especially population-based metaheuristic techniques, perform highly effective in solving the frame optimization problem. However, stochastic processes' performances included in metaheuristic techniques vary based on the problem. Accordingly, researches on the performance of novel metaheuristic techniques on challenging engineering problems continue. One of the novel metaheuristic techniques is the whale optimization algorithm (WOA) which is inspired by the bubble-net feeding behavior of humpback whales. The aim of this study is testing the performance of WOA for space steel frame optimization problems. For this purpose, WOA-cased frame optimization program will be developed. Benchmark frame structures are selected to compare optimum solutions with literature results.

Keywords Structural optimization · Metaheuristics · Whale optimization algorithm · Steel frames

B. N. Kale · İ. Aydoğdu (✉)
Akdeniz University, Antalya 07058, Turkey
e-mail: aydogdu@akdeniz.edu.tr

B. N. Kale
e-mail: baharneslisahkale@gmail.com

E. Demir
Antalya Bilim University, Antalya 07190, Turkey
e-mail: emre.demir@antalya.edu.tr

S. M. Nigdeli et al. (eds.), *Proceedings of 6th International Conference on Harmony Search, Soft Computing and Applications*, Advances in Intelligent Systems and Computing 1275, https://doi.org/10.1007/978-981-15-8603-3_13

1 Introduction

In the area of civil engineering, optimization is used for solving a variety of problems [1–7]. One of the most encountered problems is determining the positions and the dimensions of the frames. For this case, frame optimization commonly becomes the challenge in order to build an optimized structure. Because it includes challenges such as non-linearity in functions and discrete variables for the design, it develops a difficult problem to find solutions. Due to such reasons, gradient-based optimization techniques are not very adequate for achieving viable solutions, because of highly non-linear and irregular functions in problems. In order to possess reliable outcomes from the optimization, metaheuristic techniques are preferred. If there are functions including non-linearity and also discrete variables, population-based metaheuristic techniques are commonly used. They not only perform very effective and also are highly operable for having solutions of frame optimization problems. On the other hand, in metaheuristic techniques, the performances of stochastic processes show different reactions from time to time. According to the type of the problem, or the inputs of the optimization, the performance and thus the outputs can change. Therefore, the studies go on related to the performance of novel metaheuristic techniques, to overcome several engineering optimization difficulties. In this direction, recently one of the novel metaheuristic techniques is discovered. This is the whale optimization algorithm (WOA) which is nature-based and inspired by the bubble-net behavior of humpback whales. Since this up-to-date algorithm has advanced properties for solving such problems, we wonder the performance of this algorithm for frame optimization. Consequently, the main goal of this study is actually examining the performance of WOA for space steel frame optimization problems. In order to test the performance, a computer script which is based on WOA will be established with an application of frame structure optimization. Measures will be selected from the frame structures for the application, and the results will be compared by the results produced in earlier related studies. Further, the discussions will be made on the outcomes of this research. Next, size optimization techniques will be explained utilizing steel frames. After that, the novel algorithm WOA will be detailed for our study, then a design example will be indicated with its main results explored. Last but not least, a conclusion will be presented.

2 Size Optimization of Steel Frames

Ideal optimization plan of steel space structures involves picking optimum pieces for its members from accessible steel segments rundown to fulfill the workableness, strength, and geometric necessities expressed in the codes for design. Meanwhile, the structure is built inexpensively. For the most part, the optimization issues for enhancement have triple sections which are objective functions, structure factors as design variables and requirements as constraints. The scientific model of the design optimization issues reliant on the steel configuration code is taken into account by formulae and explained as follows.

2.1 *Objective and Variables*

The optimization in structures is usually proposed for limiting or minimizing the material expense of the structure. For the structures made of steel, the material expense is legitimately identified with the heaviness of the structure. Consequently, the target capacity of the optimization issue is chosen as considering the weight of the steel structure stated as follows:

$$\text{Minimize}\,W(X) = \sum_{i=1}^{N_G} m_i \cdot \left(\sum_{j=1}^{n_i} l_j \right) \tag{1}$$

In order to discover the steel parts of the optimum design, where W represents the weight of structure, m_i is the unit weight of steel segment associated with the ith group, n_i corresponds to the total member number for the group i, N_G conforms to the total number of design sets, l_j signifies the length of jth section in group i, and $X = [X_1, X_2, X_3,\ldots, X_{NG}]$ denotes the sequence number vector for steel members picked for design clusters using standards characterized as the design variable of the size optimization.

2.2 *Constraints*

2.2.1 Constraints of Strength

Every frame section in the structure has adequate strength to oppose the interior forces created because of external loads that are assigned by factors. In the analysis, the pin-jointed structures and the moment-resisting frame are considered. In the following, the constrains of strength for members are depicted in Eqs. (2) and (3).

$$g_s(X) = \frac{P_u}{\varnothing_c P_n} + \frac{8}{9}\left(\frac{M_{ux}}{\varnothing_b M_{nx}} + \frac{M_{uy}}{\varnothing_b M_{ny}} \right) - 1.0 \le 0 \; for \; \frac{P_u}{\varnothing P_n} \ge 0.2 \tag{2}$$

$$g_s(X) = \frac{P_u}{2\varnothing_c P_n} + \left(\frac{M_{ux}}{\varnothing_b M_{nx}} + \frac{M_{uy}}{\varnothing_b M_{ny}} \right) - 1.0 \le 0 \; for \frac{P_u}{\varnothing P_n} < 0.2 \tag{3}$$

In Eqs. (2) and (3), M_{nx} and M_{ny} represent the capacities of nominal moment at primary and secondary axis, respectively. Also, M_{ux} and M_{uy} signify the design moments that member exposed to, at primary and secondary axis correspondingly as well. $\emptyset_c$ and $\emptyset_b$ symbols characterize the resistance factors for compression and bending, respectively. P_n denotes the member axial load capacity, while P_u symbolizes the ultimate axial force that member exposed to. P_n and P_u are tension or compression. The moment capacities of steel frame structures are computed bearing in mind geometric non-linearity implemented as an optional method noted in Chapter C of the LRFD-AISC (LRFD 2000). To compare with literature studies, geometric non-linearity is not taken into account for the pin-jointed dome. Geometric non-linearity is not considered for the pin-jointed dome in comparison with the research completed previously.

2.2.2 Constraints for Displacement

The limitations for the inter-storey drift and top storey drift are examined for deflection of steel frame structures with the equations.

$$g_d(X) = \frac{\delta_{j,l}}{\delta_{al}} - 1.0 \leq 0 \,\text{for}\, j = 1, 2, \ldots, n_{sm} \,\text{and}\, l = 1, 2, \ldots, n_{lc} \tag{4}$$

$$g_{td}(X) = \frac{\Delta_{j,l}^{top}}{\Delta_{al}^{top}} - 1.0 \leq 0 \,\text{for}\, j = 1, 2, \ldots, n_{jtop} \,\text{and}\, l = 1, 2, \ldots, n_{lc} \tag{5}$$

$$g_{id}(X) = \frac{\Delta_{j,l}^{is}}{\Delta_{al}^{is}} - 1.0 \leq 0 \,\text{for}\, j = 1, 2, \ldots, n_{st} \,\text{and}\, l = 1, 2, \ldots, n_{lc} \tag{6}$$

In Eqs. (4), (5), and (6), $\delta_{j,l}$ and δ_{al} represent allowable and calculated deflections of the member j at loading case l, $\Delta_{j,l}{}^{top}$, and $\Delta_{al}{}^{top}$ signify permissible and calculated top storey drifts (*TSD*) on the joint j at loading case l, $\Delta_{j,l}{}^{is}$, and $\Delta_{al}{}^{is}$ indicate the occurred and allowable inter-storey drifts on the storey j at loading case l, n_{sm} implies the total number of critical members exposed to deflection, n_{lc} denotes the number of loading cases expressed in the design problem, n_{jtop} symbolizes the number of joints on the top storey, n_{st} means the storey number. In this study, limits for allowable displacement are provided from the ASCE Ad Hoc Committee report [8]. The range of drift limits accepted is as follows. 1/750 to 1/250 times the height of structure for the top storey drift (*TSD*), 1/500 to 1/200 times the storey height for the inter-storey drift (*ISD*), and 1/500 to 1/200 times the beam length for the deflection denote the drift limits. Further, only the displacement limits of nodes are examined for the pin-jointed structures with the following equation.

$$g_{di}(X) = \frac{\delta i_{j,l}}{\delta i_{al}} - 1.0 \leq 0 \,\text{for}\, j = 1, 2, \ldots, n_{nd} \,\text{and}\, l = 1, 2, \ldots, n_{lc} \tag{7}$$

In Eq. (7), $\delta i_{j,l}$ and δi_{al} represent the allowable calculated displacements of the joint j for loading case l, while n_{sm} indicates the number of joints totally on the structure.

2.2.3 Geometric Constraints

Only the frame structures, geometric constraints can be specified as follows:

$$g_{cc}(X) = \sum_{i=1}^{n_{ccj}} \left(\frac{D_i^a}{D_i^b} - 1.0 \right) + \sum_{i=1}^{n_{ccj}} \left(\frac{m_i^a}{m_i^b} - 1.0 \right) \leq 0 \tag{8}$$

$$g_{bc}(X) = \sum\nolimits_{i=1}^{n_{j1}} \left(\frac{B_f^{bi}}{D^{ci} - 2t_b^{ci}} - 1.0 \right) \leq 0 or \sum\nolimits_{i=1}^{n_{j2}} \left(\frac{B_f^{bi}}{B_f^{ci}} - 1.0 \right) \leq 0 \tag{9}$$

In Eqs. (8) and (9), $n_{cc,j}$ represent the number of column-column connection type. While $m_i{}^a$ and $m_i{}^b$ denote the upper and lower column unit weights, respectively, for the ith column-column connection, $D_i{}^a$ and $D_i{}^b$ denote the depths of upper and lower columns for the same connection type. n_{j1} signifies the number of connection type 1: beams are connected to the web of a column, and n_{j2} implies the number of connection type 2: beams connected to the flange of a column. As can be seen in Fig. 1, D^{ci} shows depth of a column, $t_b{}^{ci}$ and $B_f{}^{bi}$ are the flange thickness and flange width of a beam for the connection type 1, also $B_f{}^{ci}$ and $B_f{}^{bi}$ symbolize the flange widths of the column and beam for the connection type 2.

2.3 *Evaluation of Fitness*

Metaheuristic calculations offer arrangement of unconstrained optimization issues. The unconstrained optimization issues are required to be changed into unconstrained problems by introducing a penalty work as given in Eq. (10). On the off chance that the structure configuration does not fulfill the requirements depicted in part 2.2, the absolute penalty (*Pen*) of the structure configuration is determined as follows.

$$Pen = \sum_{i=1}^{NC} C_i; C_i = \left\{ \begin{array}{c} 0 for\ g_i(X) \leq 0 \\ g_i(X) for\ g_i(X) > 0 \end{array} \right\} \tag{10}$$

Fig. 1 A sample of space frames depicting beam and column connection

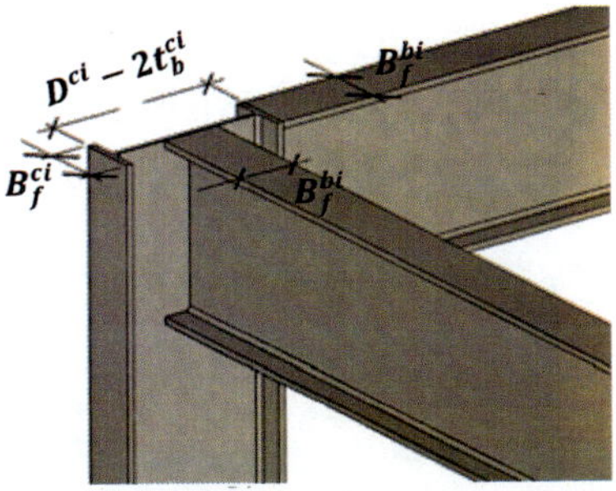

$$W_p = W \cdot (1 + Pen)^2 \tag{11}$$

$$J = \frac{1}{W_p} \tag{12}$$

C_i denotes violation of constraint in the ith constraint function, while g_i signifies the ith constraint function in Eq. (10). Moreover, depending on *Pen* value assessment, the weight of structure (W_p) can be calculated by Eq. (11), and the design fitness value (J) is acquired using Eq. (12).

3 Whale Optimization Algorithm

Whales are known as intelligent mammals. Among the kinds of whales, the humpback whales have unique behavior of hunting of prey. This special foraging behavior is called bubble-net feeding method [9].

Humpback whales mostly prefer to hunt krill and herds of small fishes close to the surface. It is observed that the humpback whales hunt in groups. They can detect how far the prey from herself and surround it. The leading whale dives and moves up with a circular or '9'-shaped path in approximately 10–15-m-deep and starts to create several various sized bubbles [10]. All created bubbles rise to the surface at the same time by forming a cylindrical or tubular bubble network. This spiral bubble net makes the prey collect toward the center of the net. So, the group of humpback whales reach the prey by following this spiral path and swallow the prey in the net [11].

In 2016, Mirjalili and Lewis [12] proposed a sole objective algorithm which is a stochastic population based, nature-inspired algorithm WOA inspired by the spiral bubble-net predation method of the humpback whales to catch their preys [13–16]. Three typical behaviors of the humpback whales are used to model WOA mathematically.

3.1 Mathematical Model

Each humpback whale is set as a search agent in order to determine the mathematical model of the searching path. Encircling prey, spiral bubble-net feeding maneuver, and search for prey are the three steps of the algorithm: encircling prey, spiral bubble-net feeding maneuver, and search for prey.

3.1.1 Encircling Prey

Humpback whales can detect the location of prey and encircle them. According to the WOA, the best current candidate solution in search space is the target prey or is close to the optimal solution. It assumes the best solution at the beginning because the position of the optimum is not known in this stage. After the best search agent is defined, along the path other search agents updates their positions according to it. The mathematical model of this is as follows:

$$\vec{X}(t+1) = \vec{X^*}(t) - \vec{A}\,\vec{D} \tag{13}$$

$$\vec{D} = |\vec{C} \cdot \vec{X^*}(t) - \vec{X}(t)| \tag{14}$$

where t is the current iteration,$\vec{X^*}$ is the optimal position vector, $\vec{X}$ is the position vector, $\vec{D}$ is the distance vector to define the distance between the prey and the whale, | | is the absolute value, and $\cdot$ is an element by element multiplication. In each iteration, $\vec{X^*}$ should be checked and if a better solution exists it should be replaced with it.

$\vec{A}$ and $\vec{C}$ are the coefficient vectors and they are calculated as follows:

$$\vec{A} = 2 \cdot \vec{a} \cdot \vec{r} + \vec{a} \tag{15}$$

$$\vec{C} = 2 \cdot \vec{r} \tag{16}$$

where $\vec{a}$ is linearly reduced from 2 to 0 over the course of iterations (including both exploration and exploitation phases) and $\vec{r}$ is a random vector in [0, 1]. The theoretical ground of the Eq. (13) in two-dimensional plane is shown in Fig. 2. As shown in the illustration, the position of the search agent (X, Y) can be changed according to the current best agent (X^*,Y^*). It is observable that if the position of (X^*,Y^*) is not selected well, the algorithm might easily fall into the local optimum. By altering the values of $\vec{A}$ and $\vec{C}$, any position close to the best agent can be obtained as demonstrated in Fig. 2.

In Fig. 3, the possible updating mechanism in three dimensions is revealed. Random vector $\vec{r}$ which exists in Eqs. (15) and (16) enables to reach any location in search space between the key points. Hence, according to the current best solution in the surrounding, the position of any search agent is updated by means of Eq. (13).

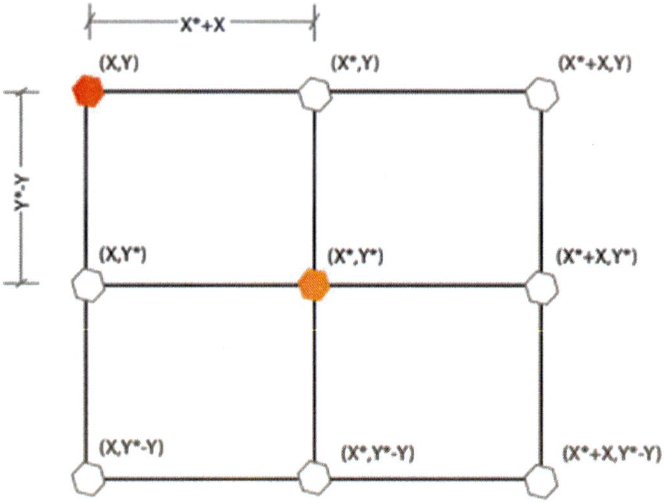

Fig. 2 The theoretical basis (position vectors with the possible following position in 2D) [12]

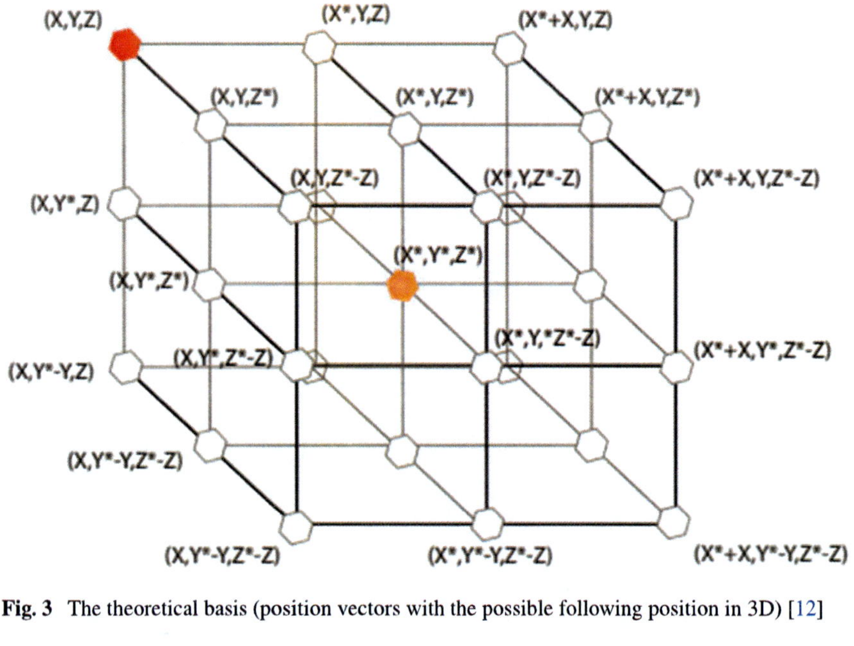

Fig. 3 The theoretical basis (position vectors with the possible following position in 3D) [12]

3.1.2 Bubble-Net Attacking Method (Exploitation Phase)

The main purpose of the bubble attacking mechanism, which is also referred to as exploitation phase, aims to use the obtained best solution as a result of exploration phase more efficiently. Two methods, which are introduced as follows, are used to mathematically model the bubble-net behavior of the Humpback whales:

Shrinking Encircling Mechanism

This mechanism occurs in case of decreasing value of "a" that reduces fluctuation range of the random variable A in [−a, a] (See Eq. 15). When random values for 'A' is set in $[-1, 1]$, the new position of a search agent is defined between the search agents at initial and the current best positions. The possible locations from (X, Y) through (X^*, Y^*) are illustrated in Fig. 4, which can be attained by $0 \leq A \leq 1$ in a 2D plane [12].

Spiral Updating Position

To achieve this approach, the distance between the position of search agent (X, Y) and the position of current best record (X^*, Y^*) is calculated at first. Then, a spiral equation is developed between the positions to update the position of whale by imitating the helix-shaped movement of whales as follows [12]:

$$\vec{X}(t+1) = e^{bl} \cdot cos(2\pi l) \cdot \vec{D} + \vec{X^*}(t) \tag{17}$$

$$\vec{D^*} = |\vec{X^*}(t) - \vec{X}(t)| \tag{18}$$

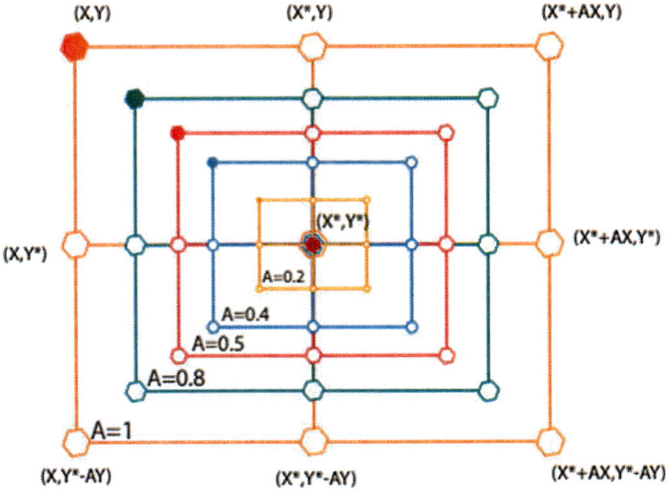

Fig. 4 Bubble-net search mechanism of WOA—shrinking encircling mechanism [12]

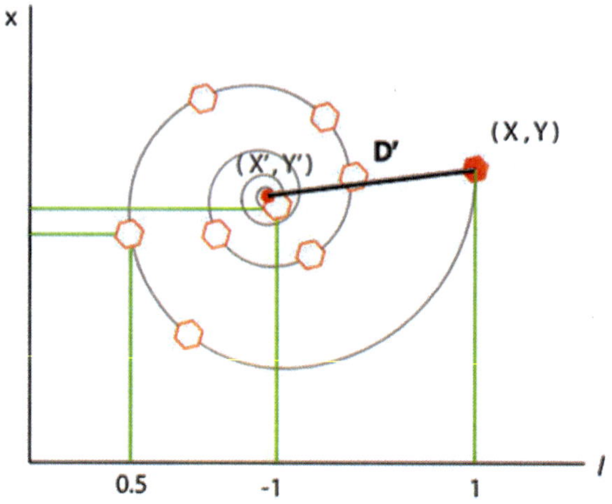

Fig. 5 Bubble-net search mechanism of WOA—spiral renewal position [12]

where b is constant which used to define the logarithmic spiral shape and l is a random value between the interval $[-1, 1]$ (Fig. 5).

The probability of that the whale either follows the shrinking encircling mechanism or spiral updating during the optimization is assumed as 50% for each, which is mathematically shown as follows:

$$\overrightarrow{X}(t+1) = \left\{ \begin{array}{l} \overrightarrow{X^*} - \overrightarrow{A}\,\overrightarrow{D},\ p < 0.5\, ise \\ e^{bl} \cdot cos(2\pi l) \cdot \overrightarrow{D} + \overrightarrow{X^*}(t),\ p \geq 0.5\, ise \end{array} \right\} \tag{19}$$

where p is a random variable in [0, 1].

3.1.3 Search for Prey (Exploration Phase)

The main object of exploration phase is to obtain a better solution. This phase allows to widen the search space by forcing search agents to move apart from each other by use of a random vector $\overrightarrow{A}$ with positive and negative values. $\overrightarrow{A}$ contains the values greater than 1 or less than -1.

$$\overrightarrow{D} = |\overrightarrow{C} \cdot \overrightarrow{X^*_{rand}}(t) - \overrightarrow{X}(t)| \tag{20}$$

$$\overrightarrow{X}(t+1) = \overrightarrow{X_{rand}} - \overrightarrow{A}\,\overrightarrow{D} \tag{21}$$

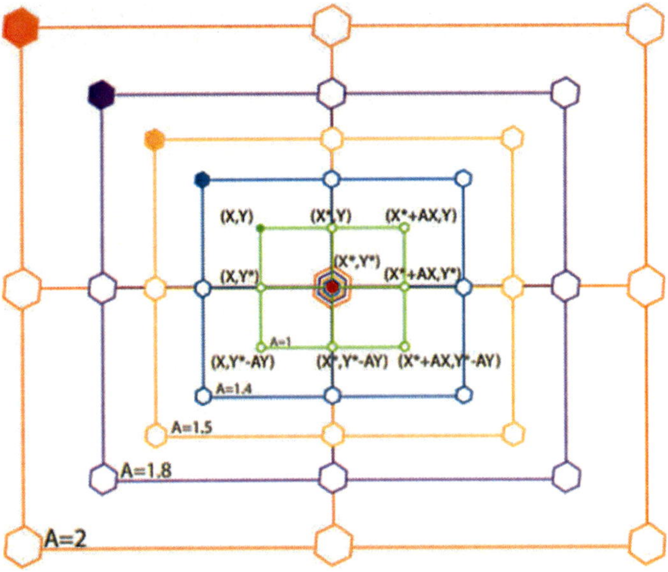

Fig. 6 Possible locations around a random search agent with A > 1 (Exploration Mechanism) [12]

where $\overrightarrow{X_{rand}}$ is random position vector that allows to explore search space wider (Fig. 6).

Vector $\overrightarrow{A}$ is used for both exploitation and exploration phases. The two conditions are concluded as follows in order to use $\overrightarrow{A}$:

1. $|\overrightarrow{A}|<1$: The location of the new search agent is updated to any location in between the current best and the starting position for shrinking the search space.
2. $|\overrightarrow{A}|>1$: The search agent chooses a random agent for exploration phase and permits the WOA to have a global search in search space.

3.1.4 Optimization Algorithm

The pseudo-code of the WOA algorithm is given below [12].

```
Generate the whale population randomly X_i (i = 1, 2, ..., n)
Calculate the fitness of each whale(solution)
x*=the best solution
Do t =1 to maximum number of iterations
        Do i =1 to each search agent
        Update a, A, C, l, and p
                If (p<0.5)
                        If (|A| < 1)
                            Update the position of the current search agent by (13)
                        else if (|A| >=1)
                            Select a random search agent (X_rand)
                            Update the position of the current search agent by (21)
                        end if
                else if (p >= 0.5)
                        Update the position of the current search by (17)
                end if
        end do
        Check if any search agent goes beyond the search space and amend it
        Calculate the fitness of each search agent
        Update x* if there is a better solution
  End do
  Return x*
```

4 Design Example

The optimum design of five-storey steel space frame which is taken from previous studies in the literature, having 105 members and 54 joints grouped into 11 independent design variables, is considered as the design example [17–20]. The related drawings of the building are illustrated in Figs. 7 and 8. Member grouping of the structure is illustrated in Table 1. In the design example, design loads are estimated according to ASCE 7–05. The combinations defined in the example are 1.2 D+1.6 L+0.5 S, 1.2 D+0.5 L+1.6 S, 1.2 D+1.6 W+0.5 L where D, L, S, and W represents dead, live, snow, and wind loads, respectively. For the given example, the steel wide flange I-profile list consisting of 272 ready sections are used to size the members of the structure.

The steel frame is optimized 50 times using the WOA with various seed values. The W-section designations, the lightest weight, and other related information are detailed in Table 2. The given table contains the results of the previous literature studies using same design circumstances as well. Among the presented results, the lightest weight obtained is 258.416 kN which is derived by SJSSSO algorithm. The estimated weight by WOA is 1.44% more than the lightest weight attained from previous studies which have the same design circumstances. The WOA has the second-best performance among given algorithms by an optimum weight of 262.13

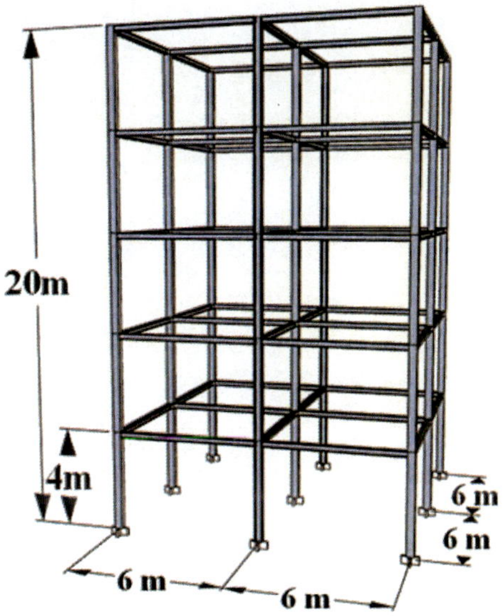

Fig. 7 3D view of the five storey, 105-member steel frame

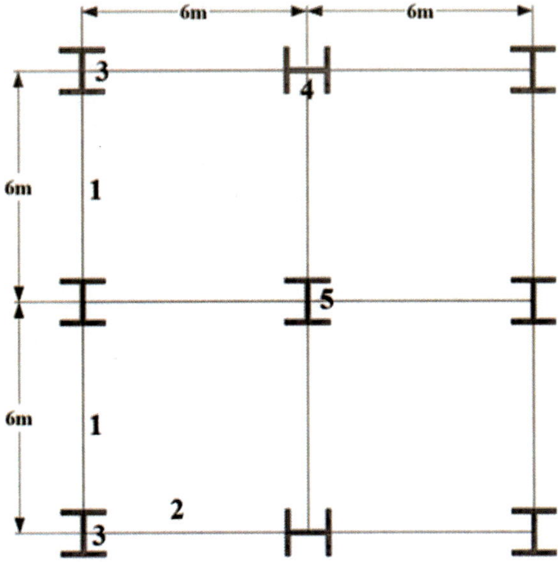

Fig. 8 Plan view of the five storey, 105-member steel frame [20]

Table 1 Group details the 105-member space frame

Definition	Story 1	Story 2–3	Story 4–5
Beams in X Dir	1	1	1
Beams in Y Dir	2	2	2
Corner columns	9	6	3
Sider columns	10	7	4
Inner columns	11	8	5

Table 2 Design sections and limit values of the optimum designs for the 105-member space frame

#	Type	ACO	HS	hTLBO-HS	SSO	SJSSSO	EFF	CSA	PSO	FF	WOA
1	Beam	W460X52	W530X66	W460X52	W460X52	W460X52	W410X46.1	W460X52	W460X52	W310X52	W410X46.1
2	Beam	W200X35.9	W310X38.7	W200X35.9	W360X44	W200X35.9	W200X41.7	W250X44.8	W200X41.7	W410X53	W310X38.7
3	Column	W200X35.9	W200X35.9	W310X38.7	W310X38.7	W310X38.7	W360X44	W610X82	W310X67	W360X101	W250X38.5
4	Column	W310X38.7	W200X35.9	W200X46.1	W200X41.7	W200X35.9	W250X49.1	W200X35.9	W460X113	W610X101	W200X46.1
5	Column	W360X57.8	W360X44	W360X44	W460X52	W410X53	W410X60	W610X125	W760X161	W310X117	W460X52
6	Column	W460X52	W310X38.7	W310X74	W310X38.7	W410X53	W410X60	W610X82	W310X67	W530X101	W360X51
7	Column	W310X86	W360X72	W250X73	W360X72	W250X73	W410X100	W250X67	W460X113	W760X134	W410X53
8	Column	W610X101	W610X92	W610X101	W760X134	W690X125	W530X123	W760X185	W760X161	W690X125	W920X201
9	Column	W530X66	W410X53	W460X74	W410X53	W410X60	W530X74	W690X170	W310X67	W690X140	W360X51
10	Column	W460X89	W360X72	W250X73	W460X74	W310X74	W460X106	W760X147	W460X113	W760X134	W460X74
11	Column	W690X170	W760X147	W760X173	W840X176	W760X134	W840X176	W920X488	W760X161	W690X170	W1100X343
Max. atrenght ratio		0.886	0.979	0.921	0.964	0.994	0.946	0.842	0.831	0.873	0.96
TopDrift (cm)		4.983	4.837	4.708	4.945	5.083	4.822	3.95	3.663	4.588	4.959
Inter storey drift (cm)		0.569	1.333	1.325	1.332	1.326	1.332	1.139	1.075	1.333	1.320
Max. deflection (cm)		0.378	0.146	–	0.188	0.28	–	–	–	–	–
Maximum iteration		50,000	50,000	50,000	50,000	50,000	50,000	50,000	50,000	50,000	50,000
Weight (kN)		265.38	278.196	269.184	270.976	258.416	284.04	336.05	338.38	391.06	262.13

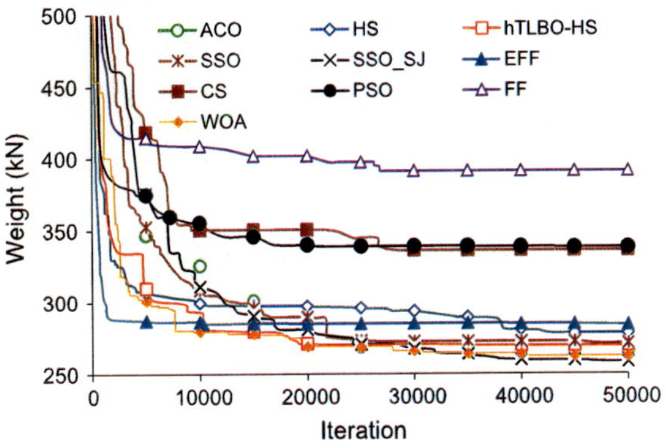

Fig. 9 Search histories of the best design for the design example

kN. Also, since maximum strength ratio has a value of 0.960 which is close to 1.00, it can be said that the third load combination has a major role in design process.

Also, Fig. 9 shows the graph of the optimum weight versus the number of iterations. As can be seen in Fig. 9, the second-best performance belongs to the WOA compared to the algorithms provided by the previous studies.

5 Conclusion

Research is carried on which focuses on the performances of up-to-date metaheuristic practices to solve a lot of puzzling engineering problems. One of them introduced recently, named WOA, is basically inspired by the hunting behavior characteristics of humpback whales. In this direction, this study examined the performance of WOA for a popular civil engineering problem of space steel frame optimization. For this reason, a WOA-cased frame optimization program was established. Accordingly, an application of benchmark frame structures was analyzed considering the measures selected from the frame structures. Then, a detailed comparison has been made for the optimum solutions inspecting the results possessed by the related studies previously. The same design circumstances as used in the earlier studies of the particular frame structure were used in our analysis. Although the outputs indicate that the lightest optimum structure weight was derived by the SJSSSO algorithm, the estimated optimum weight by WOA is only 1.44% more than the lightest weight. In this case, the methodology of WOA provided the second-best performance with an optimum weight of 262.13 kN compared to the other algorithms applied for the particular frame structure. Additionally, the maximum strength ratio derived by WOA has the value of 0.960. Therefore, the third load combination with WOA plays an important role in the design process.

References

1. T.C. Matisziw, E. Demir, Inferring network paths from point observations. Int. J. Geogr. Inf. Sci. **26**(10), 1979–1996 (2012)
2. T.C. Matisziw, E. Demir, Measuring spatial correspondence among network paths. Geogr. Anal. **48**(1), 3–17 (2016)
3. E. Demir, Assigning convenient paths by an approach of dynamic programming, in *Proceedings book of MICOPAM 2018*, Dedicated to Professor Gradimir V. Milovanović on the Occasion of his 70th Anniversary, 196–199 (2018)
4. E. Demir, Approach for siting a support facility for transporting supplies in emergency cases in the Republic of Bulgaria, in *MATEC Web of Conferences* 234. EDP SciencesE. (2018), p. 06001
5. E. Demir, N.U. Kockal, Facility location determined by an iterative technique, in *Proceedings book of MICOPAM 2018*, Dedicated to Professor Gradimir V. Milovanović on the Occasion of his 70th Anniversary, 192–195 (2018)
6. E. Demir, Havalimanlarında kalkış öncesi, acil durumlarda, yardım alınabilecek en uygun lokasyonun Weber problemine uyarlanarak belirlenmesi. Türk Coğrafya Dergisi **70**, 81–85 (2018)
7. E. Demir, N.U. Kockal, Iterative methodology on locating a cement plant. J. Inequalities Appl. **174**, 1–8 (2019)
8. B. Ellingwood, D.E. Allen, M. Elnimeiri, T.V. Galambos, H. Iyengar, L.E. Robertson, J. Stockbridge, C.J. Turkstra, Structural serviceability: a critical appraisal and research needs. J. Struct. Eng. **112**(12), 2646–2664 (1986)
9. P.R. Hof, E. Van der Gucht, Structure of the cerebral cortex of the humpback whale, Megaptera novaeangliae (Cetacea, Mysticeti, Balaenopteridae), in *The Anatomical Record: Advances in Integrative Anatomy and Evolutionary Biology: Advances in Integrative Anatomy and Evolutionary Biology*, vol. 290, no. 1 (2007), pp. 1–31
10. J.A. Goldbogen, A.S. Friedlaender, J. Calambokidis, M.F. Mckenna, M. Simon, D.P. Nowacek, Integrative approaches to the study of baleen whale diving behavior, feeding performance, and foraging ecology. Bioscience **63**(2), 90–100 (2013)
11. W.A. Watkins, W.E. Schevill, Aerial observation of feeding behavior in four baleen whales: Eubalaena glacialis, Balaenoptera borealis, Megaptera novaeangliae, and Balaenoptera physalus. J. Mammal. **60**(1), 155–163 (1979)
12. S. Mirjalili, A. Lewis, The whale optimization algorithm. Adv. Eng. Softw. **95**, 51–67 (2016)
13. A. Kaveh, M. Ilchi Ghazaan, Enhanced whale optimization algorithm for sizing optimization of skeletal structures, Mech. Based Des. Struct. Mach. **45**(3), 345–362 (2017)
14. S. Adhirai, R.P. Mahapatra, P. Singh,The whale optimization algorithm and its implementation in MATLAB. Int. J. Comput. Inf. Eng. **12**(10), 815–822 (2018)
15. W.-Z. Sun, J.-S. Wang, X. Wei, An improved whale optimization algorithm based on different searching paths and perceptual disturbance. Symmetry **10**(6), 210 (2018)
16. P. Jangir, N. Jangir, Non-dominated sorting whale optimization algorithm (NSWOA): a multi-objective optimization algorithm for solving engineering design problems. Glob. J. Res. Eng. (2017)
17. İ.Aydoğdu, Optimum design of 3-d irregular steel frames using ant colony optimization and harmony search algorithms (2010)
18. A. Akin, I. Aydogdu, Optimum design of steel space frames by hybrid teaching-learning based optimization and harmony search algorithms. World Acad. Sci., Eng. Technol. Civ. Environ. Eng. **2**(7), 739–745 (2015)
19. S. Carbas, Design optimization of steel frames using an enhanced firefly algorithm. Eng. Optim. **48**(12), 2007–2025 (2016)
20. I. Aydogdu, P. Efe, M. Yetkin, A. Akin, Optimum design of steel space structures using social spider optimization algorithm with spider jump technique. Struct. Eng. Mech. **62**(3), 259–272 (2017)

Evaluation of Metaheuristic Algorithm on Optimum Design of T-Beams

Aylin Ece Kayabekir, Gebrail Bekdaş, and Sinan Melih Nigdeli

Abstract In structural engineering and optimum design of reinforced concrete (RC) members, the existing design constraints formulated according to design regulations proposed for stress–strain capacity of members prevent the direct calculation of optimum design variables which are cross-sectional dimensions. In that case, metaheuristic methods can be easily used to find the optimum dimensions to minimize the maximum cost of an RC member. In this study, several metaheuristic algorithms such as harmony search (HS), teaching–learning-based optimization (TLBO), flower pollination algorithm (FPA) and Jaya algorithm (JA) are employed to find the optimum cross-sectional dimension of T-beams using design regulation: Eurocode 2. According to the results, TLBO has advantages comparing to other algorithms for the optimization problem.

Keywords Reinforced concrete · T-beams · Structural optimization · Metaheuristic algorithms · Eurocode 2

1 Introduction

In the engineering design of reinforced concrete (RC) members, the safety and minimum cost are both important measures. In that case, it is needed to find the optimum results via iterative techniques since the existence of design variables about the stress–strain regulations presented in design codes. In this process, metaheuristic algorithms are an effective way to find the design with the minimum cost by

A. E. Kayabekir (✉) · G. Bekdaş · S. M. Nigdeli
Department of Civil Engineering, Istanbul University-Cerrahpaşa, 34320 Avcılar, Istanbul, Turkey
e-mail: ecekayabekir@gmail.com

G. Bekdaş
e-mail: bekdas@istanbul.edu.tr

S. M. Nigdeli
e-mail: melihnig@istanbul.edu.tr

S. M. Nigdeli et al. (eds.), *Proceedings of 6th International Conference on Harmony Search, Soft Computing and Applications*, Advances in Intelligent Systems and Computing 1275, https://doi.org/10.1007/978-981-15-8603-3_14

considering design constraints. Metaheuristic algorithms include a metaphor of life to formalize different features in the generation of random design variables which are checked if these variables are optimums or not. By the elimination of weak solution in the mean of the objective function, the optimum design variables with the best objective function are found.

The restriction to use mathematical methods in the optimum design of RC members is the different stress–strain behaviors of material (steel and concrete) generating RC members. Also, these materials are needed to be used because of the different requirements. Up to now, several metaheuristics have been employed in the optimum design of RC members.

The best known evolutionary-based algorithm called Genetic Algorithm (GA) was used to design RC beams optimally [1–3]. RC continuous beams were optimized via a hybrid metaheuristic algorithm of simulated annealing and genetic algorithm by Leps and Sejnoha [4]. Harmony search (HS) algorithm is another metaheuristics, which was employed to optimize continuous beams [5] and T-beams [6, 7]. Bat algorithm was also used in the optimum design of doubly reinforced beams for different flexural moment cases [8]. The efficiency of teaching–learning-based optimization (TLBO) was also checked via the optimum design of RC beams [9]. Jaya algorithm (JA) was also employed for the optimum design of T-beams via Eurocode 2 [10, 11].

The mentioned studies about the optimum design of RC beams are related to cost optimization. CO_2 emission was also considered as the objective function by using metaheuristics [12, 13]. Different from RC beams, columns [14, 15], frames [16, 17], retaining walls [18–22] and slabs [23] are the other RC members that were optimized via metaheuristic methods.

In the present study, the optimization problem solved via GA by Fedghouche and Tiliovine [3] and JA by Kayabekir et al. [10] was investigated for the evaluation of four different algorithms. These new generation algorithms are HS [24], TLBO [25], FPA [26] and JA [27].

2 The Optimization Process

The optimization process is summarized in the flow chart given as shown in Fig. 1. After the definition of design constants (Table 1) and ranges of design variables, the initial solution matrix is generated by randomly assigning design variables as the number of population. The design variables are shown as in Table 2 with the range for a T-beam shown in Fig. 2.

For each set of design variables, the objective function given as in Eq. (1) is calculated. If the design constraints given in Table 3 with respect to Eurocode 2 [11] are not provided, the objective function is penalized with a very big value. After the generation of the initial solution matrix, it is iteratively updated by generating new sets according to algorithm rules.

$$C = b_w d + (b - b_w)h_f + (\mathrm{C}_s/\mathrm{C}_c)A_s \tag{1}$$

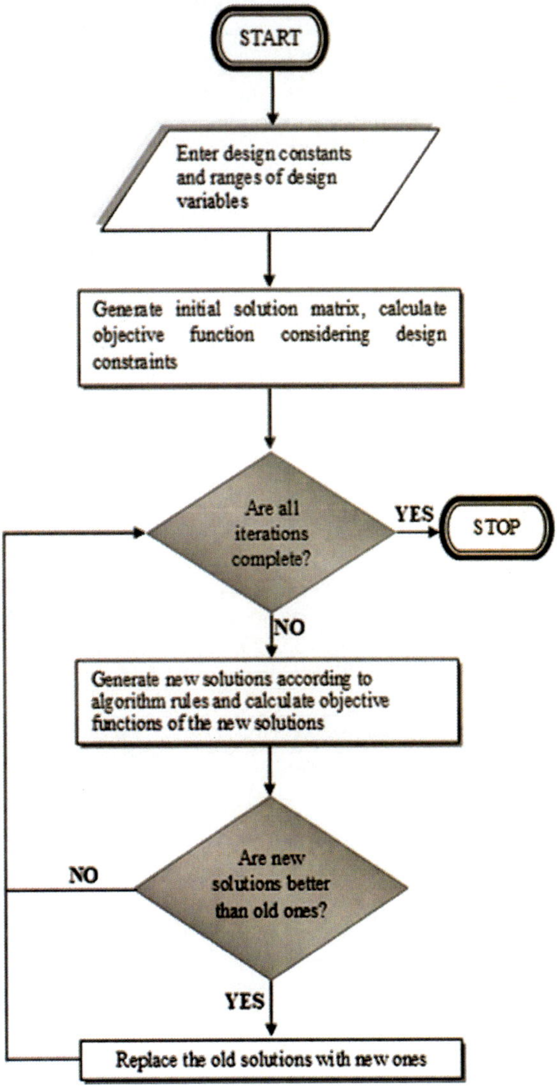

Fig. 1 The flow chart of the optimization process

The objective function is the total cost of the RC beam (C). For the design constraints, w, p, M_{Ed1} and V_{Ed1} are calculated according to formulations given as in Eqs. (2–7). The optimization code written via Matlab [28] is given in Appendix for TLBO.

$$\omega = (f_{yd}/f_{cd})(A_s/b_w d) - (b - b_w)h_f/(b_w d) \tag{2}$$

$$\rho = A_s/(b_w d) \tag{3}$$

Table 1 Design constants

Symbol	Definition
f_{ck}	Characteristic compressive strength for concrete
f_{cd}	Allowable compressive strength for concrete
f_{yd}	Characteristic yield strength of reinforcement
ρ_{max}	The maximum reinforcement ratio
ρ_{min}	The minimum reinforcement ratio
L	The length of beam
E_s	Young's elastic modulus for steel
M_{Ed}	The ultimate bending moment capacity
V_{Ed}	The ultimate bending moment capacity
C_s	The unit total cost of reinforcing steel
C_c	The unit total cost of concrete

Table 2 Design variables

Symbol	Definition	Range
b	Effective width of compressive flange (mm)	$b_w \leq b \leq \min[0.2L + b_w, 8h_f]$
b_w	Web width (mm)	$0.20d \leq b_w \leq 0.40d$
h	Height (mm)	$L/16 \leq h \leq 2.0$
h_f	Flange depth (mm)	$0.15 \leq h_f \leq d$
d	Effective depth (mm)	$d = 0.9h$
d_s	Cover of reinforcements (mm)	$d_s = 0.1h$
A_s	Area of reinforcing steel (mm^2)	$0 \leq A_s \leq 0.1$

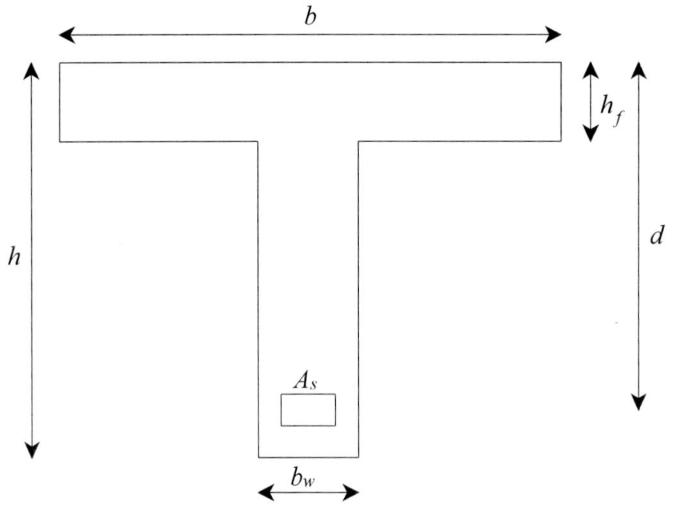

Fig. 2 T-shaped cross section

Table 3 Design constraints

Constraint number	Constraints
1	$\omega(1 - 0.5\omega) \leq 0.392$
2	$0.0035(0.8 - \omega)/\omega \geq f_{yd}/E_s$
3	$\rho_{min} \leq \rho \leq \rho_{max}$
4	$M_{Ed} \leq M_{Ed1}$
5	$V_{Ed} \leq V_{Rd\,max}$

$$M_{Ed1} = f_{cd}(b - b_w)h_f\left(d - 0.50h_f\right) + f_{cd}b_w d^2\omega(1 - 0.5\omega) \tag{4}$$

$$V_{Rd\,max} = \nu_1 f_{cd} b_w z/(\tan(45) + \cot(45)) \tag{5}$$

$$\nu_1 = 0.6(1 - f_{ck}/250) \tag{6}$$

$$z = 0.9d \tag{7}$$

3 The Numerical Example

The values of design constants for the numerical example are given in Table 4. According to the classical solution and GA approach, the objective function values, which are considered as C/C_c, are 1.117272 and 0.999221, respectively.

As seen in the optimum results given in Table 5 for 30 cycles of optimization, it is possible to find slightly better results than the GA approach in most cycles of all the algorithms.

Table 4 Design constants of numerical example [3]

Symbol	Values
f_{ck}	20 MPa
f_{cd}	11.33 MPa
f_{yd}	348 MPa
ρ_{max}	0.04
ρ_{min}	0.0013
L	20 m
E_s	200,000 MPa
M_{Ed}	4.991 N m
V_{Ed}	1.039 N
C_s/C_c	36

Table 5 Optimum design variables

Design variables	FPA	TLBO	HS	JA
b (m)	1.13879373	1.137632212	1.144541099	1.137605021
b_w (m)	0.30432074	0.304358948	0.304095428	0.304361011
h (m)	1.69067078	1.690883044	1.689419046	1.690894506
h_f (m)	1.5216037	1.52179474	1.520477141	1.521805056
d (m)	0.15	0.15	0.15	0.15
A_s (m^2)	0.01141132	0.011413083	0.011405725	0.01141303
ω	0.48660829	0.486991788	0.485020591	0.486991176
C/C_c	0.99903391	0.999033809	0.999043105	0.999033809
F_{min}	0.99903391	0.999033809	0.999043105	0.999033809
F_{mean}	1.00182853	0.999033816	1.000081551	1.000120713
F_{std}	0.00546358	5.27E−09	0.005435734	0.005917898
Mean iteration	22869.9	41595.36667	53446.31333	83928.8

3.1 Conclusions

The optimization process was done for 30 cycles for the evaluation of the employed algorithms via the minimum objective function (F_{min}), mean of the objective function (F_{mean}), standard deviation (F_{std}) and mean number of iterations. The objective function results of cycles and the iteration number at the optimum results are given as shown in Figs. 3 and 4, respectively.

All the algorithms are effective to find optimum results with a little difference. When the results are carefully investigated, it is seen that TLBO is effective to find

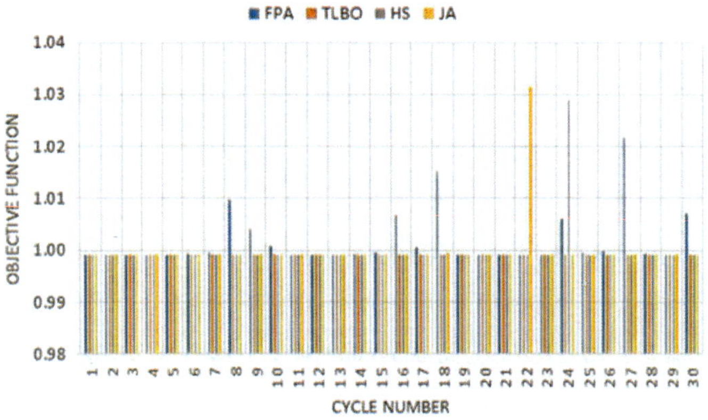

Fig. 3 The objective function

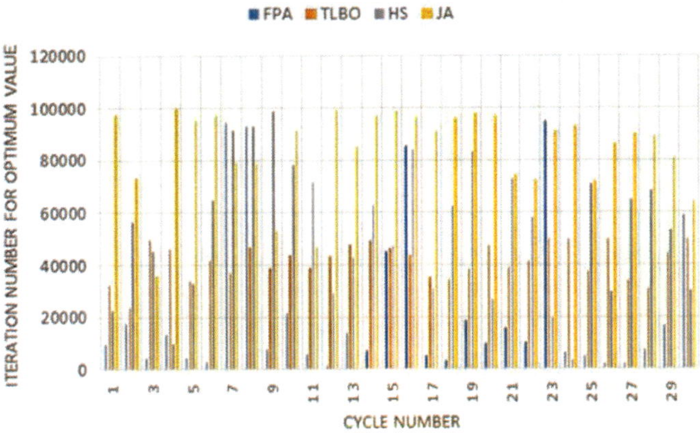

Fig. 4 Iteration number for optimum values

the best minimum cost, and also, it has a very small standard deviation value. On the other hand, FPA needs less iterations that the others to reach the optimum value in a cycle.

Appendix

```
clear all
clc

%1.ENTER DATA OF PROBLEM
fck=20;
fyd=348;
Es=200000;
fcd=11.33;
pmax=0.04; %In accordance with EC2
pmin=0.0013; %In accordance with EC2
L=20;
MEd=4.991;
VEd=1.039;
Cs=36;
Cc=1;
Cf=0;
pn=10; % pn is population number defined by user
maxiter=50000; %maximum iteration number defined by user
```

%2. GENERATING OF INITIAL SOLUTION MATRIX
for cycle=1:30
for i=1:pn % During the this loop, candidate solution vectors including design variables are generated up to pn.

% End of the this loop, candidate solution vectors provided design constraint are stored in a matrix.

%2.1. DEFINITION OF DESING VARIABLES

%NOTE: Mentioned design variables are going to be between its ultimate limits and generated randomly.

C=10^6; %C is value of objective function and 10^6 is penalization value.

while C==10^6 %With this compute, design variables which are proper to design constraints are obtained only.

```
hmin=L/16;          %hmin is lower bound of h
hmax=2.0;           %hmax is upper bound of h
```

h=hmin+rand*(hmax-hmin); %Generation of h between ultimate limits and randomly

```
d=0.9*h;  %d was generated according to h
ds=0.1*h; %ds was generated according to h

bwmin=0.2*d;         %bwmin is lower bound of bw
bwmax=0.4*d;         %bwmax is upper bound of bw
bw=bwmin+rand*(bwmax-bwmin); %Generation of bw randomly

hfmin=0.15;          %hfmin is lower bound of hf
hfmax=d;             %hfmax is upper bound of hf
hf=hfmin+rand*(hfmax-hfmin); %Generation of hf randomly

bmin=bw;             %bmin is lower bound of b
bmax=min(0.2*L+bw,8*hf);  %bmax is upper bound of b
b=bmin+rand*(bmax-bmin);  %Generation of b randomly

Asmin=0;             %As is lower bound of As
Asmax=0.1;           %As is upper bound of As
As=Asmin+rand*(Asmax-Asmin); %Generation of As randomly
```

% 2.2. OTHER CALCULATIONS ACCORDING TO EC2

w=(fyd/fcd)*(As/(bw*d))-(b-bw)*hf/(bw*d); % This function was defined with Eq.(2)
p=As/(bw*d); % The reinforcement ratio was defined with Eq.(3)

```
if p<pmin
   p=pmin;
end
```

MEd1=fcd*(b-bw)*hf*(d-0.5*hf)+fcd*bw*d^2*w*(1-0.5*w); % This function was defined with Eq.(4)
v1=0.6*(1-fck/250); % The non-dimensional coefficient was defined with Eq.(6)
z=0.9*d; % The lever arm was defined with Eq.(7)
VRdmax=v1*fcd*bw*z/(tand(45)+cotd(45)); % This function was defined with Eq.(5)

% 2.3. OBJECTIVE FUNCTON

%NOTE: The values of objective function is calculated in every cycle and through minimum value of objective function

%the candidate solution vector including values of candidate optimum design variables can be found .

C=bw*d+(b-bw)*hf+(Cs/Cc)*As+Cf/Cc*(b+2*(ds+d));

% 2.4. PART OF PENALIZATION

%NOTE: In this part,if generated any design variables do provide constraints, objective function is penalized with a large value.

%With this penaltization, selection of a solution vector including improper values of generated design variable is prevented.

```
if w*(1-0.5*w)>0.392
   C=10^6;
end
if 0.0035*(0.8-w)/w<fyd/Es
   C=10^6;
end
if p>pmax
   C=10^6;
end
if MEd>MEd1
   C=10^6;
end
if VEd>VRdmax
   C=10^6;
```

```
    end
  end

  % 2.5. INITIAL SOLUTION MATRIX

  %NOTE: In this part, all the design variables correspounding to minimum value of objetive function
  %and value of objective function were stored in a matrix named OPT.
  %Every column of this matrix is a candidate solution vector.

  OPT(1,i)=b;
  OPT(2,i)=bw;
  OPT(3,i)=h;
  OPT(4,i)=d;
  OPT(5,i)=hf;
  OPT(6,i)=As;
  OPT(7,i)=w;
  OPT(8,i)=C;
end

% 3. GENERATING OF NEW SOLUTION MATRIX

% NOTE: This part is very similar to PART 2. Differently, new solution variables generated.% and ultimate limits of variables are controlled.
% All steps during the PART3 are repeated untill all iteration is completed.

for iter=1:maxiter

  % 3.1. TEACHER PHASE

    for i=1:pn

      % 3.1.1. DEFINATION OF DESING VARIABLES

      % Values of candidate solution vector provided best/minimum value of objective function in existing solution set or matrix are assign to vector named best:
      [pe,re]=min(OPT(8,:));
      best1=OPT(1,re);
      best2=OPT(2,re);
      best3=OPT(3,re);
      best5=OPT(5,re);
      best6=OPT(6,re);


      b=OPT(1,i)+rand*(best1-round(1+rand)*mean(OPT(1,:)));
```

```
    bw=OPT(2,i)+rand*(best2-round(1+rand)*mean(OPT(2,:)));
    h=OPT(3,i)+rand*(best3-round(1+rand)*mean(OPT(3,:)));
    d=0.9*h;
    hf=OPT(5,i)+rand*(best5-round(1+rand)*mean(OPT(5,:)));
    As=OPT(6,i)+rand*(best6-round(1+rand)*mean(OPT(6,:)));
    ds=0.1*h;

    % subcode is called
    subcode_T_beam
  end

  % 3.2. LEARNER PHASE
  for i=1:pn

    % 3.2.1. DEFINATION OF DESING VARIABLES
    % New values of design variables are generated according to Eq :
    v1=ceil(rand*pn);
    v2=ceil(rand*pn);
    while v1==v2
        v1=ceil(rand*pn);
        v2=ceil(rand*pn);
  end
  if OPT(8,v1)<OPT(8,v2)
  b=OPT(1,i)+rand*(OPT(1,v1)-OPT(1,v2));
  bw=OPT(2,i)+rand*(OPT(2,v1)-OPT(2,v2));
  h=OPT(3,i)+rand*(OPT(3,v1)-OPT(3,v2));
  d=0.9*h;
  hf=OPT(5,i)+rand*(OPT(5,v1)-OPT(5,v2));
  As=OPT(6,i)+rand*(OPT(6,v1)-OPT(6,v2));
  ds=0.1*h;
  else
  b=OPT(1,i)+rand*(OPT(1,v2)-OPT(1,v1));
  bw=OPT(2,i)+rand*(OPT(2,v2)-OPT(2,v1));
  h=OPT(3,i)+rand*(OPT(3,v2)-OPT(3,v1));
  d=0.9*h;
  hf=OPT(5,i)+rand*(OPT(5,v2)-OPT(5,v1));
  As=OPT(6,i)+rand*(OPT(6,v2)-OPT(6,v1));
  ds=0.1*h;
  end

  subcode_T_beam
end
```

```
        [mi,ni]=min(OPT(8,:));
    Fmin(iter)=mi;
    mi
end
[cev,mev]=min(Fmin);
IN(cycle,1)=cev;
IN(cycle,2)=mev;
[mik,nik]=min(OPT(8,:));
Res{cycle}=OPT(:,nik);
end

[cev1,mev1]=min(IN(:,1));
INS(1,1)=min(IN(:,1));
INS(1,2)=mean(IN(:,1));
INS(1,3)=std(IN(:,1));
INS(1,4)=mean(IN(:,2));


save('TLBO')
```

subcode_T_beam.m

```
% 3.2. CONTROL OF ULTIMATE LIMITS
if h<L/16
   h=L/16;
end
if h>2
   h=2;
end
if bw<0.2*d
   bw=0.2*d;
end
if bw>0.4*d
   bw=0.4*d;
end
if hf<0.15
   hf=0.15;
end
if hf>d
   hf=d;
end
if b>0.2*L+bw
   b=0.2*L+bw;
```

```
end
if b>8*hf
   b=8*hf;
end
if b<bw
   b=bw;
end

% 3.3. OTHER CALCULATIONS
w=(fyd/fcd)*(As/(bw*d))-(b-bw)*hf/(bw*d);
p=As/(bw*d);
if p<pmin
   p=pmin;
end
MEd1=fcd*(b-bw)*hf*(d-0.5*hf)+fcd*bw*d^2*w*(1-0.5*w);
v1=0.6*(1-fck/250);
z=0.9*d;
VRdmax=v1*fcd*bw*z/(tand(45)+cotd(45));
% 3.4. OBJECTIVE FUVCTION
C=bw*d+(b-bw)*hf+(Cs/Cc)*As+Cf/Cc*(b+2*(ds+d));

% 3.5. PART OF PENALIZATION

%NOTE: In this part,if generated any design variables do providing constraint, objec-
tive function are penalized with a large value.
%With this penaltization, selection of a solution vector including improper values of
generated design variable is prevented.

if w*(1-0.5*w)>0.392
   C=10^6;
end
if 0.0035*(0.8-w)/w<fyd/Es
   C=10^6;
end
if p>pmax
   C=10^6;
end
if MEd>MEd1
   C=10^6;
end
if VEd>VRdmax
   C=10^6;
end

% 3.6. NEW SOLUTION MATRIX
OPT1(1,i)=b;
```

```
OPT1(2,i)=bw;
OPT1(3,i)=h;
OPT1(4,i)=d;
OPT1(5,i)=hf;
OPT1(6,i)=As;
OPT1(7,i)=w;
OPT1(8,i)=C;

% 3.7. PART OF COMPARISON

% NOTE: In this part, each of new candidate solution vectors in the OPT1  matrix is compared with old ones.
% If any new solution vector have a better value of objective function than old one, new values are replaced with old values.
% In this way, The matrix named OPT is updated in the every iteration.

if OPT(8,i)>OPT1(8,i)
   OPT(:,i)=OPT1(:,i);
end
```

References

1. C.C. Coello, F.S. Hernandez, F.A. Farrera, Optimal design of reinforced concrete beams using genetic algorithms. Expert Syst. Appl. **12**, 101–108 (1997)
2. V. Govindaraj, J.V. Ramasamy, Optimum detailed design of reinforced concrete continuous beams using genetic algorithms. Comput. Struct. **84**, 34–48 (2005). https://doi.org/10.1016/j.compstruc.2005.09.001
3. F. Fedghouche, B. Tiliouine, Minimum cost design of reinforced concrete T-beams at ultimate loads using Eurocode2. Eng. Struct. **42**, 43–50 (2012). https://doi.org/10.1016/j.engstruct.2012.04.008
4. M. Leps, M. Sejnoha, New approach to optimization of reinforced concrete beams. Comput. Struct. **81**, 1957–1966 (2003). https://doi.org/10.1016/S0045-7949(03)00215-3
5. A. Akin, M.P. Saka, Optimum detailed design of reinforced concrete continuous beams using the harmony search algorithm, in *The Tenth International Conference on Computational Structures Technology*, paper 131 (2010)
6. G. Bekdaş, S.M. Nigdeli, Cost optimization of t-shaped reinforced concrete beams under flexural effect according to ACI 318, in *3rd European Conference of Civil Engineering* (2012)
7. G. Bekdaş, S.M. Nigdeli, Optimization of T-shaped RC Flexural Members for Different Compressive Strengths of Concrete. Int. J. Mech. **7**, 109–119 (2013)
8. G. Bekdaş, S.M. Nigdeli, X. Yang, Metaheuristic Optimization for the Design of Reinforced Concrete Beams under Flexure Moments
9. G. Bekdaş, S.M. Nigdeli, Optimum design of reinforced concrete beams using teaching-learning-based optimization, in *3rd International Conference on Optimization Techniques in Engineering (OTENG'15)*, pp. 7–9 (2015)
10. A.E. Kayabekir, G. Bekdaş, S.M. Nigdeli, Optimum design of T-beams using Jaya algorithm, in *3rd International Conference on Engineering Technology and Innovation (ICETI)*, Belgrad, Serbia (2019)

11. EN (Veranst.): EN 1992–1–1 Eurocode 2: Design of concrete structures. Brussels: CEN (2005)
12. V.K. Koumousis, S.J. Arsenis, Genetic algorithms in optimal detailed design of reinforced concrete members. Comput-Aided Civ. Inf. **13**, 43–52 (1998)
13. V. Yepes, J.V. Martí, T. García-Segura, Cost and CO2 emission optimization of precast–prestressed concrete U-beam road bridges by a hybrid glowworm swarm algorithm. Autom. Constr. **49**, 123–134 (2015)
14. M.Y. Rafiq, C. Southcombe, Genetic algorithms in optimal design and detailing of reinforced concrete biaxial columns supported by a declarative approach for capacity checking. Comput. Struct. **69**, 443–457 (1998)
15. L.M. Gil-Martin, E. Hernandez-Montes, M. Aschheim, Optimal reinforcement of RC columns for biaxial bending. Mater. Struct. **43**, 1245–1256 (2010)
16. C.V. Camp, S. Pezeshk, H. Hansson, Flexural design of reinforced concrete frames using a genetic algorithm. J. Struct. Eng.-ASCE **129**, 105–11 (2003)
17. V. Govindaraj, J.V. Ramasamy, Optimum detailed design of reinforced concrete frames using genetic algorithms. Eng. Optimiz. **39**(4), 471–494 (2007)
18. B. Ceranic, C. Fryer, R.W. Baines, An application of simulated annealing to the optimum design of reinforced concrete retaining structures. Comput. Struct. **79**, 1569–1581 (2001)
19. C.V. Camp, A. Akin, Design of retaining walls using Big Bang–Big crunch optimization. J. Struct. Eng.-ASCE **138**(3), 438–448 (2012)
20. A. Kaveh, A.S.M. Abadi, Harmony search based algorithms for the optimum cost design of reinforced concrete cantilever retaining walls. Int. J. Civ. Eng. **9**(1), 1–8 (2011)
21. S. Talatahari, R. Sheikholeslami, M. Shadfaran, M. Pourbaba, Optimum design of gravity retaining walls using charged system search algorithm. Math. Probl. Eng. **2012**, Article ID 301628
22. R. Temur, G. Bekdaş, Teaching learning-based optimization for design of cantilever retaining walls. Struct. Eng. Mech. **57**(4), 763–783 (2016). https://doi.org/10.12989/sem.2016.57.4.763
23. M.G. Sahab, A.F. Ashour, V.V. Toropov, Cost optimisation of reinforced concrete flat slab buildings. Eng. Struct. **27**, 313–322 (2005). https://doi.org/10.1016/j.engstruct.2004.10.002
24. Z.W. Geem, J.H. Kim, G.V. Loganathan, A new heuristic optimization algorithm: harmony search. Simulation **76**, 60–68 (2001)
25. R.V. Rao, V.J. Savsani, D.P. Vakharia, Teaching–learning-based optimization: a novel method for constrained mechanical design optimization problems. Comput. Aided Des. **43**(3), 303–315 (2011)
26. X.S. Yang, Flower pollination algorithm for global optimization, in *International Conference on Unconventional Computing and Natural Computation* (Springer, Berlin, Heidelberg, 2012), pp. 240–249
27. R. Rao, Jaya: A simple and new optimization algorithm for solving constrained and unconstrained optimization problems. Int. J. Ind. Eng. Comput. **7**(1), 19–34 (2016)
28. T. MathWorks, *Matlab R2018a* (Natick, MA, 2018)

Multi-objective Optimization of the Reinforced Concrete Beam

Zhi-Yu Zhang, Zulkarnaen Gifari, Young-Kyu Ju, and Joong Hoon Kim

Abstract This paper introduces two kinds of multi-objective optimization algorithms. The optimal values are determined through multi-objective functions and various equality and inequality constraints. The optimal value results of the two algorithms with different parameters are discussed. A simplified optimization case of reinforced concrete beam was discussed that minimizes the total cost of reinforced concrete beams while complying with all strength and serviceability requirements for a given level of the applied load. This paper focuses on the differences between Multi-objective Harmony Search Algorithm (MOHSA) and Multi-objective Genetic Algorithm (MOGA) for reinforced concrete beam design subjected to a specified set of constraints by considering aspects of the Harmony Memory Considering Rate (HMCR) parameters in HSA and Population Mutation (Pm) parameters in GA. Through HSA and GA for RC beam problem, with same reference strength, the result using GA has a lower cost than using HSA.

Keywords Reinforced concrete beam · Multi-objective optimization · Harmony search algorithm · Genetic algorithm

Z.-Y. Zhang · Z. Gifari · Y.-K. Ju (✉) · J. H. Kim (✉)
Korea University, Anam-ro, Seongbuk-gu, Seoul, Korea
e-mail: tallsite@korea.ac.kr

J. H. Kim
e-mail: jaykim@korea.ac.kr

Z.-Y. Zhang
e-mail: Zhiyu199576@gmail.com

Z. Gifari
e-mail: gifariz@korea.ac.kr

S. M. Nigdeli et al. (eds.), *Proceedings of 6th International Conference on Harmony Search, Soft Computing and Applications*, Advances in Intelligent Systems and Computing 1275, https://doi.org/10.1007/978-981-15-8603-3_15

1 Introduction

1.1 Background

For the optimization design of reinforced concrete, the design method is to propose a design plan, and through mathematical analysis, to verify whether the requirements of the problem are satisfied.

The design of reinforced concrete beams is usually an iterative process. Firstly, the designer will assume the weight of the beam, determine the section of the beam, and then determine the resistance moment of the part in order to check whether it is suitable for a given applied bending moment. Repeat the process until finding the most suitable section. Due to the weight of the beam, the resistance moment of the section and the applied bending moment cannot satisfy the requirements in this process, and the above problems also exist in many practical cases [1]. However, this paper discussed the differences between Multi-objective Optimization HSA (MOHSA) and Multi-objective Optimization GA (MOGA) for reinforced concrete beams by considering aspects of the HMCR parameters in HSA and Pm parameters in GA.

1.2 Multi-objective Optimization

Multi-objective optimization problems show that the actual optimization problem involves not only optimizing one objective but simultaneously optimizing several conflicting objectives, include multiple objective functions to be optimized simultaneously and various equality and inequality constraints.

Generally, multi-objective optimization problems include multiple objective functions to be optimized simultaneously and various equality and inequality constraints. This can be generally formulated as follows:

$$Min\ fi(x) \quad i = 1, 2, 3, 4, \ldots, N \tag{1}$$

$$S.t.gj(x) = 0 \quad j = 1, 2, 3, 4, \ldots, M \tag{2}$$

$$hk(x) \leq 0 \quad k = 1, 2, 3, 4, \ldots, K \tag{3}$$

(Adopted from [2])

where N, M, and K are, respectively, the number of objective functions, the number of equality constraints, and the number of inequality constraints; and *fi* is the *i*th objective function, x is a variable representing the objective function. For the results of the multi-objective optimization problem, any two solutions $\times$ 1 and $\times$ 2 are decided with the dominant position. In this paper, the price and strength of the concrete are optimized and the dominant position of price or strength changes with different needs.

Harmony Search Algorithm and Genetic Algorithm. Harmony search algorithm has been proposed by Geem et al. [3]. The pitch of each musical instrument determines the esthetic quality, just as the fitness function value determines the quality of the decision variables. The HS algorithm proceeds as follows [4, 5]:

1. Initialize the optimization problem and algorithm parameters.
2. Harmony memory initialization.
3. New harmony improvisation.
4. Update the harmony memory.
5. Repetition of steps 3 and 4 until satisfied the termination criterion.

(Adopted from [4])

Genetic Algorithm (GA) was introduced by John Holland in 1960 under the concept of Darwin's theory of evolution. In general, GA is to search the solution of the problem in multiple directions and exchange and fuse information between different directions [1, 6]. This population has undergone an evolution of simulation. The GA for a particular problem proceeds as follows [7]:

1. A representation for potential solutions to the problem.
2. A way to create an initial population of potential solutions.
3. An evaluation function that plays the role of the environment, rating solutions in terms of their "fitness."
4. Genetic operators that alter the composition of children.
5. Values for various parameters that the genetic algorithm uses (population sizes, probabilities of applying genetic operators, etc.).

(Adopted from Michalwicz 1991)

2 Methodology

2.1 Case Study

An optimization case that minimizes the total cost of reinforced concrete beams while achieving maximum strength, the illustration of RC beam shows in Fig. 1.

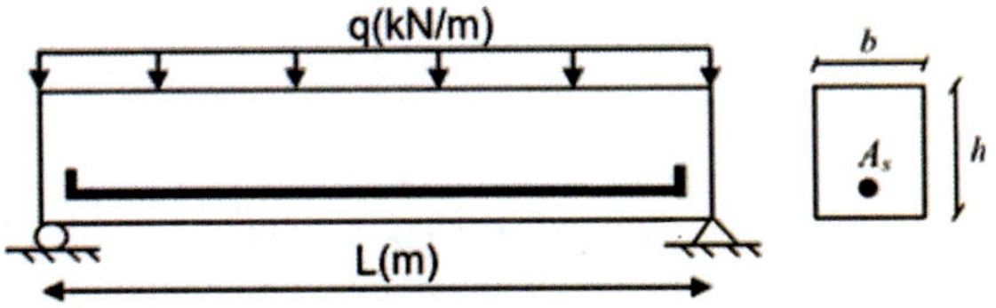

Fig. 1 Illustration of RC beam Adopted from [8]

This model follows the model of [8], and some modifications are made to it (revised some constraints) to make it suitable for practical problems.

As shown in Fig. 1, the beam span is L (m) and the load including the beam weight is q (kN/m) are constant. The purpose of the design is to confirm the cross-sectional area of the reinforcement (As), the width of the beam(d), the depth of the beam(h), and the compressive strength of the concrete (fc). The cross-sectional area of the reinforcing bar (As) is a discrete variable that should be chosen from possible reinforcing bars by ACI as shown in Table 1 [9].

The basic assumptions for the design of reinforced concrete are as following [11]:

1. The yield stress of the reinforcing steel is constant (355 MPa).
2. The average tensile stress in the reinforcement does not exceed yield stress.
3. The average compressive stress in the concrete is 0.85fc.
4. The effective depth is assumed to be $de = 0.8\ h$.
5. The bounds of the variables are $100 \leq b \leq 600$ mm, $150 \leq h \leq 1200$ mm, $20 \leq fc \leq 80$ MPa.

2.2 *Objective Function*

According to the price of concrete and steel, the defined first objective function and the structure should satisfy the ACI building code with a bending strength to the defined second objective function [10, 12].

$$Min : f(As, b, h, fc) = 0.0059 * As + (1.0971 * fc + 56.891) * b * h/10^6 \quad (4)$$

$$Max : 1/M(u) = 1/0.9 * As * fy * (d - a/2)/10^6 \quad (5)$$

The first constraint is the depth to width ratio of the beam is restricted to be less than, or equal, to 4 [9, 13].

$$g(x) = h/b - 4 \leq 0 \quad (6)$$

Defined the Parameters of HSA and GA. Compare the output codes by changing the HMCR parameter in the HSA, and Pm in the GA to compare the output data. The HSA parameters need to be defined:

1. Harmony memory size $HMS = 50$,
2. Harmony memory considering rate $HMCR = 0.7, 0.8, 0.9$.
3. Pitch adjusting rate $PAR = 0.2$.
4. Bandwidth $BW = 1$.
5. Maximum Pareto size $Ps = 20$.
6. Maximum iteration $MaxIter = 1000$.

Table 1 Design space for reinforcing bar. Adopted from [8, 10]

As	Bar type	Area (sq in.)	As	Bar type	Area (sq in.)
1	1#4	0.2	40	5#8	3 95
2	1#5	0.31	41	9#6	3.96
3	2#4	0.4	42	4#9	4
4	1#6	0.44	43	13#5	4.03
5	3#4, 1#7	0.6	44	7#7	4.2
6	2#5	0.62	45	14#5	4.34
7	1#8	0.79	46	10#6	4.4
8	4#4	0.8	47	15#5	4.65
9	2#6	0.88	48	6#8	4.74
10	3#5	0.93	49	8#7	4.8
11	5#4, 1#9	1	50	11#6	4.84
12	6#4, 2#7	1.2	51	5#9	5
13	4#5	1.24	52	12#6	5 28
14	3#6	1.32	53	9#7	5.4
15	7#4	1.4	54	7#8	5.53
16	5#5	1.55	55	13#8	5.72
17	2#8	1.58	56	10#7, 6#9	6
18	8#4	1.6	57	14#6	6.16
19	4#6	1.76	58	8#8	6.32
20	9#4, 3#7	1.8	59	15#6, 11#7	6 6
21	6#5	1.86	60	7#9	7
22	10#4, 2#9	2	61	9#8	7.11
23	7#5	2.17	62	12#7	7.2
24	11#4, 5#6	2.2	63	13#7	7.8
25	3#8	2.37	64	10#8	7.9
26	12#4, 4#7	2.4	65	8#9	8
27	8#5	2.48	66	14#7	8.4
28	13#1	2.6	67	11#8	8.69
29	6##	2.64	68	15#7	9
30	9#5	2.79	69	12#8	9.48
31	14#4	2.8	70	13#8	10.27
32	15#4, 5#7, 3#9	3	71	11#9	11
33	7#6	3.08	72	14#8	11.06
34	10#5	3.1	73	15#8	11.85
35	4#8	3.16	74	12#9	12
36	11#5	3.41	75	13#9	13

(continued)

Table 1 (continued)

As	Bar type	Area (sq in.)	As	Bar type	Area (sq in.)
37	8#6	3.52	76	14#9	14
38	6##	3.6	77	15#9	15
39	12#5	3.72			

Note 1 in, = 25.4 mm; 1 sq in = 645.16 mm^2; 1 cu in = 16,400 mm^3; 1 in ^{4} = 416,200 mm^4

HSA first generates M initial solutions and puts them into the harmony memory HM, where HMCR is the probability of selecting a harmony from the HM [14, 15].

The GA parameters also need to be defined:

1. Population Size $Ps = 50$.
2. Population Crossover $Pc = 0.8$.
3. Population Mutation $Pm = 0.1$ & 0.2.
4. Maximum iteration $MaxIter = 1000$.

The mutation in GA is an auxiliary method to generate new individuals. It determines the local search ability of GA while maintaining the diversity of the population [1].

3 Result of HSA and GA

3.1 *Result of Coding*

Coding for the problem using MATLAB and obtain multiple sets of data by modifying parameters of HSA and GA [16].

Figures 2 and 3 represent $HMCR = 0.7, 0.8,$ *and* 0.9 in HSA and $Pm = 0.1$ *and* 0.2 in GA, respectively. And combined the three sets of data in HSA and two sets of data in GA to get Fig. 4.

Delete wrong data and find the optimization point with MOHSA and MOGA from Fig. 4 in each algorithm. In the MOHSA, the beam size $B = 365$ mm, $H =$

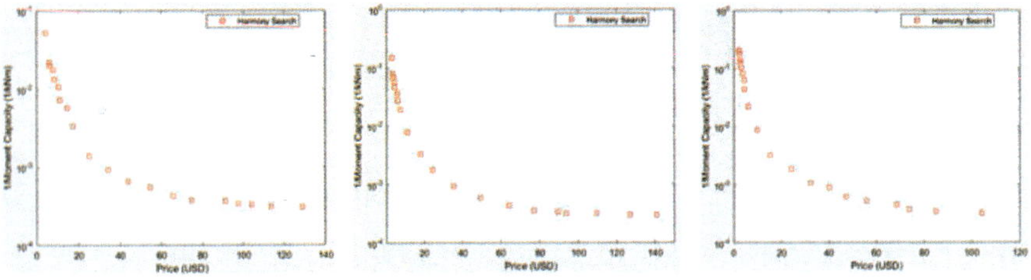

Fig. 2 Result of HSA coding $HMCR = 0.7$ (left), $HMCR = 0.8$ (medium) and $HMCR = 0.9$ (right)

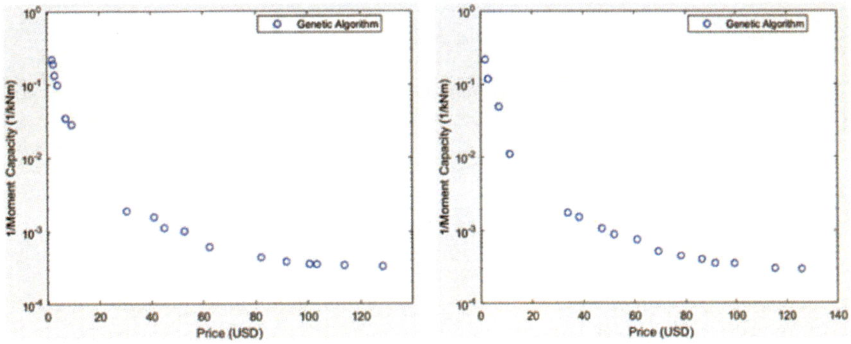

Fig. 3 Result of GA coding *Pm* = *0.1* (left) and *Pm* = *0.2* (right)

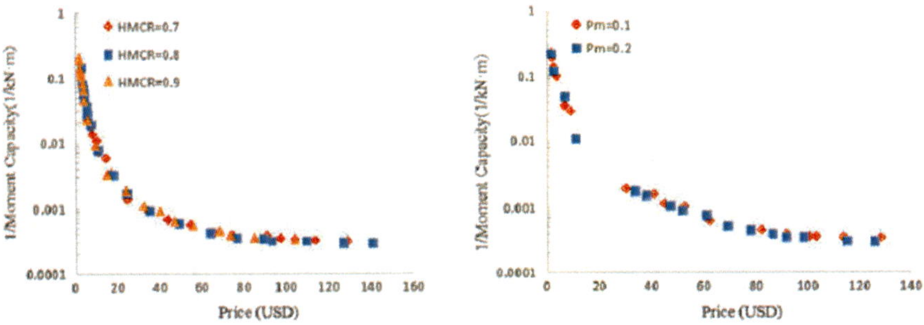

Fig. 4 Result of MOHSA (left) and result of MOGA (right)

1116 mm, bar Type 15#9, *As* = *9677* mm^2, *fc* = *53* MPa, and *Price* = *$104* and the optimization beam size in the MOGA *B* = *356*, *H* = *1118*, bar Type 14#9, *As* = *9032* mm^2, *fc* = *53* MPa, and *Price* = *$99*.

4 Conclusion

This paper focuses on the optimized design of RC beam using MOHSA and MOGA, data are obtained by changing the parameters of HMCR in HSA and Pm in GA, respectively. According to data analysis and comparison, with the same reference strength, the result using GA has a 5% lower cost than using HSA.

This paper is limited to the analysis of HSA and GA. Therefore, multiple algorithms and different parameters in each algorithm are required to improve the optimal design of RC beams.

References

1. V. Govindaraj, J.V. Ramasamy, Optimum detailed design of reinforced concrete continuous beams using genetic algorithms. Comput. Struct. **84**(1–2), 34–48 (2005)
2. S. Sivasubramani, K.S. Swarup, Multi-objective harmony search algorithm for optimal power flow problem. Int. J. Electr. Power Energy Syst. **33**(3), 745–752 (2011)
3. Z.W. Geem, J.H. Kim, G.V. Loganathan, A new heuristic optimization algorithm: harmony search. Simulation **76**(2), 60–68 (2001)
4. P. Chakraborty, et al., An improved harmony search algorithm with differential mutation operator. Fundamenta Informaticae **95**(4), 401–426 (2009)
5. A. Vasebi, M. Fesanghary, S.M.T. Bathaee, Combined heat and power economic dispatch by harmony search algorithm. Int. J. Electr. Power Energy Syst. **29**(10), 713–719 (2007)
6. C.A. Coello, A.D. Christiansen, F. Santos Hernandez, A simple genetic algorithm for the design of reinforced concrete beams. Eng. Comput. **13**(4), 185–196 (1997)
7. G.A. Vignaux, Z. Michalewicz, A genetic algorithm for the linear transportation problem. IEEE Trans. Syst. Man Cybern. B Cybern. **21**(2), 445–452 (1991)
8. H.M. Amir, T. Hasegawa, Nonlinear mixed-discrete structural optimization. J. Struct. Eng. **115**(3), 626–646 (1989)
9. S. Koziel, X.-S. Yang (eds.) Computational Optimization, Methods and Algorithms, vol. 356 s(Springer, 2011)
10. ACI Committee, Commentary on building code requirements for reinforced concrete (ACI 318–77). American Concrete Institute (1977)
11. B. Saini, V.K. Sehgal, M.L. Gambhir, Genetically optimized artificial neural network based optimum design of singly and doubly reinforced concrete beams (2006), pp. 603–619
12. A. Kaveh, R.A. Izadifard, L. Mottaghi, Optimal design of planar RC frames considering CO_2 emissions using ECBO, EVPS and PSO metaheuristic algorithms. J. Build. Eng. **28**, 101014 (2020)
13. A. Kaveh, R.A. Izadifard, L. Mottaghi, COST optimization of RC frames using automated member grouping. Iran Univ. Sci. Technol. **10**(1), 91–100 (2020)
14. A. Verma, B.K. Panigrahi, P.R. Bijwe, Harmony search algorithm for transmission network expansion planning. IET Gener. Transm. Distrib. **4**(6), 663–673 (2010)
15. A. Kaveh, O. Sabzi, A comparative study of two meta-heuristic algorithms for optimum design of reinforced concrete frames (2011), pp. 193–206
16. A. Kaveh, T. Bakhshpoori, S.M. Hamze-Ziabari, GMDH-based prediction of shear strength of FRP-RC beams with and without stirrups. Comput. Concr. **22**(2), 197–207 (2018)

Optimal Cost Design of Single-Story Reinforced Concrete Frames Using Jaya Algorithm

Elmas Rakıcı, Gebrail Bekdaş, and Sinan Melih Nigdeli

Abstract A method has been presented for the design of reinforced concrete plane frame systems at minimum cost by using the Jaya algorithm. The total material cost is at the objective function, and the cross-sectional dimensions were taken as design variables. These design variables were assigned with candidate solutions according to the rules of the algorithm in the numerical iterations. The total material cost was calculated according to the amount of concrete and reinforcements, and the matrix displacement method was used to analyze structures. The reinforced concrete design was made according to ACI 318-05 (Building code requirements for structural concrete and commentary) rules published by American Concrete Institute. These rules are taken as design constraints. The developed method has been applied to a single-story structure for different loading cases. Since the results have a direct match with the expected optimum results, the method is feasible for the optimization problem.

Keywords Optimum design · Reinforced concrete plane frame system · Jaya algorithm · Optimization

1 Introduction

The design of structures is done according to analysis results. During the numerical optimization process, the employed code of metaheuristic-based optimization must contain the combination of structural analysis code and the calculation of the design

E. Rakıcı · G. Bekdaş · S. M. Nigdeli (✉)
Department of Civil Engineering, Istanbul University - Cerrahpaşa, 34320, Avcılar Istanbul, Turkey
e-mail: melihnig@istanbul.edu.tr

E. Rakıcı
e-mail: elmasrakici@hotmail.com

G. Bekdaş
e-mail: bekdas@istanbul.edu.tr

S. M. Nigdeli et al. (eds.), *Proceedings of 6th International Conference on Harmony Search, Soft Computing and Applications*, Advances in Intelligent Systems and Computing 1275, https://doi.org/10.1007/978-981-15-8603-3_16

of structures. After the structural analysis, the internal forces on the critical sections are used in the design constraints to provide stress–strain requirements of structures. Then, the assigned candidate design variables given for dimensions are also used in the design constraints for providing additional constructive rules of design codes and requirements.

The optimum design of structures includes steel, reinforced concrete and composite structures. In the present study, reinforced concrete (RC) frames were optimized. Metaheuristic methods have been employed to single components of RC structures or frames, which are combinations of beams and columns.

For the optimum design of RC columns, harmony search algorithm (HS) [1], artificial bee colony algorithm (ABC) [2], bat algorithm (BA) [3] and teaching–learning-based optimization (TLBO) [4] were employed. The optimum design of the beams was done by employing genetic algorithm (GA) [5, 6], ant colony optimization (ACO) [7], particle swarm optimization (PSO) [8, 9] and ABC [10, 11].

The recent studies about optimization of RC frames includes the methods employing BA [12], GA [13, 14], PSO [15, 16], HS [17–21], artificial neural network (ANN) [22] and big bang–big crunch (BB-BC) [23].

In the present study, the total material cost of RC plane frame systems is minimized by optimizing dimension design variables. It is presented by employing Jaya algorithm (JA), which is a parameter-free metaheuristic algorithm. The internal forces are found via analysis according to the matrix displacement method, and the design constraints are considered according to ACI 318: Building code requirements for structural concrete and commentary [24].

2 The Design Methodology

2.1 The Optimization Problem

In every loop of the numerical optimization, the structural analysis is done according to the loading condition of the structure. In the matrix displacement method, element stiffness matrices are provided for all members. Then, the system stiffness matrix is found by combining the element matrices transformed into global coordinates. Since the values of system stiffness matrices are the forces needed for the unit displacements of the structure, the nodal displacements under the system load vector are found. According to displacement values, the internal forces are found by multiplying stiffness with the corresponding displacements. For these calculations, the rigidity of members must be known. The rigidity of members is related to the cross-sectional dimensions of members. Due to that, the design variables must be assigned with a candidate value before the analysis of the system.

After the internal forces of beams and columns are known, the design of RC members is done to find the required reinforcements.

In the design of beams, an equivalent rectangular pressure block is considered in the balance of compressive forces of concrete and tensile forces of steel reinforcement bars. The minimum and maximum values of reinforcement are checked as design constraints, and the maximum value is found according to the balanced reinforcement ratio.

Shear reinforcements are also calculated according to shear forces. The design rules of shear reinforcements are taken as design constraints.

For RC columns, the second-order effects are considered according to the slenderness of the members. A moment magnification factor is calculated to multiply with the first-order moments. The required reinforcements are found according to the flexural moment–axial force relationship.

The cost of a member is calculated according to Eq. (1).

$$C_m = C_v * C_c + C_w * C_s \tag{1}$$

In Eq. (1), C_c and C_s are the cost of concrete per unit volume ($/m^3) and steel per unit weight ($/ton), respectively. C_w is the total weight of steel and C_v is the total volume of the concrete.

The objective function (f(x)) is calculated by summing the cost of all N elements of the frame system as seen in Eq. (2).

$$f(X) = \sum_{i=1}^{n} (C_m)_i \tag{2}$$

In application, the cross-sections are taken to be rectangular. Height (h) and breadth (b) of all members are the design variables. The design constraints are given in Table 1.

2.2 The Optimization Process

There are many types of metaheuristic algorithms used in the optimization process. In 2016, Rao developed an algorithm called the Jaya algorithm (JA) [25], where the algorithm was based on the concept that the solution obtained with randomly defined design variables to accomplish the objective function converging to the best solution and diverging from the worst solution. This optimization process can be defined in 4 steps as follows.

Step 1: In this step, the objective function of the optimization problem is determined first. The algorithm parameters are defined as number of population, termination criterion (maximum number of iterations) and number of design variables. Design variables are produced randomly to provide design constraints. Solution vectors containing design variables are generated and recorded in the initial solution matrix. The number of solution vectors is the number of population entered by

Table 1 Design constraints

Explanation	Symbol	Unit	Value or formula
Minimum section breadth	b_{min}	mm	250
Maximum section breadth	b_{maks}	mm	450
Minimum section height	h_{min}	mm	300
Maximum section height	h_{maks}	mm	600
Minimum diameter of shear reinforcement	ϕ_{vmin}	mm	8
Maximum diameter of shear reinforcement	ϕ_{vmaks}	mm	16
Minimum reinforcement ratio for column	ρ_{min}	–	0.01
Maximum reinforcement ratio for column	ρ_{max}	–	0.06
Maximum reinforcement ratio for beam	ρ_{max}	–	$0.75\rho_b$
Minimum reinforcement area for beam	$A_{s,min}$	mm^2	$\frac{\sqrt{f'_c}}{4f_y}bd$ or $\frac{1.4}{f_y}bd$
Minimum horizontal reinforcement area	$(A_v)_{min}$	mm^2	$\frac{1}{3}\frac{bs}{f_y}$
Compressive strength of concrete	f_c	MPa	30
Yield strength of concrete	f_y	MPa	420
Specific gravity of steel	γ_s	t/m^3	7.86
Specific gravity of concrete	γ_c	t/m^3	2.5
Clear cover	c_c	mm	40
Cost of concrete per unit volume	C_c	\$/m^3	45
Cost of steel per unit weight	C_s	\$/ton	400
Live disturbed load	L	kN/m	25, 37.5, 50
Dead disturbed load	D	kN/m	50, 75, 100

the user. Analysis and design are done for each solution vector and the results are recorded in the objective function vector (f(x)) after the objective function (total cost) is calculated.

Step 2: After the analysis, the values that give the best solution ($f(x)_{best}$) and the values that give the worst solution ($f(x)_{worst}$) (Eq. 3) are saved separately.

$$f(x)_{best} = \text{The best solution from } f(x) \text{ and } f(x)_{worst} = \text{The worst solution from } f(x) \quad (3)$$

According to the design variables in $f(x)_{best}$ and $f(x)_{worst}$, new design variables are produced according to Eq. (4).

'i' is an iterative number, 'vn' is the number of the variable (j = 1, 2, …, vn), pn is the number of population (k = 1, 2, …, pn) and $X_{j,k,i}$ is the value of *j*th variable for the *k*th population at *i*th iteration.

The updated value of $X_{j,k,i}$ is $X'_{j,k,i}$ and is defined with

$$X'_{j,k,i} = X_{j,k,i} + r_{1,j,i}\left(X_{j,best,i} - \left|X_{j,k,i}\right|\right) - r_{2,j,i}\left(X_{j,worst,i} - \left|X_{j,k,i}\right|\right) \quad (4)$$

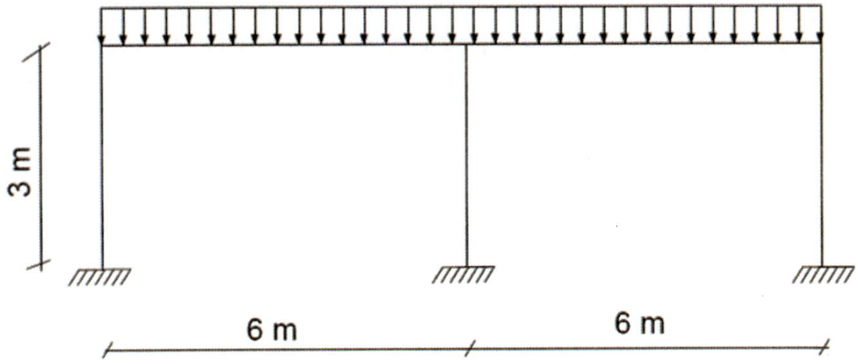

Fig. 1 Two-span single-story reinforced concrete frame system

In Eq. 4, $r_{1,j,i}$ and $r_{2,j,i}$ are the two random numbers for the jth variable during the ith iterations. $X_{j,best,i}$ is the best value for the jth variable and $X_{j,worst,i}$ is the worst value for the jth variable.

Using new design variables, new solution vectors are generated, and finally new objective function vectors are obtained.

Step 3: If the newly produced objective function is better than the worst function in the solution matrix where the existing vectors are present, the existing vector is deleted from the solution matrix, and the values of the newly produced solution vector are written instead, otherwise the solution matrix is retained.

Step 4: The stop condition with the maximum number of iterations is checked. It is repeated from step 2 until the stop condition is achieved. In case of stopping condition, the optimization process is ended.

3 Numerical Examples

The two-span single-story reinforced concrete frame design example was optimized using the Jaya algorithm. The frame system used is shown in Fig. 1. Figure 2 shows the numeration of the elements and nodes in the frame system for structural analysis. For optimization, the stop criterion (maximum number of iterations) is 100,000 and the population number is 20. The optimum results are given in Tables 2, 3 and 4 for different loading cases.

4 Conclusion

In this study, by using the Jaya algorithm, design optimization of reinforced concrete plane frames according to the provisions of ACI 318-05 and the matrix displacement

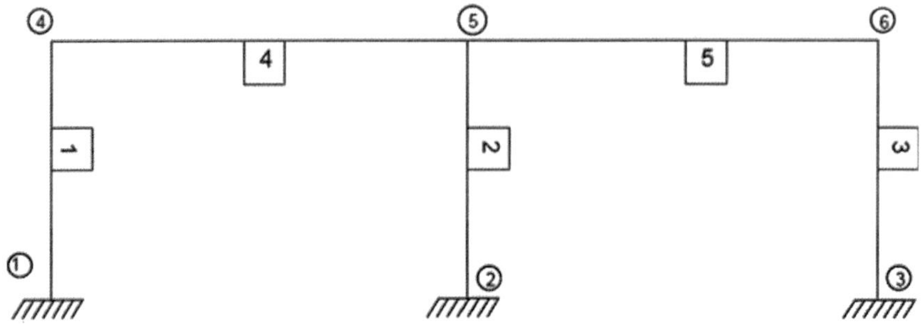

Fig. 2 Element and node point numbering for example 1

Table 2 Optimum results for D = 50 kN/m and L = 25 kN/m

Element number	1	2	3	4	5
b (mm)	250	250	250	250	250
h (mm)	300	300	300	600	600
Cost ($)	21.0531	21.0531	21.0531	84.9025	84.9025
Total cost ($)	232.9644				

Table 3 Optimum results for D = 75 kN/m and L = 37.5 kN/m

Element number	1	2	3	4	5
b (mm)	400	400	400	350	350
h (mm)	300	300	300	600	600
Cost ($)	31.6905	31.6905	31.6905	123.826	123.826
Total cost ($)	342.7244				

Table 4 Optimum results for D = 100 kN/m and L = 50 kN/m

Element number	1	2	3	4	5
b (mm)	450	450	450	450	450
h (mm)	600	350	600	600	600
Cost ($)	83.5052	47.5201	83.5052	154.824	154.824
Total cost ($)	524.1780				

method is examined. The optimum cost was aimed under all design constraints and the optimization process was tested on a two-span single-story reinforced concrete plane frame. The optimum cost of the frame system is between 232.9644 $ and 524.1780 $ according to the loading cases.

The cross-sectional dimensions of the columns in the reinforced concrete plane frame were obtained as the minimum value for the lowest distributed loads. For the first two cases, all columns have the same optimum design. It is good proof of the success of the algorithm since the symmetric columns 1 and 3 have the same results. For the highest loads, the symmetric columns 1 and 3 have bigger dimensions than the middle column. By the increase of the loads, the flexural moments are increasing for the side columns.

In the beams, the cross-sectional breadth was minimized, and the cross-sectional height was maximized for the first case. By increasing the loads, the optimum breadth values start to increase. In all cases, it is thought that the cross-sectional height of the beam is maximized to ensure safety conditions.

According to the results, the optimum values support the expected results, and show a direct match with the theory. The proposed method employing JA for the optimum design of RC frames is a feasible method. This study can be updated and used for multi-story structures and three-dimensional reinforced concrete frame systems.

References

1. G.F. De Medeiros, M. Kripka, Optimization of reinforced concrete columns according to different environmental impact assessment parameters. Eng. Struct. **59**, 185–194 (2014)
2. H.T. Ozturk, A. Durmus, Optimum cost design of RC columns using artificial bee colony algorithm. Struct. Eng. Mech. **45**(5), 643–654 (2013)
3. G. Bekdaş, S.M. Nigdeli, Bat algorithm for optimization of reinforced concrete columns. PAMM **16**(1), 681–682 (2016)
4. G. Bekdaş, S.M. Nigdeli, Optimum design of reinforced concrete columns employing teaching-learning based optimization. CHALLENGE **2**(4), 216–219 (2016)
5. M. Leps, M. Sejnoha, New approach to optimization of reinforced concrete beams. Comput. Struct. **81**(18–19), 1957–1966 (2003)
6. S.T. Yousif, R.M. Najem, Optimum cost design of reinforced concrete continuous beams using Genetic Algorithms. Int. J. Appl. Sci. Eng. Res. **2**(1), 79–92 (2013)
7. P. Sharafi, M.N.S. Hadi, L.H. Teh, Geometric design optimization for dynamic response problems of continuous reinforced concrete beams. J. Comput. Civ. Eng. **28**(2), 202–209 (2012)
8. S. Chutani, J. Singh, Design optimization of reinforced concrete beams. J. Inst. Eng. (India): Ser. A **98**(4), 429–435 (2017)
9. A.N. Hanoon et al., Energy absorption evaluation of reinforced concrete beams under various loading rates based on particle swarm optimization technique. Eng. Optim. **49**(9), 1483–1501 (2017)
10. M.M. Jahjouh, M.H. Arafa, M.A. AlqedrA, Artificial Bee Colony (ABC) algorithm in the design optimization of RC continuous beams. Struct. Multidiscip. Optim. **47**(6), 963–979 (2013)
11. H.T. Ozturk, Ay. Durmus, Ah. Durmus, Optimum design of a reinforced concrete beam using artificial bee colony algorithm. Comput. Concr. **10**(3), 295–306 (2012)
12. S. Gholizadeh, V. Aligholizadeh, Optimum design of reinforced concrete frames using bat meta-heuristic algorithm (2013)
13. S. Rajeev, C.S. Krishnamoorthy, Genetic algorithm–based methodology for design optimization of reinforced concrete frames. Comput. Aided Civ. Infrastruct. Eng. **13**(1), 63–74 (1998)
14. V. Govindaraj, J.V. Ramasamy, Optimum detailed design of reinforced concrete frames using genetic algorithms. Eng. Optim. **39**(4), 471–494 (2007)

15. M.J. Esfandiary et al., A combination of particle swarm optimization and multi-criterion decision-making for optimum design of reinforced concrete frames. Int. J. Optim. Civil Eng **6**(2), 245–268 (2016)
16. S. Gharehbaghia, M.J. Fadaee, Design optimization of RC frames under earthquake loads. Int. J. Optim. Civ. Eng. **2**, 459–477 (2012)
17. G. Bekdaş, S.M. Nigdeli, Optimization of RC frame structures subjected to static loading, in *11th World Congress on Computational Mechanics* (2014), pp. 20–25
18. M. Kripka et al., Optimization of reinforced concrete frames by harmony search method, in *11th World Congress on Structural and Multidisciplinary Optimisation* (2015)
19. G. Bekdaş, S.M. Nigdeli, X. Yang, *Optimum Reinforced Concrete Design by Harmony Search Algorithm* (Metaheuristics and Optimization in Civil Engineering, Springer, Cham, 2016), pp. 165–180
20. G. Bekdaş, S.M. Nigdeli, Modified harmony search for optimization of reinforced concrete frames, in *International Conference on Harmony Search Algorithm* (Springer, Singapore, 2017), pp. 213–221
21. A. Akin, M.P. Saka, Harmony search algorithm based optimum detailed design of reinforced concrete plane frames subject to ACI 318–05 provisions. Comput. Struct. **147**, 79–95 (2015)
22. A.A.A. Aga, F.M. Adam, Design optimization of reinforced concrete frames. Open J. Civ. Eng. **5**(01), 74 (2015)
23. A. Kaveh, O. Sabzi, Optimal design of reinforced concrete frames using big bang-big crunch algorithm. Int. J. Civ. Eng. **10**(3), 189–200 (2012)
24. ACI Committee, American Concrete Institute, & International Organization for Standardization, Building code requirements for structural concrete (ACI 318-05) and commentary. American Concrete Institute (2008)
25. R. Rao, Jaya: a simple and new optimization algorithm for solving constrained and unconstrained optimization problems. Int. J. Ind. Eng. Comput. **7**(1), 19–34 (2016)

Optimum Design of Reinforced Concrete Retaining Walls Under Static and Dynamic Loads Using Jaya Algorithm

Nur Yılmaz, Sena Aral, Sinan Melih Nigdeli, and Gebrail Bekdaş

Abstract In this study; a reinforced concrete retaining wall was dimensioned under static and dynamic loads at optimum cost using the Jaya algorithm, which is one of the metaheuristic algorithms. As an objective function, the total cost for the unit length of the retaining wall and the cross-sectional dimensions are defined as design variables. Thanks to the single phase of the Jaya algorithm, the solution was reached quickly, and the best design variables were obtained with the minimal solution in the objective function compared to regular calculations. In addition to achieving optimum dimensioning results in terms of safety and cost, the relationship between earthquake and cost has been examined with the optimization method used as a result of the reinforced concrete design made by applying the regulations on the Buildings to be built in Earthquake Zones (DBYBHY 2007).

Keywords Cantilever retaining wall · Reinforced concrete structures · Jaya algorithm · Optimum design · Optimization

1 Introduction

Structural design, which is one of the specialized areas of civil engineering, is the process of fabricating a building safely, economically and aesthetically. Structural engineers design the structure by prioritizing strength, durability and safety. In this

N. Yılmaz (✉) · S. Aral · S. M. Nigdeli · G. Bekdaş
Department of Civil Engineering, Istanbul University, 34320 Cerrahpaşa, Avcılar, Istanbul, Turkey
e-mail: nuryilmaz19@ogr.iu.edu.tr

S. Aral
e-mail: senaaral@ogr.iu.edu.tr

S. M. Nigdeli
e-mail: melihnig@istanbul.edu.tr

G. Bekdaş
e-mail: bekdas@istanbul.edu.tr

S. M. Nigdeli et al. (eds.), *Proceedings of 6th International Conference on Harmony Search, Soft Computing and Applications*, Advances in Intelligent Systems and Computing 1275, https://doi.org/10.1007/978-981-15-8603-3_17

process, engineers should be able to solve many complex problems such as choosing the appropriate material, analyzing the material properties and analyzing different types of forces affecting the structure. Engineers rely on accepted rules and their own experience to solve these problems. After the design of the structure is defined in this way, they analyze the structure according to the predicted critical situations. If the initial variables are insufficient or excessive, the structure is redesigned by trial and error. Structural optimization methods are very beneficial at this stage, because it is a great advantage for an engineer to perform more trials than the trial and error method in a certain period of time, as well as obtaining the most optimum solution among them.

Retaining walls are the rigid structures that engineers have built to hold the soil volume at two different levels. These structures designed according to the lateral ground pressures (active, passive) should also securely transmit the soil pressures that will occur during earthquakes to the ground. According to TS 7944 [1], retaining walls are divided into groups as rigid, semi-rigid and flexible.

While pre-dimensioning reinforced concrete cantilever retaining walls, one of the rigid retaining wall types is associated with the wall height and the initial values of the other zone lengths of the wall are determined [2]. Static and dynamic lateral soil pressures that will affect the wall are calculated with certain initial dimensions, and sliding, overturning and bearing capacity tests are performed. If the results of the controls are negative, these operations are repeated by changing the dimensions. In addition to these investigations, structural optimization methods have been used in the design of the retaining walls due to the steps that cannot be solved by many mathematical equations such as reinforcement areas varying according to the internal forces and reinforcement ratios that do not exceed the limit values.

The optimum design studies of reinforced concrete cantilever retaining walls that started in the 1980s focused on the availability of metaheuristic algorithms in optimization after the 2000s. The optimum design of the reinforced concrete retaining wall under static loads has been carried out on the parameters affecting the minimum cost, and economic design using the Simulated Annealing (SA) algorithm [3, 4]. In the study conducted for the safe optimization of the reinforced concrete cantilever retaining wall, Target Reliability Approach (TRA) was used [5]. Other metaheuristic algorithms used in the optimum design of the reinforced concrete retaining wall; Particle Swarm Optimization (PSO) [6], Ant Colony Optimization (ACO) [7], Harmony Search (HS) Algorithm [8], Big Bang–Big Crunch (BP-BC) Optimization [9], Genetic Algorithm (GA) [10], Firefly Algorithm (FA) [11], Bat Algorithm (BA) [12], Colliding Object (CO) Optimization and Democratic Particle Swarm Optimization (DPSO) [13] have been applied for the optimum design of reinforced concrete retaining walls. The optimum design of the reinforced concrete retaining wall under the influence of static and dynamic loads was made by Temür and Bekdaş [14] by applying a Teaching–Learning-Based Optimization algorithm (TLBO). The optimum design of the retaining wall based on the Jaya algorithm is a study in which the effect of the magnitude of the surcharge load on the cost and CO_2 emission value of the wall with the parameters of the ground was investigated [15]. However, in the literature, the optimum design of the reinforced concrete retaining wall under both static and earthquake loads has not been found with the Jaya algorithm.

In this study, the economic design of the reinforced concrete cantilever retaining wall in accordance with the Regulation on Buildings to be constructed in Earthquake Zones (DBYBHY) [16] using the Jaya algorithm under the influence of static and dynamic loads is aimed. In addition, the design of a retaining wall was designed for two different situations under static and dynamic loads acting on the retaining wall, and the effect of the two states on the cost was examined. The compatibility of the Jaya algorithm with the commonly used methods in the literature has been determined. In addition to calculating the dimensions of the reinforced concrete cantilever retaining wall with the lowest cost and safety constraints using the Jaya algorithm, the relationship of the earthquake to the retaining wall cost was also investigated.

2 The Design Methodology

2.1 The Optimization Problem

When designing retaining walls, damages that may occur should be considered. Vertical and horizontal loads that can affect a cantilever retaining wall are shown in Fig. 1. External forces (mass of the wall, soil pressures, etc.) acting under static conditions are in equilibrium. The conveniently planned retaining wall prevents the shear stresses from reaching the shear strength of the ground, ensuring the balance of these forces. On the other hand, dynamic loads (inertial forces and ground resistance changes) that occur during an earthquake can cause permanent deformations. As a result of excessive deformations, collapse types arise from rotation problems, stability of the structure and structural damages. In order to design the retaining wall safely, bearing capacity, sliding and overturning tests are performed.

For ground stress analysis, the maximum and minimum ground stresses under the base plate of the wall must first be calculated Eq. (1) according to axial load (N), flexural moment (M_o), area (A_t) and section modulus (W_t) of the base. It is desired

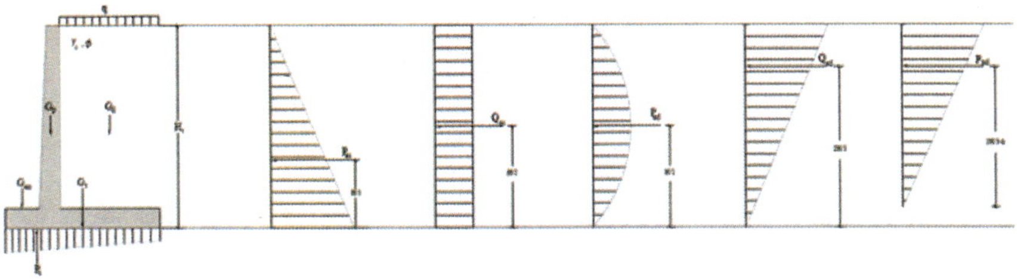

Fig. 1 Static and dynamic loads that can affect a cantilever retaining wall

that the largest ground stress ($\sigma_{\max}$) formed at the base is smaller than the ground safety stress (σ_u) (Eq. 2). In order to avoid undesired tensile stresses at the smallest ground stress ($\sigma_{\min}$) formed at the base, the value of $\sigma_{\min}$ must be positively designed Eq. (3).

$$\sigma_{\max,\min} = \frac{N}{A_t} \pm \frac{M_o}{W_t} \tag{1}$$

$$\sigma_{\max} < \sigma_u \tag{2}$$

$$\sigma_{\min} > 0 \tag{3}$$

In case of earthquake loading, ground safety stress can be increased by 50% at most (Eq. 4).

$$\sigma_{\max} < 1,5\sigma_u \tag{4}$$

The sliding test is determined by the ratio of the forces (F_R) against the sliding to the forces (F_O) that cause the wall to slide (Eq. 5). The number of safety (SF_S) is considered to be at least 1.5 on granular floors and at least 1.1 under earthquake loads [17].

$$SF_S = \frac{\sum F_R}{\sum F_O} \tag{5}$$

The safety coefficient (SF_O) for the overturning test is expressed by the ratio of the moments taken according to the lower end of the pre-amplification, to the moments that resist for overturning (M_R) to the forces (M_O) trying to overturn the system (Eq. 6). SF_O should provide at least 2 without earthquake and 1.2 in the earthquake state. The passive state at the front can be neglected, as more torque will be required to create a passive state according to the active state.

$$SF_O = \frac{\sum M_R}{\sum M_O} \tag{6}$$

In the optimization problem examined in this study, five design variables were used, namely stem thickness at the top of the wall (b1), toe projection length (b2), stem thickness at the bottom of the wall (b3), heel projection length (b4) and base slab thickness (h) (Fig. 2). In the calculation of these variables, Eqs. 2–6 are considered as design constraints.

In the cantilever reinforced concrete retaining walls in Fig. 3, height-dependent front dimensions are also used as design constraints.

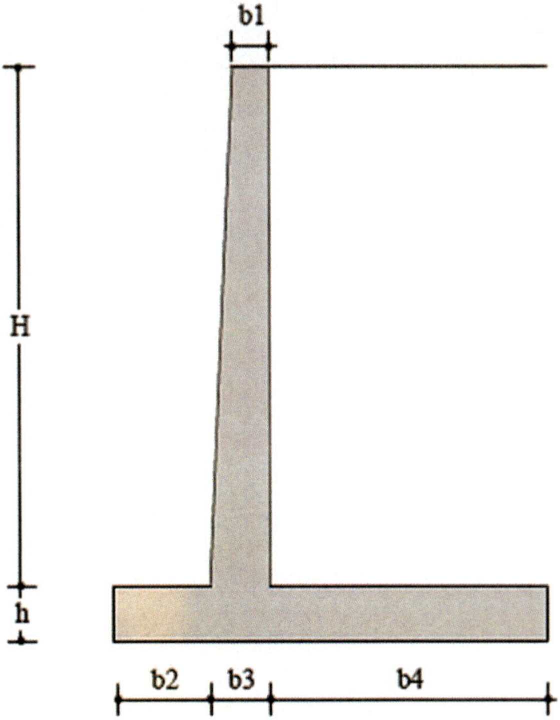

Fig. 2 Design variables of a cantilever retaining wall

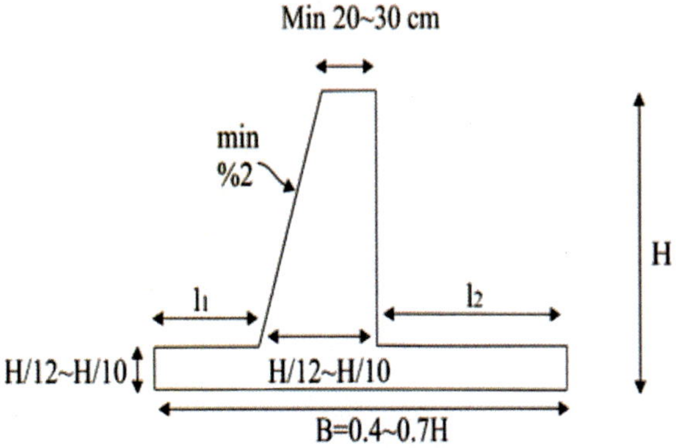

Fig. 3 Pre-sizing constraints on the cantilever reinforced concrete retaining walls [2]

In addition to these; the critical internal force (shear and bending) calculation was made for the earthquake and earthquake-free cases in the console, heel and toe projections. The design constraints shown in Table 1 are taken into consideration.

The objective function in the optimization problem is to minimize the cost obtained as a result of the volume of concrete (V_c) and the weight of steel (W_s) and the sum of the products of the unit the cost of concrete and steel (C_c, C_s) (Eq. 7). V_c and W_s values vary according to the design variables.

Table 1 Restrictions on the strength of the retaining wall

Explanation	Restrictions	
	Under static loads	Under static and dynamics loads
Safety for bearing capacity	$\sigma_{max,design} < \sigma_u$ $\sigma_{min,design} > 0$	$\sigma_{max,design} < 1,5\sigma_u$ $\sigma_{min,design} > 0$
Safety for sliding stability	$SF_{S,design} > 1,5$	$SF_{S,design} > 1,1$
Safety for overturning stability	$SF_{O,design} > 1,5$	$SF_{O,design} > 1,2$
Maximum reinforcement area of console section, $A_{s,max}$	$A_{s,design} < A_{s,max}$	$A_{sd,design} < A_{s,max}$
Minimum reinforcement area of console section, $A_{s,min}$	$A_{s,design} \geq A_{s,min}$	$A_{sd,design} \geq A_{s,min}$
Maximum reinforcement area of toe projection, $A_{so,max}$	$A_{so,design} < A_{so,max}$	$A_{sod,design} < A_{so,max}$
Minimum reinforcement area of toe projection, $A_{so,min}$	$A_{so,design} \geq A_{so,min}$	$A_{sod,design} \geq A_{so,min}$
Maximum reinforcement area of heel projection, $A_{sa,max}$	$A_{sa,design} < A_{sa,max}$	$A_{sad,design} < A_{sa,max}$
Minimum reinforcement area of heel projection, $A_{sa,min}$	$A_{sa,design} \geq A_{sa,min}$	$A_{sad,design} \geq A_{sa,min}$
Shear strength capacities of critical sections, V_{max}	$V_{max,design} \leq V_{cr}/2$	$V_{max,design} \leq V_{cr}/2$

$$\min\ f(X) = C_c.V_c + C_s.W_s\ldots \tag{7}$$

2.2 *The Optimization Process*

In the literature, many types of metaheuristic algorithms used by engineers have been found. The Jaya algorithm is an algorithm developed by Rao [18] that works to get the solution obtained by multiplying the design variables defined to realize the objective function by random coefficients and to get away from the bad solutions.

First, the objective function ($f(x)$) is determined for the targeted maximization or minimization problem. Population size, design variables and termination criteria (maximum number of iterations) are the parameters required for the Jaya algorithm. The initial solution matrix is generated and saved by randomly obtaining design variables that provide design constraints. The number of the population determined by the user determines the size of the solution matrix.

In the iterative process using the Jaya algorithm, the best solution ($f(x)_{best}$) and worst solution ($f(x)_{worst}$) values from the solution matrix created are used. According to these values in Eq. 8, the updated value of the jth variable of the kth population is generated. The candidate solution in the ith iteration is shown as $X'_{i,k,i}$.

$$X'_{j,k,i} = X_{j,k,i} + r_{1,j,i}\left(X_{j,best,i} - \left|X_{j,k,i}\right|\right) - r_{2,j,i}\left(X_{j,worst,i} - \left|X_{j,k,i}\right|\right) \quad (8)$$

In Eq. 8, $r_{1,j,i}$ and $r_{2,j,i}$ are random numbers that change in the range [0,1]. While $X_{j,k,i}$ is the previous design variable value, $X_{j,best,i}$ represents the design variable values in the best solution and $X_{j,worst,i}$ in the worst solution. If the updated variable $X'_{j,k,i}$ gives a better purpose function, the new value is accepted by replacing it with $X_{j,k,i}$. If it gives a worse purpose function, $X_{j,k,i}$ value is accepted. After checking compliance with the restriction criteria, it is recorded in the optimum solution matrix and used as input data in the next iteration.

This process is repeated until the condition of stopping is met with the specified number of termination criteria. In case of stopping condition, optimization ends. The flow diagram of the Jaya algorithm is presented in Fig. 4.

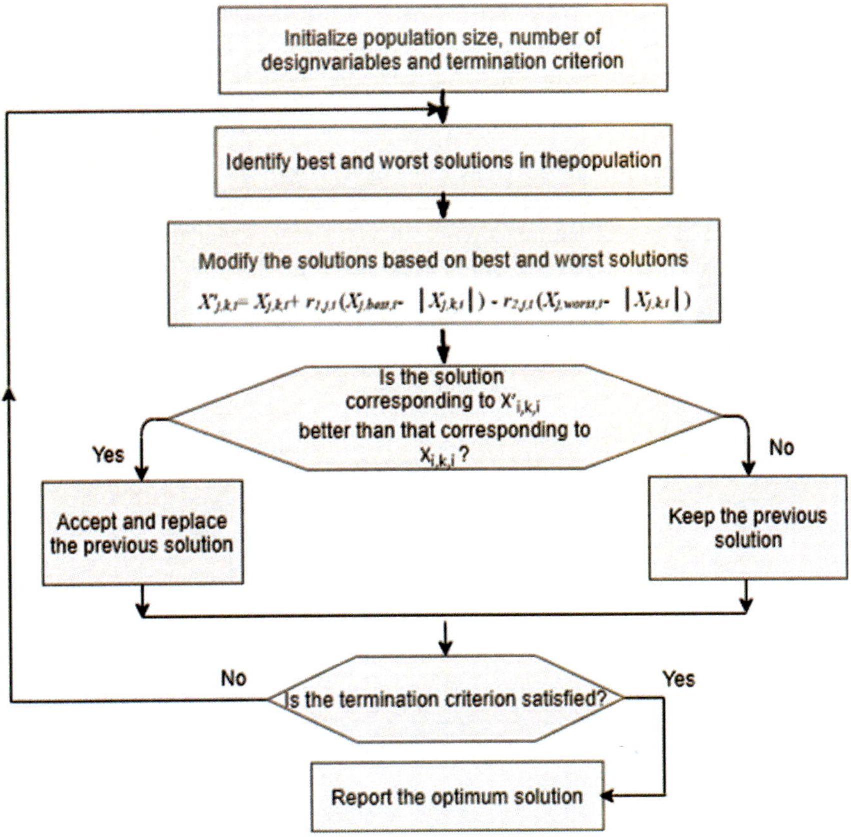

Fig. 4 Jaya algorithm flow chart [18]

3 Numerical Examples

The methodology has been studied for two different situations. The design of the cantilever retaining wall made by Celep [19] (Solution 1) was optimized under static loads only in the first case using the Jaya algorithm and named as Solution 2. In the second case, the same design of the cantilever retaining wall is minimized by dimensioning under dynamic loads (Solution 3) in addition to static loads, and the results are expressed as the third solution.

The design constants are defined at the same values so that the comparison of both cases is more effective. The design of Celep [19] and the Jaya algorithm were compared with the optimization results of the two states.

Calculations were made for the length of the cantilever retaining wall 1 m. The design parameters of the cantilever retaining wall in the second earthquake zone and the internal friction angle of the infill ground is 30°, and the unit volume weight is 18 kN/m^2, ground safety stress 200 kN/m^2, material class C30/S420, the charge load is 5 kN/m^2 and the height of the wall 5.4 m excluding the base slab thickness (h) were taken. The unit price of steel is defined as 1400 TL/ton and the unit price of concrete is 111 TL/m^3.

The design variables are as shown in Fig. 2. The objective function changes according to the design constants and design variables. The stop criterion (maximum number of iterations) for optimization is 100,000 and the population number is 20. All dimensions, reinforcements and total costs are given as in Tables 2, 3 and 4, respectively.

Table 2 Geometric size comparison

Variables (m)	Solution 1	Solution 2	Solution 3
h	0,6000	0,4480	0,4537
b1	0,2500	0,2000	0,2000
b2	0,9000	1,2303	1,3076
b3	0,6000	0,6993	0,7123
b4	3,0000	1,9760	2,1296

Table 3 Reinforcement area comparisons on console (A_s), toe projection ($A_{s,o}$) and heel projection ($A_{s,a}$)

Reinforcement areas (mm^2)	Solution 1	Solution 2	Solution 3
A_s	1578,6	1278,6	1304,5
$A_{s,o}$	1080,0	776,0	787,4
$A_{s,a}$	1298,9	1156,5	1146,3

Table 4 Total cost comparison

	Solution 1	Solution 2	Solution 3
Total cost (TL/m)	2286,9	1789,7	1854,0

Table 5 Comparison of 10 runs solved by Jaya algorithm for cases with earthquake and without earthquake

Cases		h (m)	b1 (m)	b2 (m)	b3 (m)	b4 (m)	A_s (m)	$A_{s,o}$ (m)	$A_{s,a}$ (m)	f(x) Cost (TL/m)
Without earthquake	Avg	0,448	0,2	1,230	0,699	1,976	1278	775,7	1156	1789,7
	Std	0	0	0,0004	0,0001	0	0,0001	0,0005	0,0003	0
With earthquake	Avg	0,4536	0,2	1,307	0,712	2,129	1304	787,3	1146	1854
	Std	0,0001	0	0,0004	0	0	0	0,0001	0,0003	0

The optimization process of the Jaya algorithm-based methodology was repeated 10 times and the average results are presented in Table 5 with the standard deviation results. Since the standard deviation results are zero, the proposed methodology is robust.

4 Conclusion

In this study, using the Jaya algorithm, the optimum design of the cantilever reinforced concrete retaining wall according to the rules of DBYBHY [16] was investigated.

In order to test the performance and success of the method, analyses aiming the lowest cost under design constraints were made using the Jaya algorithm and compared with a sample retaining wall problem [19] solved using the same design constants. The design under the static loads presented by Celep [19] was calculated as 2255 TL/m according to the unit costs used in this study. It was observed that the design without seismic condition obtained by optimization was approximately 21% (465.3 TL/m) more economical. In addition, the optimization of the cantilever reinforced concrete retaining wall was carried out under both static loads and dynamic loads. The design under the static and dynamic loads presented by Celep [19] was calculated as 2286.9 TL/m according to the unit costs used in this study. The earthquake design obtained by optimization was found to be approximately 19% (432.9 TL/m) more economical. In both cases, it is understood that optimization with the Jaya algorithm gives more efficient results.

In the study, the analysis of the reinforced concrete retaining wall under earthquake-free and earthquake conditions was made and optimum design results were obtained. When compared, according to these results, it was found that earthquake-free conditions were approximately 4% (64.9 TL/m) of lower cost. Since dynamic loads additional to static loads increase the load on the wall, sizes increase by approximately 1.0259% to provide the necessary conditions to ensure the stability and required reinforcement area. Thus, it is understood that the reinforced concrete retaining wall's dynamic forces caused an increase in the cost. Thanks to the study, the dimensions of the reinforced concrete cantilever retaining wall that provide economic and safety constraints were calculated with the Jaya algorithm, and a more efficient design was made by examining the relationship of the earthquake to the retaining wall cost.

References

1. TS7944, *Soil Bearing Structures: Classification, Properties and Design Principles.* (Turkish Standards Institute, ITU. Central Library, 1990)
2. C.R.I. Clayton, *Earth Pressures and Earth Retaining Structures*, 3rd edn. (CRC Press, London and New York, 2014)
3. B. Ceranic, C. Freyer, R.W. Baines, An application of simulated annealing to the optimum design reinforced concrete retaining. Struct. Comput. Struct. **79**, 1569–1581 (2001)
4. V. Yepes, J. Alcala, C. Perea, F. Gonzalez-Vidosa, A parametric study of optimum earth-retaining walls by simulated annealing. Eng. Struct. **30**, 821–830 (2008)
5. G.L.S. Babu, B.M. Basha, Optimum design of cantilever retaining walls using target reliability approach. Int. J. Geomech. **8**(4), 240–252 (2008)
6. B. Ahmadi-Nedushan, H. Varaee, *Optimal Design of Reinforced Concrete Retaining Walls Using a Swarm Intelligence Technique.* In: Proceeding of the First International Conference on Soft Computing Technology in Civil, Structural and Environmental Engineering (Civil-Comp Press, Scotland, 2009)
7. M. Ghazavi, S.B. Bonab, *Optimization of Reinforced Concrete Retaining Walls Using ant Colony Method*, Schuppener, Straub, Bräu (Eds.), (2011) pp 297–305
8. A. Kaveh, A.S.M. Abadi, Harmony search based algorithms for the optimum cost design of reinforced concrete cantilever retaining walls. Int. J. Civ. Eng. **9**(1), 1–8 (2011)
9. C.V. Camp, A. Akin, Design of retaining walls using big bang-big crunch optimization. J. Struct. Eng. **138**(3), 438–448 (2012)
10. Y. Pei, Y. Xia, Design of reinforced cantilever retaining walls using heuristic optimization algortihms. Procedia Earth Planet. Sci. **5**, 32–36 (2012)
11. A. Akin, Aydogdu I, *Optimum Design of Retaining Walls Using Adaptive Firefly Algorithm.* International Civil Engineering and Architecture Symposium for Academicians, Geotechnical Engineering (2014), pp 57–67
12. S. Talatahari, R. Sheikholeslami, Optimum design of gravity and reinforced retaining walls using enhanced charged system search algorithm. KSCE J. Civil Eng. **18**(5), 1464–1469 (2014)
13. A. Kaveh, N. Soleimani, CBO and DPSO for optimum design of reinforced concrete cantilever retaining walls. Asian J Civil Eng. **16**(6), 751–774 (2015)
14. R. Temur, G. Bekdaş, Teaching learning-based optimization for design of cantilever retaining walls. Struct. Eng. Mech. **57**(4), 763–783 (2016)
15. H.T. Öztürk, E. ve Türkeli, Optimum design of reinforced concrete retaining walls with a key section at the base with the algorithm of Jaya. Polytechn. Mag. **22**(2), 283–291 (2019)
16. Ministry of Public Works and Settlement, *Principles about Buildings to be made in Earthquake Zones* (Ankara, Turkey, 2007)
17. S. Yıldırım, *Soil Investigation and Foundation Design* (Birsen Publishing, Istanbul, 2009)
18. R. Rao, Jaya: a simple and new optimization algorithm for solving constrained and unconstrained optimization problems. Int. J. Ind. Eng. Comput. **7**(1), 19–34 (2016)
19. Z. Celep, *Introduction to Earthquake Engineering and Earthquake Resistant Building Design.* (Istanbul, 2014), pp 299–305

Jaya Optimization for the Design of Cantilever Retaining Walls with Toe Projection Restriction

Sena Aral, Nur Yılmaz, Gebrail Bekdaş, and Sinan Melih Nigdeli

Abstract In this study, optimum dimensioning of a reinforced concrete cantilever retaining wall under static loads was made by developing a method using the Jaya algorithm. During the optimization process, the retaining wall stability checks, cross-section capacities and controls including reinforced concrete structural design rules were made. With the optimum design, it is aimed that the retaining wall cross-section dimensions reach the optimum value. The total material cost is defined as the objective function and the cross-section dimensions are defined to be variable. In order to determine the effect of the front encasement (toe projection), different design limits are entered into the program, and program outputs are examined. The results of the analysis showed that the standard deviation value increases as the design limits get narrow. The dimensioning of the retaining wall was made according to TS 7994 (Soil Resistance Structures; Classification, Properties and Project Design Principles) and the reinforced concrete design was made according to TS 500 (Design and Construction Rules of Reinforced Concrete Structures) rules. The Jaya optimization method has been found to be superior to the existing methods, and it has been understood that it is appropriate to use the method in the optimum design of the retaining wall.

Keywords Cantilever retaining wall · Reinforced concrete structures · Jaya algorithm · Optimum design · Optimization

S. Aral (✉) · N. Yılmaz · G. Bekdaş · S. M. Nigdeli
Department of Civil Engineering, Istanbul University-Cerrahpaşa, 34320 Avcılar, İstanbul, Turkey
e-mail: senaaral@ogr.iu.edu.tr

N. Yılmaz
e-mail: nuryilmaz19@ogr.iu.edu.tr

G. Bekdaş
e-mail: bekdas@istanbul.edu.tr

S. M. Nigdeli
e-mail: melihnig@istanbul.edu.tr

S. M. Nigdeli et al. (eds.), *Proceedings of 6th International Conference on Harmony Search, Soft Computing and Applications*, Advances in Intelligent Systems and Computing 1275, https://doi.org/10.1007/978-981-15-8603-3_18

1 Introduction

The main purpose of structural engineering is to provide adequate security to withstand the loads and forces that the structures will be exposed to throughout their service life and to ensure that the structures remain within the safety limits predicted for cracking and displacement during use. In addition to providing definite security conditions for buildings, depleted natural resources and increases in environmental pollution require a design with minimum cost and minimum CO_2 emission. The structural system dimensions are changed for several times to determine the design with the lowest cost in traditional design. Besides these processes being boring and monotonous, the most economical solution may not be achieved. It is necessary to take advantage of the optimization process, which is mathematical programming, to reach the design that provides all security requirements. At the same time, metaheuristic optimization methods are effective in finding the exact results.

For this problem, metaheuristic algorithms such as Simulated Annealing algorithm (SA) [1], Particle Swarm Optimization (PSO) [2], Genetic Algorithm (GA) [3] and Teaching–Learning-Based Optimization (TLBO) [4] have been also used by designers. Reinforced concrete retaining wall optimization studies started in the 1980s. In the literature, studies using an object-oriented visual programming language provide opportunities for continuous monitoring, evaluation and modification of control parameters using Simulated Annealing algorithm (SA) related to the minimum cost design of the retaining wall [1]. Structural optimization was performed by controlling all geotechnical and structural design constraints while reducing the total cost of retaining walls using particle herd optimization (PSO) [2]. With the genetic algorithm (GA), the strength structure design variable values that minimize the cost function were found [3]. Also, in recent years, the retaining walls of the console have been optimized under static and dynamic loads with the TLBO algorithm [4].

The Jaya algorithm used in the optimization process of the study is a single-phase metaheuristic algorithm developed by Rao [5]. It is a metaheuristic algorithm that can solve both constrained and unconstrained optimization problems. In this metaheuristic optimization process, design constants, population number, maximum number of iterations, ranges of design variables, design constraints and objectives are defined.

In this process, the Jaya algorithm can find the minimum or maximum values of the objective function by trying to approach the best and get away from the worst among the solutions that are created and renewed in every iterations. The purpose of the Jaya algorithm is to converge to the best solution and deviate from the worst solution in a single phase. It is not an algorithm affected by parameters since it does not contain any certain parameter other than the population number [5]. Jaya has been used to find the optimum reinforcement design for reinforcing reinforced concrete beams in carbon fiber [6]. In addition, the study to minimize vertical displacement of beams is also included in the literature [7].

In this study, unlike the studies in the literature, the Jaya algorithm, which is a parameterless algorithm, was used in the optimization of the retaining wall minimum cost and minimum cross-section design. In this paper, console-type retaining wall optimization taken from Celep [8] has been made by considering static loads according to TS 7994 (Soil Resistance Structures; Classification, Properties and Design Principles) [9] and reinforced concrete design according to TS 500 (Design and Construction Rules of Reinforced Concrete Structures) [10]. Retaining wall optimization has been defined as the objective function to minimize concrete and reinforcing steel costs. Different design limits are entered into the program, and program outputs are examined for the toe projection.

2 The Design Methodology

2.1 *The Optimization Problem*

Jaya means victory in Sanskrit. The Jaya algorithm, which is one of the metaheuristic algorithms, aims to achieve victory by reaching the optimum solution. Numerical optimization consists of three phases: structural modeling, optimum design modeling and optimization algorithm. In this structural modeling, the basic design condition is to maintain the stability of the retaining wall structure against sliding and overturning under lateral ground pressures and to remain within the specified safety limits. Therefore, the optimization problem can also be expressed as the stability analysis of the structure. The objective function used in the optimum design can be defined as in Eq. 1:

$$\text{Min } f(x) = C_c \cdot V_c + C_s \cdot W_s \tag{1}$$

In Eq. 1, C_c refers to unit concrete cost (TL/m^3), C_s refers to unit reinforcement cost (TL/ton), V_c refers to unit length concrete volume and W_s refers to unit length reinforcement steel weight.

The retaining wall design was made under the loads shown in Fig. 1. G_d is the weight of the retaining wall; G_T is the weight of the ground on the rear of the wall; q is the surcharge load; P_{AS} is the active ground pressure, P_T is the base shear force and the Q_{AS} is the active ground pressure consisting of the surcharge load. Also, the definition of these loads is given in Table 1.

In the retaining wall optimization process, it is aimed to create a structural system that provides overturning, shear and stress safety conditions. The first one is the safety coefficient (SF_O) against overturning, and it is defined as in Eq. 2 as the ratio of the sum of moments trying to overturn the wall that is taken from the front end of the encasement.

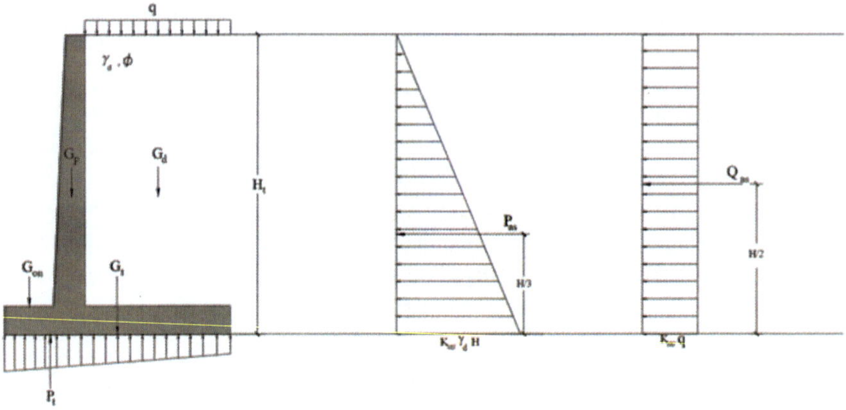

Fig. 1 Loads acting on a cantilever retaining wall

Table 1 Definition of wall loads

P_{AS} (Static load of active ground)	$K_a \cdot \gamma_{(ground)} \cdot H^2/2$
Q_{AS} (Static load of surcharge load)	$K_a \cdot q \cdot H$
W_w (Weight of the wall)	Area · thickness · $\gamma_{(concrete)}$
W_r (Ground weight)	$H \cdot b4 \cdot \gamma_{(concrete)}$
$W_{surcharge}$ (resultant surcharge load)	$q \cdot b4$

$$SF_o = \frac{\sum M_r}{\sum M_o} \tag{2}$$

In the process of calculating static soil pressures, active soil pressure was calculated as in Eq. 3 according to the Rankine theory and the positive effect of passive soil pressure was neglected.

$$k_a = cos\beta \cdot \frac{cos\beta \cdot \sqrt{cos2 \cdot \beta \cdot cos2\theta}}{cos\beta \cdot +\sqrt{cos2 \cdot \beta \cdot cos2\theta}} \tag{3}$$

The second one of the examinations is the shear safety coefficient (SFs) that is defined as the ratio of the resisting forces (FR) and the overturning forces (FO) that are defined as follows.

$$SF_S = \frac{\sum F_R}{\sum F_O} \tag{4}$$

The definition of $\sum F_R$ (resisting forces) and $\sum F_O$ (overturning forces) static charges are as follows.

Table 2 Design variables

Variables	Explanation	Design variable
With geometry related variables	Stem thickness at the top of the wall	b1
	Toe projection length	b2
	Stem thickness at the bottom of the wall	b3
	Heel projection length	b4
	Base slab thickness	h

$$\sum F_R = W_w + W_R + W_{surcharge} \tag{5}$$

$$\sum F_O = P_{AS} + Q_{AS} \tag{6}$$

The stresses on the retaining wall base should be checked to ensure that they does not exceed the existing ground safety stress and meet safety requirements. Control of stresses can be done as in Eq. 7. W_t expresses the moment of resistance. The total area A_t, the axial force N and the moments that occur according to the base midpoint M_o are the other parameters in Eq. (7).

$$\sigma_{1,2} = \frac{N}{A_t} \pm \frac{M_o}{W_t} \tag{7}$$

During the retaining wall optimization process, 5 variables related to geometry were defined. The definition of these variables is as in Table 2 and Fig. 2.

Design constraints are determined according to TS 7994 (Soil Resistance Structures; Classification, Properties and Project Design Principles) [9]. The reinforced concrete design has been checked under the maximum and minimum boundary conditions specified in the regulation according to TS 500 (Design and Construction Rules of Reinforced Concrete Structures) [10]; reinforcement area, shear and tensile safety controls, and reinforcement area calculated according to internal forces. In the analysis, it is aimed to reach the best solution by considering situations that are not suitable for limit conditions. In Table 3, definition constraints and formulations are given.

2.2 *The Optimization Process*

The Jaya algorithm, which is a metaheuristic algorithm, can find the minimum or maximum values of the optimum objective function determined by trying to approach the best and get away from the worst among the candidate solutions created and renewed in every iteration. The flow diagram that defines this working process, where

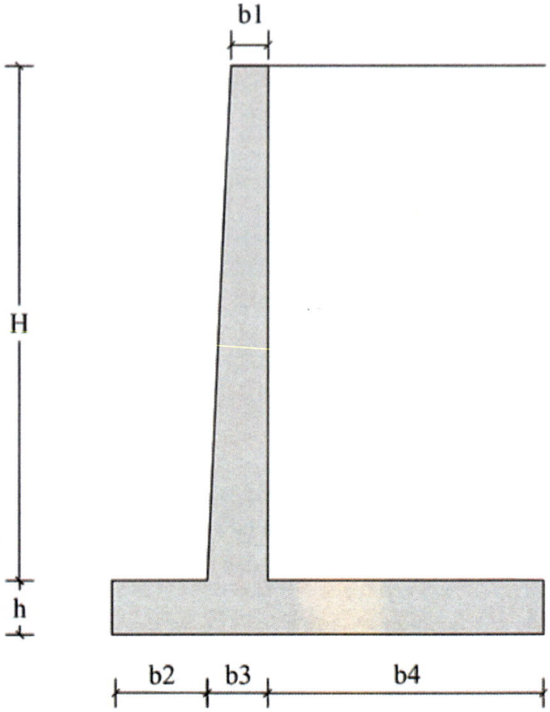

Fig. 2 A cantilever retaining wall cross section

Table 3 Design constraints

Explanation	Design variable
Overturn safety	g1(X): SFO, design ≥ SFO
Shear safety	g2(X): SFS, design ≥ SFS
Ground stress	g3(X): SFB, design ≥ SFB
Minimum bearing tension	g4(X): qmin ≥ 0
Minimum reinforcement area of critical sections, Asmin	g5-6(X): As ≥ Asmin
Maximum reinforcement area of critical sections, Asmax	g7-8(X): As ≤ Asmax

the number of iterations determined by the user is realized and enables achieving the optimum solution, is as in Fig. 3.

The optimization process with Jaya consists of 4 steps. These stages can be defined as follows.

At this stage, the number of populations, the maximum number of iterations, the number of design variables are determined by the designer. For these design variables, the target range is defined and in order to select the optimum values of the variables, the program is defined in order to multiply it by random coefficients provided that it is in the maximum and minimum ranges. The number of solution vectors is the

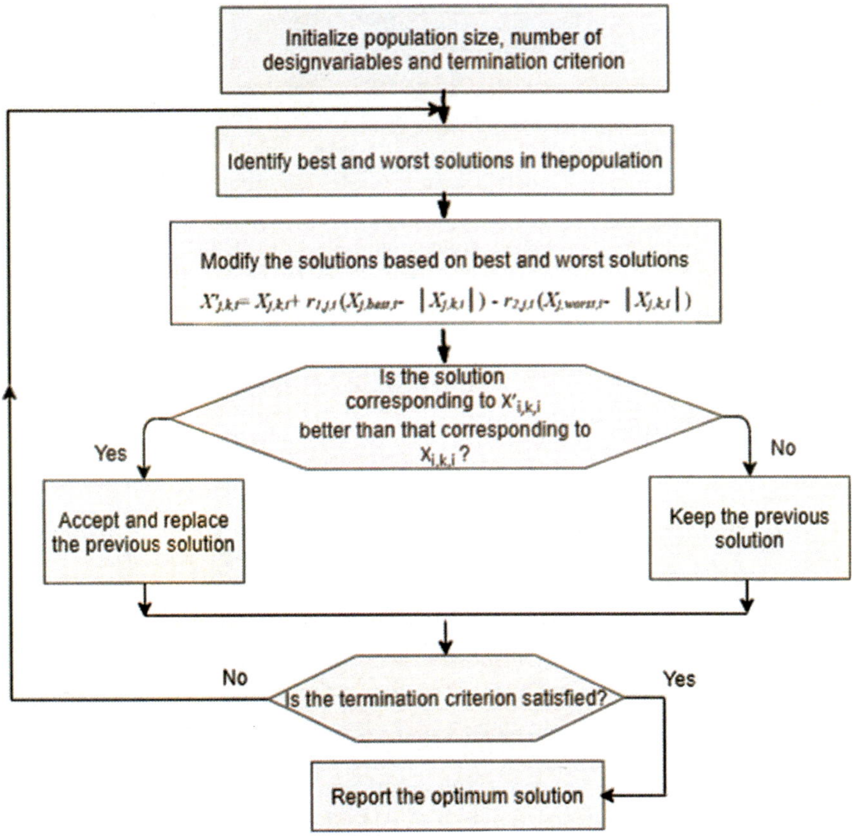

Fig. 3 Flow chart of Jaya algorithm [5]

number of populations entered by the designer. Analysis and design are made for each solution, and the results are recorded in the objective function vector f (x) after the objective function is calculated.

After the analysis, values containing design variables that give the best and worst objective functions [f(x)] are recorded separately in the optimum and optimum1 matrix. These functions are defined in Eq. 8. The objective function is the total cost of the retaining wall.

$$\begin{aligned} \mathrm{f(x)\ best} &= \mathrm{Best\ solution\ function\ f(x)\ and} \\ \mathrm{f(x)\ worst} &= \mathrm{Worst\ solution\ function\ f(x)} \end{aligned} \tag{8}$$

In addition, the general Jaya equation is defined in Eq. 9.

$$X'_{j,k,i} = X_{j,k,i} + r_{1,j,i}\left(X_{j,best,i} - \left|X_{j,k,i}\right|\right) - r_{2,j,i}\left(X_{j,worst,i} - \left|X_{j,k,i}\right|\right) \tag{9}$$

'I' is the number of iterations and i shows the ith iteration. 'v_n' is the number of variables and j = 1, 2, ... is defined as v_n. P_n is the population number (k = 1, 2, ..., pn). In this iterative process, the updated $X_{j,k,i}$ value is expressed as $X_{j,k,i}'$. In Eq. 9, $r_{1,j,i}$ and $r_{2,j,i}$ are two random numbers for the variable j during iterations. $X_{j,best,i}$ is the best value and $X_{j,worst,i}$ is the worst value. New solution vectors are produced according to the new design variables, and the resulting values of the objective function are stored. The new function value is compared with the function in the existing solution matrix. The worst solution is deleted, and the better solution is continued. The cycle is repeated until the maximum number of iterations defined in the program is completed. After these stages, the optimization process is completed.

3 Numerical Examples

An example of a retaining wall design made by Celep [8] has been optimized using the Jaya algorithm. The reinforced concrete design was made for three separate sections of the retaining wall, including the console, front and rear encasements. Also, the effect of the front-encasement design boundary on the cost was checked for 6 cases. In addition, in the absence of toe projection, the values of the cost were examined by comparing the classical calculations of Celep [8]. In Table 4, the section sizes obtained as a result of the optimization and the section sizes used in the example by Celep [8] were compared.

Comparison of analysis results and costs are given in Tables 5 and 6. In Table 7, the effect of variation of the toe projection limits on optimization is given with the average and standard deviation variables of 10 cycles of optimization runs.

Parameters used in the design of the retaining wall under static loads are material class as C30-S420, ground safety stress as 200 kN/m^3, internal friction angle of the floor (tight sand) as 30°, Internal friction angle between the floor and the floor as μ

Table 4 Geometric size comparisons

Variables	Celep [8] (m)	Jaya algorithm results (m)
b_1	0.25	0.2
b_2	0.9	1.23
b_3	0.6	0.7
b_4	3	1.98
h	0.6	0.44

Table 5 Reinforcement design comparisons

Variables	Celep [8] (mm^2)	Jaya algorithm results (mm^2)
As (console)	1530	1278
As (back)	1298	1156
As (front)	1080	776

Table 6 Total cost comparisons

Cost	Celep [8]	Jaya algorithm results
Total cost (TL/m)	2255	1789

Table 7 The effect of toe projection range on optimization

Variables	Case	b2 limits	Cost $*10^3$	b1 (m)	b2 (m)	b3 (m)	b4 (m)	h (m)	A_s $*10^3$	A_{sa} $*10^3$	A_{so} $*10^3$
Average	1	0.2–10	1.79	0.2	1.23	0.7	1.98	0.45	1.28	1.16	0.78
Std. deviation			0	0	0	0	0	0	0	0	0
Average	2	0.2–8	1.79	0.2	1.23	0.7	1.98	0.45	1.28	1.16	0.78
Std. deviation			0	0	0	0	0	0	0	0	0
Average	3	0.2–6	2.06	0.21	1.25	0.76	2.48	0.44	1.4	1.27	0.88
Std. deviation			0.55	0.04	0.21	0.16	1.67	0.1	0.33	0.35	0.21
Average	4	0.2–4	2.11	0.2	1.15	0.64	2.15	0.44	1.53	1.34	0.97
Std. deviation			0.77	0	0.32	0.12	1.24	0.03	0.5	0.71	0.63
Average	5	0.2–2	1.91	0.23	1.16	0.66	1.9	0.42	1.45	1.33	0.92
Std. deviation			0.28	0.09	0.18	0.1	0.4	0.07	0.43	0.41	0.42
Average	6	0	2.79	0.23	0	0.9	3.5	0.53	1.89	1.66	0
Std. deviation			1.01	0.07	0	0.5	2.39	0.19	0.96	0.9	0

= 0.55, unit volume weight of the filled soil as 18 kN/m^3 and console wall height as H = 5.6 m, surcharge load as q = 5 kN/m^2. The unit price of concrete is 111 TL/m^3 and the unit price of steel is 1400 TL/ton.

4 Conclusion

In this study, the optimization of a reinforced concrete console retaining wall under static loads is done by the Jaya algorithm according to TS 7994 and TS 500 rules. It is aimed to achieve the optimum design with minimum cross-section dimensions that maintain its stability under external loads, and this optimization problem is adapted to the example presented by Celep [8]. According to this study, the cost obtained by Celep [8] was found to be 466 TL/m higher. In addition, when the analyses made with 6 different design ranges for the toe projection are compared, it was found that as this interval narrowed, the standard deviation increased and the design without pre-encasement increased the cost by 55%. Construction costs were calculated as 1790 TL/m and 2790 TL/m, respectively, in the case where the toe projection ranges are maximum and without toe projection.

Compared to the design made by Celep [8], the retaining wall design shows that a 21% lower construction cost can be achieved. According to these data, the design made with the Jaya optimization method was found to be superior to the existing methods, and it was understood that it was appropriate to use the method in the optimum design of the retaining wall.

References

1. B. Ceranic, C. Freyer, R.W. Baines, An application of simulated annealing to the optimum design reinforced concrete retaining structures. Comput. Struct. **79**, 1569–1581 (2001)
2. M. Khajehzadeh, M.R. Taha, A. El-Shafie, M. Eslami, Economic design of retaining wall using particle swarm optimization with passive congregation. Struct. Eng. Electr. Eng. **4**, 5500–5507 (2010)
3. N.A. Jasim, A.M. Yaqoobi, Optimum design of tied back retaining wall. Civ. Eng. **6**, 139–155 (2016)
4. R. Temur, G. Bekdaş, Teaching learning-based optimization for design of cantilever retaining walls. Struct. Eng. Mech. **57**, 763–783 (2016)
5. R. Rao, Jaya: a simple and new optimization algorithm for solving constrained and unconstrained optimization problems. Int. J. Ind. Eng. Comput. **7**(1), 19–34 (2016)
6. A.E. Kayabekir, B. Sayın, S.M. Nigdeli, G. Bekdas, Jaya algorithm based optimum carbon fiber reinforced polymer design for reinforced concrete beams. AIP Conf. Proc. **1978**, 260006 (2018). https://doi.org/10.1063/1.5043891
7. G. Bekdas, S.M. Nigdeli, A.E. Kayabekir, Y.C. Toklu, Minimization of vertical deflection of an optimum I-beam by Jaya algorithm. AIP Conf. Proc. **1978**, 260002 (2018). https://doi.org/10.1063/1.5043887
8. Z. Celep, Deprem Mühendisliğine Giriş ve Depreme Dayanıklı Yapı Tasarımı, İstanbul, sayfa (2014), pp. 299–305
9. Turkish Standards Institute, Soil Retaining Structures; Properties and Guidlines for Design, Ankara, TS7994 (1990)
10. Turkish Standardization Institute, Design and Construction of Concrete Structures, Ankara, Turkey, TS500 (2000)

Transportation Path Assignment Within the Airports in Turkey

Emre Demir and İbrahim Aydoğdu

Abstract To speed up the development of civil aviation, many academic researchers are being carried out. One of the most encountered and important issues is the necessity for handling the visiting order of the inspections or audits of all airports. In this study, the evaluation of the most suitable transportation path takes place, and the assignment issue is resolved with the help of a traveling salesman problem. However, finding the best solution to this problem is considerably difficult. Stochastic-Based Optimization (SBO) techniques are efficient tools to overcome this difficulty. In the study, the optimization program will be developed to find the best transportation path Assignment within the airports in Turkey. The Tree Seed Algorithm (TSA) and Symbiotic Organisms Search (SOS) optimization techniques are utilized to find the best path. In addition, performances of the TSA and SOS techniques will be investigated for the current optimization problem.

Keywords Transportation optimization · Civil aviation · Path assignment

1 Introduction

For many decades, civil aviation in Turkey has advanced and this evolving rate stays. Lots of passengers or travelers, as well as cargo has been transferred by the means of aviation every day. In order to organize the high demand for aviation carriage, the management of civil aviation is very crucial, but still evolving [1]. As many authorities would appreciate, this growth is fast and definitely requires several kinds of inspections or audits that must be conducted at every airport. To keep up the speed of the development or increase the rate of advance, many academic researches have

E. Demir (✉)
Department of Civil Engineering, Antalya Bilim University, 07190 Antalya, Turkey
e-mail: emre.demir@antalya.edu.tr

İ. Aydoğdu
Department of Civil Engineering, Akdeniz University, 07058 Antalya, Turkey
e-mail: aydogdu@akdeniz.edu.tr

S. M. Nigdeli et al. (eds.), *Proceedings of 6th International Conference on Harmony Search, Soft Computing and Applications*, Advances in Intelligent Systems and Computing 1275, https://doi.org/10.1007/978-981-15-8603-3_19

been completed and many are being carried out [2, 3]. One of the most encountered and important issues is the necessity for handling the visiting order of the inspections or audits at airports. The required visits or seminars of audits must be arranged very systematically so that the financial loss of not only the airport companies but also the local government can be minimized. In this case, the data is important and must be taken from a reliable source. With the help of the data delivered from the civil aviation airports in Turkey, the assessment of the most appropriate transportation path for visiting each airport takes place. Also, the assignment path problem in this article is resolved with the support of commonly used assignment problems.

Although conceptually assignment problems such as traveling salesman problem seem easy to handle, the finding best solution to this problem is considerably complicated as the number of joints increases. To overcome such challenges, for instance, several linear integer programming techniques are applied for determining the paths, their similarities, as well as the locations of facilities [4–9]. Likewise, Stochastic-Based Optimization (SBO) techniques are great tools to prevail against this difficulty. In recent years, several effective SBO techniques have been developed and successfully applied to challenge engineering and planning problems [10–14]. The Tree Seed Algorithm (TSA) and Symbiotic Organisms Search (SOS) are two of the newest SBO techniques that are developed. They are adopted from the natural phenomena of raising the trees and seeds, and the relationship between two organisms in ecosystems, respectively [15–19]. These methodologies and algorithms were utilized to solve different aspects of engineering problems. However, in this study, they are first time applied for optimizing and determining the route and its cost as path inference at the industry of civil aviation.

2 Optimization Problem of Airport Transport

Companies in global try to find ways in order to advance more and gain more benefits. This is almost similar in the sector of civil aviation. Aviation companies would like to develop to be in a high-ranking position in the tough competition. In this case, airport hubs are important for aviation companies to deliver service or open a new branch at an airport or increase the service capability at a particular airport is crucial. For increasing the capability of the airports, some necessary or compulsory audits or inspections for those companies should be completed [20–22]. Generally, an authorized group of professionals are assigned to visit airports or clients at the airports. They commonly conduct seminars or make inspections and usually proximity is important to the clients for success [23]. Further most of the time, inspections lead to decrease the cost of the service at the airports [24].

The problem statement in this study is inferring the path of those specialists' travel. Especially if the visits include all of the airports of the country and if the visits are needed to be done within a limited budget, finding the route of the visits is getting complex to solve. Shortly, optimization problem for airport transportation can be described as finding optimal flight routes by minimizing total cost (i.e., monetary flight cost) including take-off, landing, and cruise. This problem definition is formulated as follows.

$$\min f(\overrightarrow{x}) = \sum_{i=1}^{N_{AP}+1} \left(c_{takeoff} + c_{cruise} \cdot (d_{cruise})_{i,i+1}\right) \quad (1)$$

In Eq. (1), $\overrightarrow{x}$ defines the airports (i.e., the nodes in the network given) where the final path visits only once which is the design variable vector of the optimization. Also, N_{AP} represents the number of airports that must be visited. $c_{takeoff}$ and c_{cruise} illustrate the cost of flight during take-off and the cost of flight during the cruise, respectively. $(d_{cruise})_{i,i+1}$ is the total distance during the cruise flight from airport i to airport $i + 1$. The value of d_{cruise} is computed as in Eq. (2).

$$d_{cruise} = d_{total} - d_{takeoff} - d_{landing} \quad (2)$$

d_{total} is the total distance between an airport to another airport, while $d_{takeoff}$ and $d_{landing}$ are the take-off horizontal distance and the landing horizontal distance, respectively. In this case, d_{total} can be defined as in the following formulation as well.

$$d_{total} = \sqrt{(X_{i+1} - X_i)^2 + (Y_{i+1} - Y_i)^2} \quad (3)$$

Equation (3) provides d_{total} which is the total distance between an airport to another airport in the network as mentioned before. To gain this variable, the coordinates (e.g., X and Y) of the airports are required and the Euclidean distance can be computed by Eq. (3).

In this research, there are a couple of constraints. Every airport must be visited only once and the visitor must come back to the initial airport after completing one tour among the airports. Additionally, take-off cost is assumed as a constant due to the fact that the particular airplane climbs to the same altitude at the beginning of every flight.

3 Metaheuristic Techniques

3.1 Tree Seed Algorithm (TSA)

In nature, trees proliferate and become more widespread, through their seeds. The seeds can reach to regions that are kilometers away with stochastic effects such as wind and animals. However, all seeds cannot grow and only seeds that are in optimal conditions can become trees. The TSA algorithm was developed by imitating this natural phenomenon. In the method, tree, seed, and location terms are defined as solution vector, candidate solution, and design variables, respectively. Besides, the method has a Search Tendency (ST) parameter that adjusts the stochastic spread rate of seeds. The main steps of the TSA study for the current study are as follows:

Step 1, Initialization: First, Search and problem parameters are defined in this step. These are Number of trees (N_{pop}), minimum and maximum number of seeds produced by a tree ($seed_{\min}$ and $seed_{\max}$), ST, and maximum iteration number ($MaxIter$). Then, initial solutions (flight routes) are generated randomly using Eq. (4). These evaluated and their fitness values are stored into algorithm memory.

$$X_{i,j} = round(1 + (N_{AP} - 1) \cdot rnd(0, 1)); i = 1, 2, \ldots, N_{pop}; j = 1, 2, \ldots, N_{AP} - 1 \tag{4}$$

where X represents matrix that contains design variable values of solutions in the algorithm memory, $rnd(a, b)$ is mathematical function to generate random real number between interval (a, b), $round$ is a function that rounds real value to the nearest integer.

Step 2, Search with seeds: In this step, seeds (candidate solutions) are produced for each tree in the solution pool. Number of seeds (N_{seed}) is between $seed_{\min}$ and $seed_{\max}$. The values of the seeds are determined as follows.

$$\begin{aligned} &if\, rnd < ST\, then\, S_{k,j} = X_{i,j} + rnd(-1,\ 1) \cdot \left(X_{best,j} - X_{r,j}\right) \\ &if\, rnd \geq ST\, then\, S_{i,j} = X_{i,j} + rnd(-1,\ 1) \cdot \left(X_{i,j} - X_{r,j}\right) \\ &i,\ r \in 1, 2, \ldots, N_{pop};\ i \neq r;\ k = 1, 2, \ldots, N_{seed};\ j = 1, 2, \ldots, N_{AP} - 1 \end{aligned} \tag{5}$$

where i, r, respectively, are the indexes of current and randomly chosen trees; S is the seed matrix of the tree and X_{best} is the best solution in the solution pool.

Step 3, Growth of seeds: The seeds produced for each tree are evaluated and the best seeds replace their belonged trees.

TSA uses processes between Steps 2 and 3 until the total number of evaluations reach to the $MaxIter$.

3.2 Symbiotic Organisms Search (SOS)

A new and powerful metaheuristic optimization method called Symbiotic Organism Search (SOS) simulates the interaction strategies that organisms use to survive in the ecosystem. Organisms rarely live in isolation due to relying on other species for feeding themselves and even survival. This dependency or relationship is called symbiosis. The most common symbiotic interactions in nature are mutualism, commensalism, and parasitism. Mutualism is a symbiotic interaction between two different species which benefit; Commensalism is a symbiotic interaction between two different species, one of which benefits and the other is unaffected or neutral;

Parasitism, on the other hand, refers to the symbiotic interaction between two different species, one of which benefits and the other is actively damaged. The SOS method is based on these interactions when creating a candidate solution, and these interactions construct the main stages of the optimization technique. These definitions, mentioned in the SOS method, are adapted to the current optimization problem as follows:

- Organism: candidate flight route (candidate solution),
- The fitness and survival power of the organism: 1/total route cost,
- Commensalism: Changing of a randomly selected solution according to the another randomly selected solution,
- Mutualism: Changing of the two solutions selected in the solution pool by affecting in a positive way each other,
- Parasitism: Changing some dimensions of a solution (parasitic organism) that is randomly selected from the solution pool.

Accordingly, the main steps of the SOS method can be described as follows:

Step 1, Initialization: Initial solutions are generated randomly and evaluated in the same way with the first step of the TSA algorithm.

Step 2, Mutualism phase: Solutions of randomly selected two organisms having i and k indices are updated according to the following formula:

$$\begin{aligned} X_{i,j}^{new} &= round\left(X_{i,j} + rnd(0,1)\cdot X_{best,j} - \left(\frac{X_{i,j}+X_{k,j}}{2}\right)\cdot rint(1,2)\right) \\ X_{k,j}^{new} &= round\left(X_{k,j} + rnd(0,1)\cdot X_{best,j} - \left(\frac{X_{i,j}+X_{k,j}}{2}\right)\cdot rint(1,2)\right) \\ & i,k \in 1,2,\ldots,N_{pop};\ i \neq k;\ j = 1,2,\ldots,N_{AP}-1 \end{aligned} \tag{6}$$

where $rint(a, b)$ randomly generates integer number a or b. Updated solutions are evaluated, and their fitness are compared with their previous versions. If new solutions have better performance, new solutions replace old versions.

Step 3, Commensalism phase: In this phase, one of the randomly selected organisms (having i index) updates its solution with respect to the other randomly selected organism (having k index) as in the formula below.

$$\begin{aligned} & X_{i,j}^{new} = round\left(X_{i,j} + rnd(-1,1)\cdot\left(X_{best,j} - X_{k,j}\right)\right) \\ & i,k \in 1,2,\ldots,N_{pop};\ i \neq k;\ j = 1,2,\ldots,N_{AP}-1 \end{aligned} \tag{7}$$

As a similar way in Step 2, if the new solutions have better fitness, it replaces its previous version.

Step 4, Parasitism phase: dimensions of randomly selected solution (having i index) change randomly as follows:

$$X_{i,j}^{new} = round(1 + (N_{AP} - 1) \cdot rnd(0,\ 1)) \\ i \in 1, 2, \ldots, N_{pop}; i \neq k; j \in 1, 2, \ldots, N_{AP} - 1 \tag{8}$$

The modified solution called parasite solution is compared to another randomly selected solution (having k index). If parasite solution has better fitness, it replaces with a solution having index k.

Processes among Steps 2 and 5 repeats by the time that $MaxIter$ is reached.

4 Design Example

In the study, the airports in Turkey are used to test the performance of TSA and SOS techniques on the optimal airport transportation path assignment. All the public airports in Turkey (totally fifty-five airports) were considered. In the journey, each airport is used only once, and the plane should come back to the initial location. For this problem, the airport for the starting journey is defined as "Istanbul Airport (IST)". The latitude and longitude of the airports are taken from the google maps application [25]. The latitude and longitude values have been converted to the bird's flight (Euclidean) distances of the airports with the help of the Haversine formula [26] and stored in the program memory. Cessna 208 Caravan plane is used for the journey. In the calculation of monetary flight cost of Cessna 208 Caravan, $c_{takeoff}$, c_{cruise}, $d_{takeoff}$, and $d_{landing}$ parameters, respectively, are taken as 80.8 \$, 2.03369 \$/km, 87.5 km, and 87.5 km. Search parameters of TSA and SOS techniques, determined according to literature studies [13, 14], are shown in Table 1.

Each optimization technique is tested ten times using different random_seed values. Statistical information of these tests is given in Table 2. According to the results of these tests, the best flight cost has been found from TSA. The best cost of the SOS technique is 32% higher than the TSA's best cost. Map views of the best optimum routes are shown in Figs. 1 and 2. In addition, route details of the best optimum solutions are given in Table 3. It is clear in figures and tables that there is not much difference in optimum routes except for the Marmara region and the southeastern Anatolia region. Search histories of TSA and SOS techniques for their best run (processes) are given in Fig. 3. According to this figure convergence rate of TSA is better than the SOS method.

Table 1 Search parameters of optimization methods

Parameter	Value
N_{pop}	100
ST	0.5
$seed_{\min}$	10
$seed_{\max}$	25
$MaxIter$	5000

Table 2 Statistical information about the optimization tests

	TSA	SOS
Minimum cost ($)	4813.99	6350.27
Average cost ($)	5049.39	6612.63
Standard deviation	221.08	230.52
Maximum cost ($)	5537.41	6994.09
Median cost ($)	5023.06	6507.19

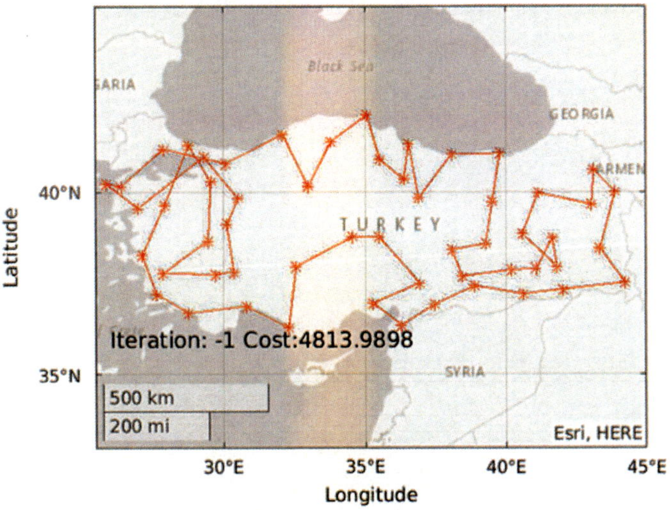

Fig. 1 Map view of the best optimum routes of tests obtained from TSA algorithm

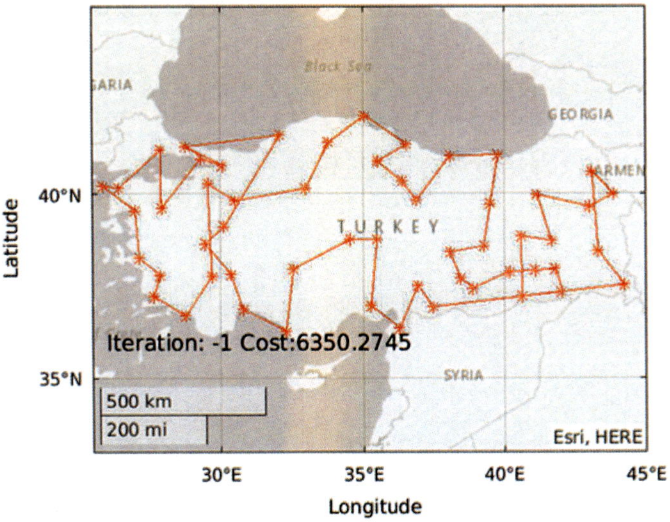

Fig. 2 Map view of the best optimum routes of tests obtained from SOS algorithm

Table 3 Details of the best optimum route in tests for the design problem (A.: Airport)

	TSA		SOS			TSA		SOS	
#	A. Code	A. Name	A. Code	A. Name	#	A. Code	A. Name	A. Code	A. Name
1	IST	Istanbul	IST	Istanbul	29	DIY	Diyarbakir	GNY	Sanliurfa
2	YEI	Bursa	KCO	Kocaeli	30	BAL	Batman	DIY	Diyarbakir
3	USQ	Usak	SAW	SabihaGokcen	31	MSR	Mus	BAL	Batman
4	CII	Aydin	BZI	BKMerkez	32	SXZ	Siirt	SXZ	Siirt
5	DNZ	Denizli	TEQ	Corlu	33	BGG	Bingol	NKT	Sirnak
6	ISE	Isparta	CKZ	Canakkale	34	ERZ	Erzurum	MQM	Mardin
7	KZR	Zafer	GKD	Gokceada	35	AJI	Agri	BGG	Bingol
8	AOE	Eskisehir	EDO	BKKocaSeyit	36	KSY	Kars	MSR	Mus
9	SAW	SabihaGokcen	ADB	Izmir	37	IGD	Igdir	ERZ	Erzurum
10	EDO	BKKocaSeyit	CII	Aydin	38	VAN	Van	AJI	Agri
11	GKD	Gokceada	BJV	Bodrum	39	YKO	Hakkari	IGD	Igdir
12	CKZ	Canakkale	DLM	Dalaman	40	NKT	Sirnak	KSY	Kars
13	TEQ	Corlu	DNZ	Denizli	41	MQM	Mardin	VAN	Van
14	KCO	Kocaeli	YEI	Bursa	42	GNY	Sanliurfa	YKO	Hakkari
15	ONQ	Zonguldak	AOE	Eskisehir	43	GZT	Gaziantep	GZT	Gaziantep
16	ESB	Ankara	ESB	Ankara	44	HTY	Hatay	KCM	KMaras
17	KFS	Kastamonu	KFS	Kastamonu	45	ADA	Adana	HTY	Hatay
18	NOP	Sinop	NOP	Sinop	46	KCM	KMaras	ADA	Adana
19	MZH	Amasya	SZF	Samsun	47	ASR	Kayseri	ASR	Kayseri
20	TJK	Tokat	MZH	Amasya	48	NAV	Kapadokya	NAV	Kapadokya

(continued)

Table 3 (continued)

	TSA		SOS			TSA		SOS	
#	A. Code	A. Name	A. Code	A. Name	#	A. Code	A. Name	A. Code	A. Name
21	SZF	Samsun	TJK	Tokat	49	KYA	Konya	KYA	Konya
22	VAS	Sivas	VAS	Sivas	50	GZP	Gazipasa	GZP	Gazipasa
23	OGU	Ordu-Giresun	OGU	Ordu-Giresun	51	AYT	Antalya	AYT	Antalya
24	TZX	Trabzon	TZX	Trabzon	52	DLM	Dalaman	ISE	Isparta
25	ERC	Erzincan	ERC	Erzincan	53	BJV	Bodrum	USQ	Usak
26	EZS	Elazig	EZS	Elazig	54	ADB	Izmir	KZR	Zafer
27	MLX	Malatya	MLX	Malatya	55	BZI	BKMerkez	ONQ	Zonguldak
28	ADF	Adiyaman	ADF	Adiyaman	56	IST	Istanbul	IST	Istanbul

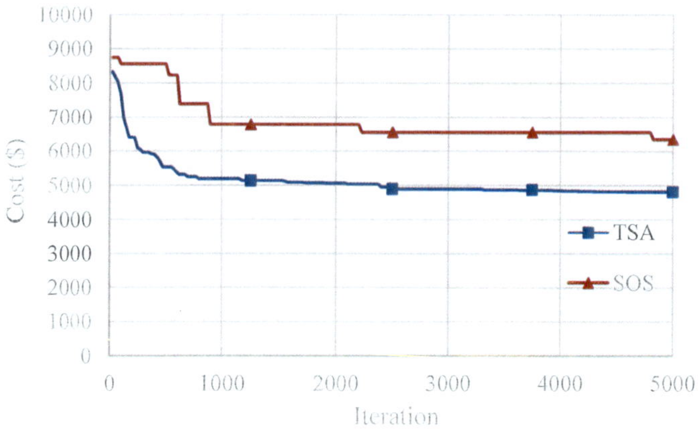

Fig. 3 Search histories of the best solutions

5 Conclusion

In the study, a novel network-based optimization problem, called airport path assignment according to the monetary flight cost, has been introduced. Two new meta-heuristic optimization algorithms named TSA and SOS were utilized and their performances were examined in the presented optimization problem. Public airport locations in Turkey were used as the network points of the design example.

Results obtained from the test example clearly show that although both optimization methods performed sufficiently, TSA has better performance than SOS method according to its convergence speed of reaching the optimum and final optimization value which is monetary cost of flights.

Additionally, route details (shape of the route, monetary flight cost, and visiting order of the network nodes) of the best optimum solutions are similar to each other. Only, for the Marmara and the southeastern Anatolia regions, the assignment path seems to change within the airports of the regions.

References

1. S. Albers, B. Herbert, A. Stefan, D.Werner, *Strategic Management in the Aviation Industry* (Routledge, 2017)
2. M. Dougan, *A Political Economy Analysis of China's Civil Aviation Industry* (Routledge, 2016)
3. AJ Stolzer, *Safety Management Systems in Aviation* (Routledge, 2017)
4. J. Current, C. ReVelle, J. Cohon, The shortest covering path problem: an application of locational constraints to network design. J. Reg. Sci. **24**(2), 161–183 (1984)
5. T.C. Matisziw, E. Demir, Inferring network paths from point observations. Int. J. Geogr. Inf. Sci. **26**(10), 1979–1996 (2012)
6. T.C. Matisziw, E. Demir, Measuring spatial correspondence among network paths. Geogr. Anal. **48**(1), 3–17 (2016)

7. E. Demir, Assigning convenient paths by an approach of dynamic programming, in *Proceedings Book of MICOPAM 2018*, Dedicated to Professor Gradimir V. Milovanović on the Occasion of his 70th Anniversary (2018), pp. 196–199
8. E. Demir, Approach for siting a support facility for transporting supplies in emergency cases in the Republic of Bulgaria, in *MATEC Web of Conferences*, vol. 234 (EDP Sciences 2018), p. 06001
9. E. Demir, N.U. Kockal, Iterative methodology on locating a cement plant. J. Inequalities Appl. article no: 174, 1–8 (2019)
10. İ Aydoğdu, P. Efe, M. Yetkin, A. Akın, Optimum design of steel space structures using social spider optimization algorithm with spider jump technique. Struct. Eng. Mech. **62**(3), 259–272 (2017)
11. İ Aydoğdu, A. Akın, Teaching and learning-based optimization algorithm for optimum design of steel buildings. Comput. Civ. Build. Eng. **2014**, 2167–2175 (2014)
12. C. Yıldırım, İ Aydoğdu, Artificial bee colony algorithm for thermohydraulic optimization of flat plate solar air heaters. J. Mech. Sci. Technol. **31**(7), 3593–3602 (2017)
13. İ. Aydoğdu, Comparison of metaheuristics on multi objective (cost&C02) optimization of RC cantilever retaining walls. Pamukkale Univ. J. Eng. Sci. **23**(3) (2017)
14. B. Uzun, Ö. Civalek, İ. Aydoğdu, Optimum design of nano-scaled beam using the social spider optimization (SSO) algorithm. J. Appl. Comput. Mech.
15. G.G. Tejani, J.S. Vimal, K.P. Vivek, Adaptive symbiotic organisms search (SOS) algorithm for structural design optimization. J. Comput. Des. Eng. **3**(3), 226–249 (2016)
16. M.S. Kıran, TSA: tree-seed algorithm for continuous optimization. Expert Syst. Appl. **42**, 6686–6698 (2015)
17. Z. Ding, J. Li, H. Hao, Z.R. Lu, Nonlinear hysteretic parameter identification using an improved tree-seed algorithm. Swarm Evol. Comput. **46**, 69–83 (2019)
18. J. Jiang, S. Jiang, X. Meng, C. Qiu, EST-TSA: an effective search tendency based to tree seed algorithm. Phys. A: Stat. Mech. Appl. **534**, 122323 (2019)
19. A.E. Ezugwu, D. Prayogo, Symbiotic organisms search algorithm: theory, recent advances and applications. Expert Syst. Appl. **119**, 184–209 (2019)
20. M.B. Salter (ed.), *Politics at the Airport* (University of Minnesota Press, 2008)
21. M.C. Leva, D. Del Sordo, F. Mattei, Day-to-day performance management in a small regional airport and management of change for safer operations. Cogn. Technol. Work **17**(2), 237–248 (2015)
22. J. Kolesár, L. Melníková, D. Heralová, P. Daňko, Methodology of the auditing measures to civil airport security and protection. MAD Mag. Aviat. Dev. **4**(20), 38–44 (2016)
23. J.R. Francis, N. Golshan, N.J. Hallman, Does audit partner proximity to clients matter. Working Paper, University of Missouri and the University of Texas at Austin (2018)
24. D.A. da Cunha, R. Macário, V. Reis, Keeping cargo security costs down: a risk-based approach to air cargo airport security in small and medium airports. J. Air Transp. Manag. **61**, 115–122 (2017)
25. Google Maps Homepage, https://www.google.com/maps. Accessed 08 Mar 2020
26. N.R. Chopde, M. Nichat, Landmark based shortest path detection by using A* and Haversine formula. Int. J. Innov. Res. Comput. Commun. Eng. **1**, 298–302 (2013)

Structural Strengthening of RC Buildings for Enhanced Seismic Resistance

Baris Gunes, Turgay Cosgun, Atakan Mangir, and Baris Sayin

Abstract This study aims to present a strengthening technique for the seismic resistance of existing reinforced concrete (RC) buildings subjected to earthquake effects. The entire process is illustrated with a case study of an existing RC building. In this study, the building is examined with field surveys and numerical analyses. In the first stage, the damage assessment of the building was done with visual inspection and observation of structural members, and then the seismic performance of the examined RC building was determined using a nonlinear analysis method. Accordingly, a strengthening technique is comprehensively presented for this building, which was determined to be seismically insufficient. The enhancement achieved for the sufficient seismic strength is discussed by numerical comparisons considering the analysis results of the structural model that includes the strengthening members proposed in the presented technique. The presented research is important in terms of the numerical determination of the current state of similar RC buildings against the seismic effects, and to verify the feasibility of the proposed strengthening technique.

Keywords RC building · Seismic performance · Field study · Numerical analysis · Strengthening

1 Introduction

The seismic design and analysis of structures located in the earthquake zones of Turkey, which has significant geographical importance, must be taken into consideration seriously. The acceptance of the possibility that the buildings may be exposed to the mentioned seismic activity at least once during their lifetime is an indication that the seismic effects are one of the most critical loading conditions. Besides, it

B. Gunes · T. Cosgun · B. Sayin (✉)
Department of Civil Engineering, Istanbul University-Cerrahpasa, 34320 İstanbul, Turkey
e-mail: barsayin@istanbul.edu.tr

A. Mangir
School of Engineering and Natural Sciences, Istanbul Medipol University, 34810 İstanbul, Turkey

S. M. Nigdeli et al. (eds.), *Proceedings of 6th International Conference on Harmony Search, Soft Computing and Applications*, Advances in Intelligent Systems and Computing 1275, https://doi.org/10.1007/978-981-15-8603-3_20

is necessary to eliminate adverse conditions and provide the structural safety of a building when there is a possibility of damage or structural deficiency due to various reasons. For this reason, it may be necessary to repair or strengthen the structure or inadequate structural elements. Modification of the building function, project design or improper manufacturing that is not suitable for the intended design loads, and damages due to earthquakes or any changes in the earthquake specifications may be the reason for strengthening [1]. In order to decide that a building needs to be strengthened, it is vital to determine the current state of the structure under study. While determining the current state of the building, it is necessary to examine with many parameters such as concrete properties, possible irregularities, workmanship quality, number of floors, residential area, and current status of structural elements. Currently, the addition of new structural members is generally preferred in strengthening procedures. Thus, in order to avoid restricting the usable area of the building, new structural elements are located on the outer facades of the building in most cases.

There are many studies in the literature where various strengthening proposals are presented by determining the performance level of existing RC structures. Yüzbaşı and Yerli [2] analyzed two different RC structures and assessed the performance level of the building. In addition to this, a proposal was made for the strengthening or reconstruction of buildings, taking into account the cost parameter for the structures examined. Similarly, Keleşoğlu et al. [3] analyzed a five-storey RC structure built in 1995. In the first stage, considering the analysis results, it is examined whether the building provides the life safety performance level. In the second stage, as the existing building did not meet the necessary conditions, it was proposed to strengthen it in accordance with the seismic regulations. Arı et al. [4] analyzed an RC structure damaged in the Afyon earthquake in 2002. As a result of the analysis, the most appropriate reinforcement method was determined for the building, and a strengthening proposal was presented, and the issues encountered during the application phases were discussed. Severcan and Sınanı [5] determined the damage levels and performance level of the structural elements by the nonlinear pushover analysis method for an eight-storey RC building. Olbak and Naimi [6] examined two different RC structures defined as risky structures within the scope of urban transformation. In this context, an analysis was performed using incremental equivalent earthquake load and incremental mode superposition methods, which are nonlinear analysis methods. As a result of the analysis, strengthening techniques were proposed for the existing two buildings. Tekin et al. [7] applied nonlinear analysis methods for an existing reinforced concrete structure in the study. The results obtained by using displacement coefficients and dynamic analysis methods from nonlinear analysis methods were compared. Chaulagain et al. [8] have presented a strengthening proposal for an examined RC building, which includes reinforced concrete wall attachment, adding steel members and increasing section dimensions. Yalçıner and Hedayat [9] numerically investigated an existing RC building. The appropriate methods of increasing section dimensions, strengthening with steel elements, and adding shear walls are discussed. Kap et al. [10] determined the structural system and material strength of a reinforced concrete school building and then analyzed the structure. As a result of the analysis, the performances of the building and the elements to be strengthened were determined. Hueste and Bai [11] examined the performance level of a five-storey RC building built in the United States in 1980. In order to strengthen the building, which did not have sufficient performance level, shear wall addition, increase of cross-section dimensions, and retrofitting methods with steel sections were examined. It has been determined that the construction of new RC shear walls is more suitable in terms of increasing the strength of the structural system compared to other approaches. Ghobarah et al. [12] investigated the effect of strengthening strategies on the structures for RC columns using the nonlinear pushover analysis method. It was determined that increasing the strength of the columns is the most suitable strengthening technique considering the data obtained from the analysis results. Korkmaz [13] examined the performance of RC structures which are strengthened by steel cross members. Tama et al. [14] examined the current state of an RC structure. As a result of the analysis, it is proposed to add RC shear walls or external steel walls, which are among the strengthening methods for the building. At the last stage, they compared the two techniques in terms of structural performance. Inoue and Youcef [15] examined three different reinforced concrete structures (apartments, schools, and hospitals). As a result of the investigations, the seismic behavior of the buildings was determined, and a strengthening plan was presented. Bal and Kılıç [16] analyzed an RC industrial building built in 1966. By determining the damage condition of the structural elements and two different strengthening methods (steel jacketing and CFRP), the effects of these methods on the structure behavior were compared in terms of economy and ductility. Rocha et al. [17] evaluated various strengthening practices in terms of efficiency and applicability. For this purpose, the results obtained from different numerical methods in the strengthening of RC frames were compared with experimental studies. Mutlu [18] evaluated what factors should be taken into consideration during the decision to strengthen or reconstruct reinforced concrete buildings by considering the economic factors. Cicen [19] analyzed a reinforced concrete public building and determined its performance. Accordingly, the strengthening of the building with seismic isolation was investigated.

In this study, the seismic performance of a strengthened RC building based on a strengthening proposal was investigated. The scope of this work includes field study, laboratory tests, and numerical analysis. The damages in the structure were determined first, and the reinforcement properties and configuration were determined. In the second stage, the compressive strength of the concrete samples taken from the building was determined in the laboratory, and the characteristics of the soil samples taken by drilling in the examination area were determined by laboratory tests. In the last stage, based on the data obtained in the field and laboratory studies, numerical analysis of the 3D finite element model of the building was performed using the nonlinear method, and the seismic performance of the structure was determined.

2 Examination of the Building

The building is located in Istanbul, Kucukcekmece district (Figs. 1 and 2). The building has a reinforced concrete frame structural system and an area of 800 m^2 which consists of four storeys including basement, ground, and two standard storeys. The storey heights of the building are 4 m at the basement storey and 3.5 m at the other storeys. The slab system of the building is constructed as two-way RC plates, and the foundation system is RC strip footing. The building has a regular plan geometry with 23.2 and 26.7 m dimensions in the x-direction and 32 m in the y-direction. There

Fig. 1 Aerial photo of the examined building

Fig. 2 The facade views of the examined building

is no RC shear wall in the existing building. No visible structural damage (crack, crush, etc.) was observed in the on-site inspections of RC members.

The rebar determination of the structural elements in the building was performed in accordance with the principles specified in the seismic code [20], using two different methods presented in TS708 [21]. In the first method, steel rebar positions were detected by X-ray imaging, while in the second method, steel rebar properties were determined by stripping the cover concrete.

Within the scope of the undamaged rebar detection approach, steel rebars of 34 structural members were examined. Rebar features (dimension, class) are checked with the other rebar determination method by stripping the cover concrete in 3 of structural members on all storeys. An example of some results obtained from both methods is given in Table 1. As a result of the two methods, it is understood that S220 class plain bars are used as longitudinal rebars and stirrups in the beams and columns. In addition, no damage due to corrosion effect was observed in the rebars.

In accordance with the relevant codes [22, 23], the laboratory studies are carried on seven cylinder core samples taken from the columns of the building. The concrete strength of the building is determined by concrete compressive strength tests performed on these samples. Obtained strength values are given in Table 2. As a result of the laboratory tests on the samples gathered from the building and the evaluation made according to the conditions of the related code [20], the unconfined concrete compressive strength of the building was determined as 13.06 MPa (Table 3).

Table 1 Rebar properties of some structural members in the building

Structural member	Dimensions (cm)	Number of examined member	Longitudinal rebar	Cover concrete (mm)	Stirrup
Columns	25 × 60	S01	8ϕ14	40	ϕ8/270
	25 × 60	S02	8ϕ14	40	ϕ8/270
	25 × 60	S03	8ϕ14	40	ϕ8/270

Table 2 The results of compressive tests of the samples

Storey	Core sample	*h/D*	f_c (MPa)	$f_{c,cube,av.}$ (MPa)
Ground	01	1.0	14.20	30.17
	02		15.14	
	03		13.50	
First	04	1.0	30.54	
	05		41.68	
	06		48.83	
	07		47.33	

h height, *D* diameter (100 mm), f_c compressive strength

Table 3 The evaluation of concrete compressive strength test results

$f_{c,cube,av}$	σ	$f_{c,cube} = f_{c,cube,av} - \sigma$	$f_{c,cyl} = 0.85 \cdot f_{c,cube}$	E_c
30.17	14.80	15.37	13.06	25747

All units: MPa, σ standard deviation, E_c Young's modulus of concrete

3 Strengthening Proposal

Within the scope of strengthening, new RC elements are added to the existing strip foundations. Dimensions of the new elements are 1.5 m in height, 1 and 1.2 m in width. ϕ22 ribbed bars are used as the top and bottom rebars with 15 cm spacing. In addition, in the web of strip foundations, ϕ16 web rebars and ϕ16 stirrups are used with 20 cm spacing. The foundation plan of the building is given in Fig. 3. Moreover, RC columns were jacketed from two, three, or four sides by the RC layer with 15–20 cm thickness. In these jacketing applications, ϕ16 longitudinal reinforcements and ϕ10 stirrups are used with 10 cm spacing. New RC shear walls with 30 cm thickness have been added between some columns. In these hear walls, ϕ12 and ϕ20 longitudinal reinforcements, and ϕ14 and ϕ16 stirrups are used with 10 cm spacing. In Fig. 3, jacketed members and added shear walls for strengthening purposes are also presented.

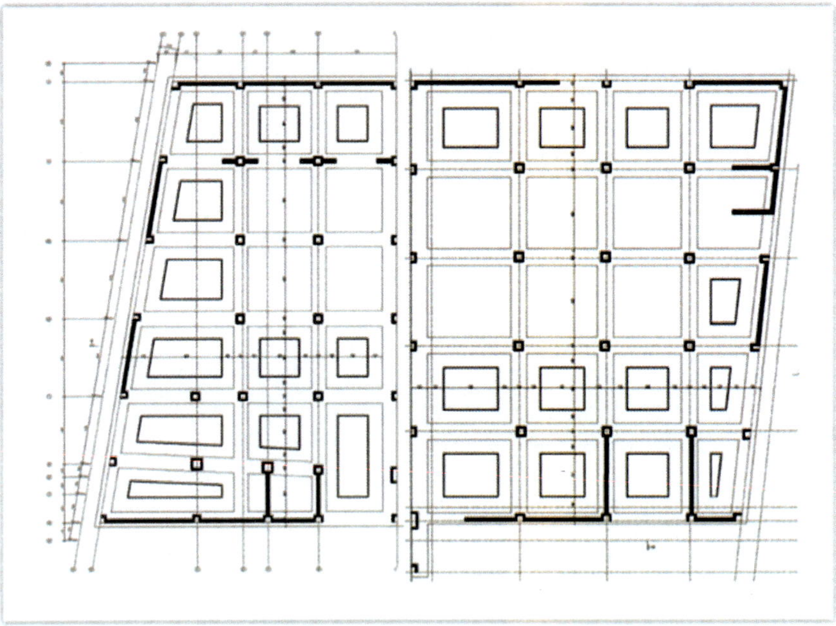

Fig. 3 Strengthened state of the building

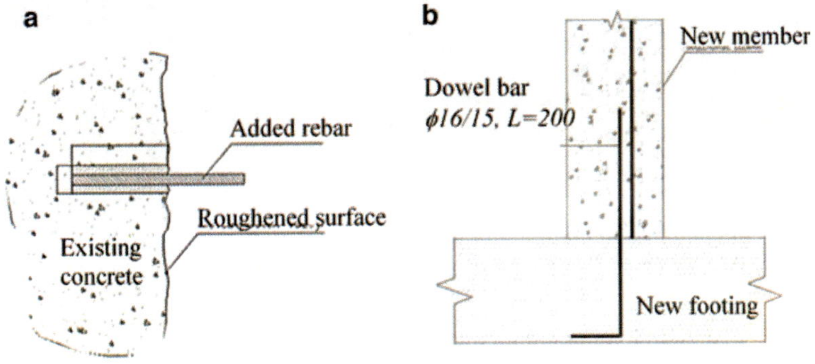

Fig. 4 **a** Epoxy anchorage application detail. **b** Shear wall–new foundation connection detail

The details related to the anchorages, dowel bars, and studs that are used in the strengthening applications are presented in Fig. 4a. ϕ16 anchors are used with 15 cm spacing in the shear wall—new foundation dowel connection. The details are shown in Fig. 4b. ϕ16 studs between existing and new elements are used with 30 cm spacing. Between the existing concrete and new elements, ϕ20 chemical anchors are used with 30 cm spacing (Fig. 5).

Anchorage detailing of the new shear walls include the connection of the walls to the existing columns and beams. Accordingly, ϕ20 chemical anchors are used with 30 cm spacing. In the anchorage details of the existing foundations and jackets/walls, ϕ20 chemical anchors are used with 15 cm spacing in the concrete jackets and ϕ20 anchors with 30 cm spacing in the walls (Fig. 6). ϕ10 bars are used as tie spacers in the concrete jackets with 25 cm horizontal and 20 cm vertical spacings. In the walls, ϕ10 tie spacers are used with 30 cm horizontal and 20 cm vertical spacings.

4 Numerical Analyses

Considering the findings within the scope of the field and laboratory studies, the strengthened state of the building was modeled using a finite element software [24], and nonlinear performance analyses of the building were carried out. The finite element analysis model of the building is given in Fig. 7. The parameters used in the analyses are presented in Table 4.

Mode superposition method is used in the seismic analysis. Based on the code, existing material strength values were used in the calculation of the member capacity calculations. In order to determine the performance level of the building, nonlinear multi-mode incremental pushover analysis was selected. Performance analysis was performed for the earthquake level, which has a 10% probability of exceedance in 50 years. In accordance with the purpose of use of the existing building, it is checked

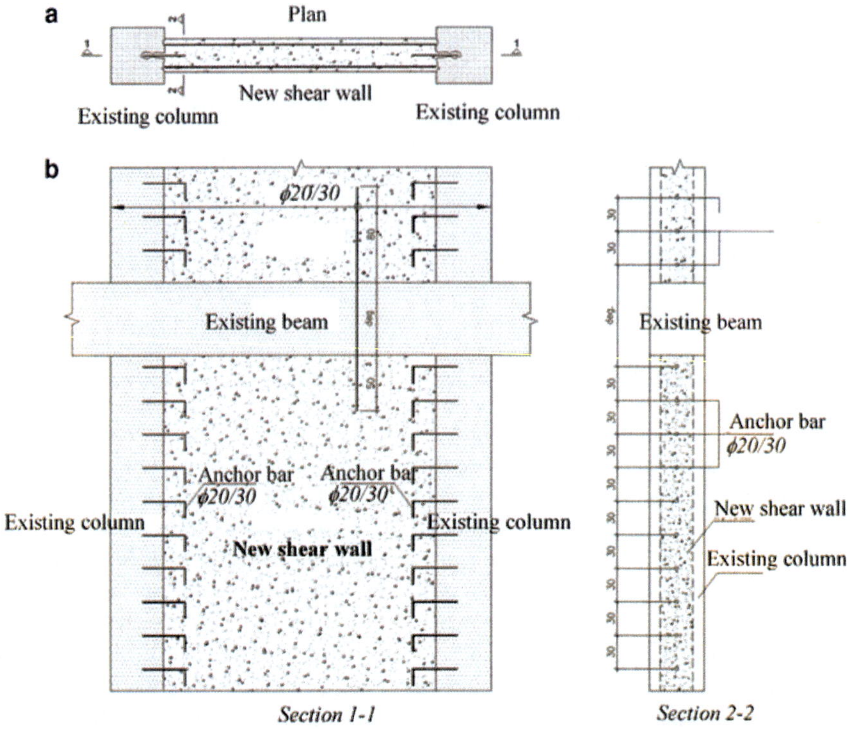

Fig. 5 **a** Typical column anchorage detail. **b** Anchorage detail for new shear wall element

whether the performance of the building satisfies "Life Safety" conditions (Table 5). Four damage regions are defined in the code for the ductile structural members (Fig. 8).

5 Results and Discussion

The results of the strengthened state are presented under this chapter. Base shear—roof displacement and demand spectrum—capacity curves were plotted for the "Life Safety" performance level considering the target earthquake (Fig. 9). The columns and shear walls are sufficient in terms of shear strength under the seismic effects in both orthogonal directions in all storeys. On each storey, the ratio of the shear forces acting on these members to the storey shear is less than the limit value presented in the code for "Life Safety" performance level (<20%). There are no vertical structural elements in this storey that have damages classified in the advanced damage region. Under the effects of the earthquake in the x and y directions, all of the damages in the beams are classified in the minimum or significant damage regions. There are no

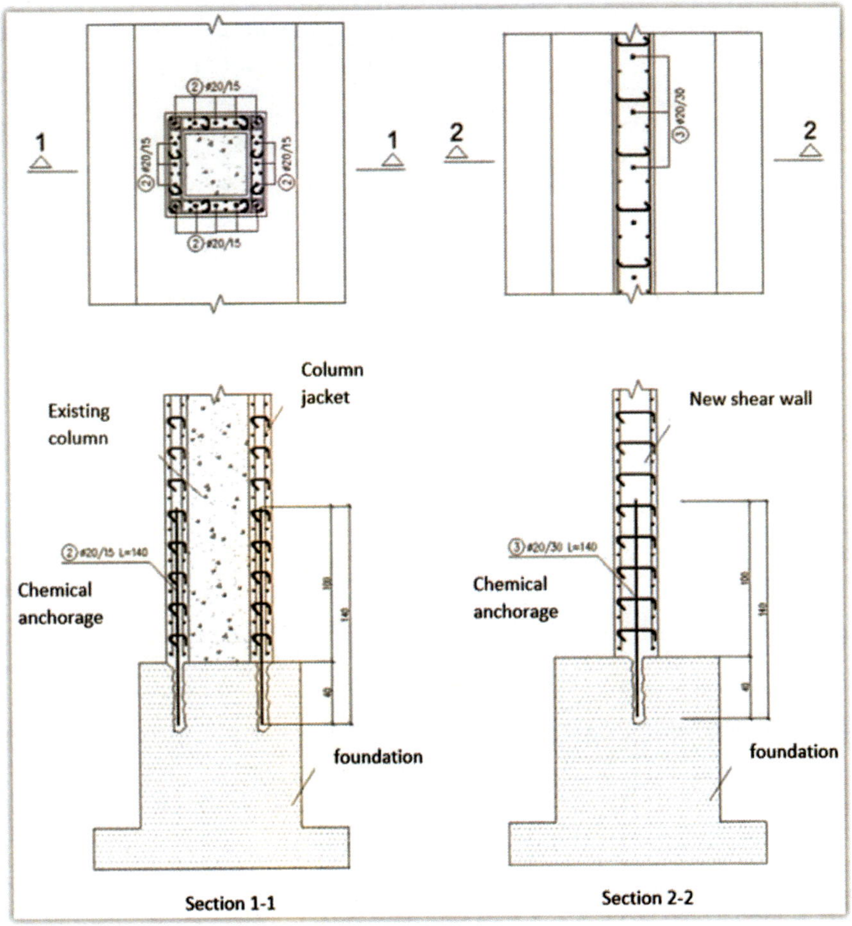

Fig. 6 Details of the connections between jacketing members/new shear walls and footings

beams that pass to the advanced damage region. In the code, the elements that may face with a brittle failure are not allowed for "Life Safety" performance level. The checks show that there are no brittle damaged elements under the seismic effects acting in both directions. The strengthened foundation was also modeled and then analyzed to determine whether the strength of the foundation is sufficient (Fig. 10). The envelope moment graphs of the foundation are given in Figs. 11 and 12, and the vertical displacement envelope diagrams of the foundation are presented in Fig. 13. The results show that the foundation has sufficient strength considering the capacity of the strengthened RC members and allowable stress of the soil.

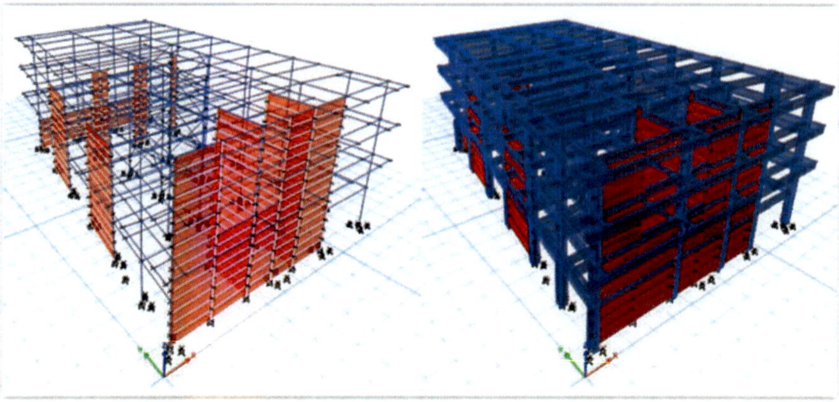

Fig. 7 Finite element model of the examined building

Table 4 Analysis parameters

Parameter	Definition	Value
A_o[a]	Effective ground acceleration coefficient	0.30
I	Building importance factor	1.0
$T_{x,y}$[b]	Natural vibration periods of building in x and y directions	$T_x = 0.23$ s, $T_y = 0.25$ s
T_A, T_B[c]	Spectrum characteristic periods	$T_A = 0.15$ s, $T_B = 0.40$ s
$S(T_1)$	Spectrum coefficient	2.5
$R_a(T_1)$	Seismic load reduction factor	2.0
R	Structural system behavior factor	1.0

[a]For seismic zone I
[b]Analysis result
[c]For Z2 local site class

Table 5 Minimum building performance targets anticipated for different earthquake levels [20]

The purpose of use and the type of building	Probability for the earthquake to be exceeded in 50 years		
	50%	10%	2%
The buildings that should be used after earthquakes		RU	LS
The buildings that people stay in for a long time period		RU	LS
The buildings that people visit densely and stay in for a short time period	RU	LS	
Buildings containing hazardous materials		RU	PC
Other buildings		LS	

RU Ready for usage; *LS* Life safety; *PC* Pre-collapse

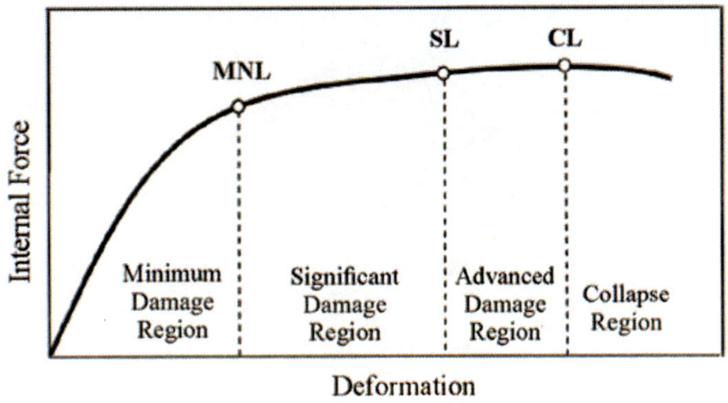

Fig. 8 Cross-section damage regions [20]

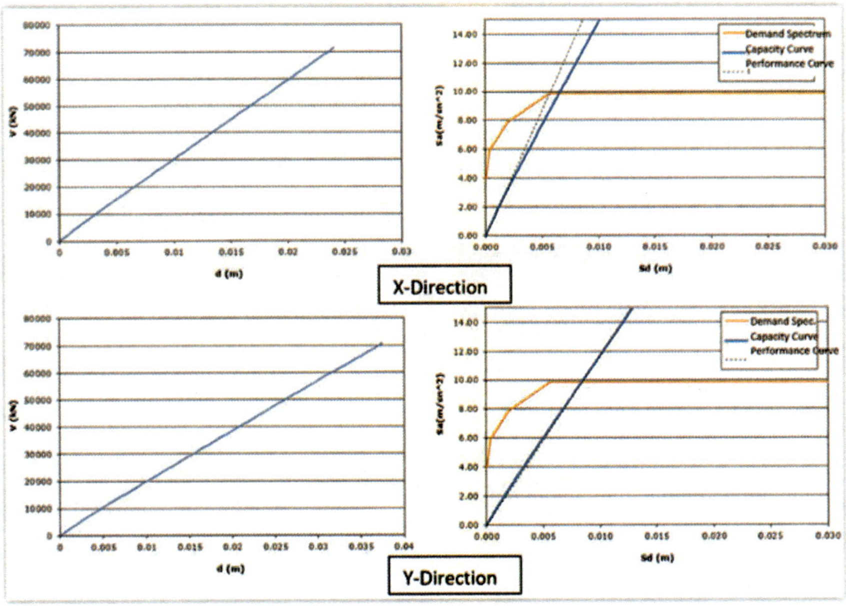

Fig. 9 Base shear versus roof displacement and demand spectrum versus capacity curve relationships

6 Conclusions

In this study, seismic performance analyses were carried out for an existing building with the nonlinear analysis method. The work consists of numerical analyses based on the data obtained by field study and laboratory test results. A strengthening proposal was presented for the building. The proposed strengthening method was applied

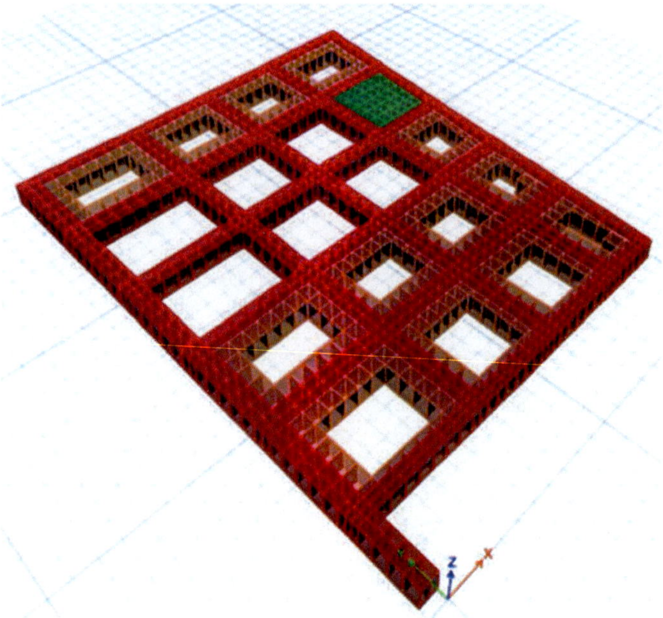

Fig. 10 Finite element model of the strengthened state of the foundation

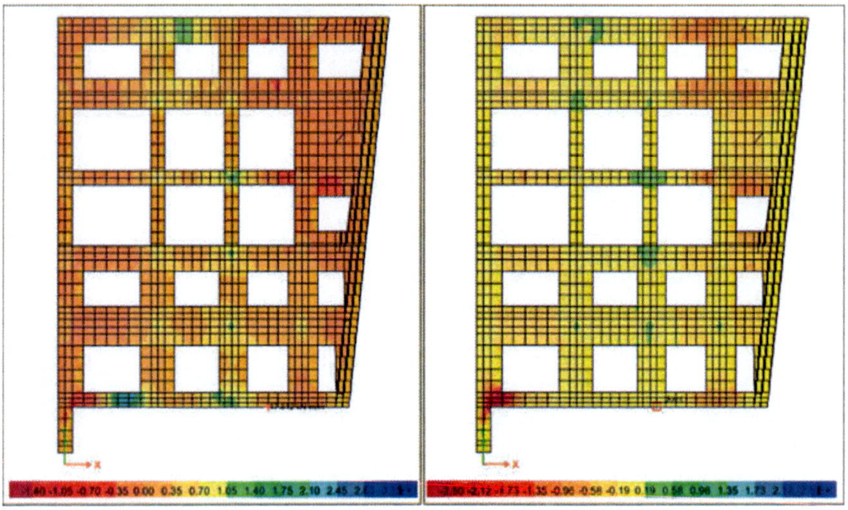

Fig. 11 M11 moment distribution for the building

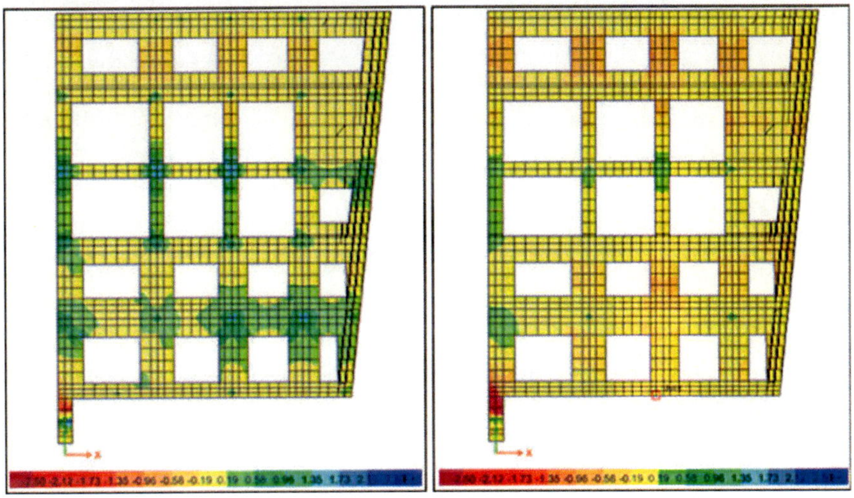

Fig. 12 M22 moment distribution for the building

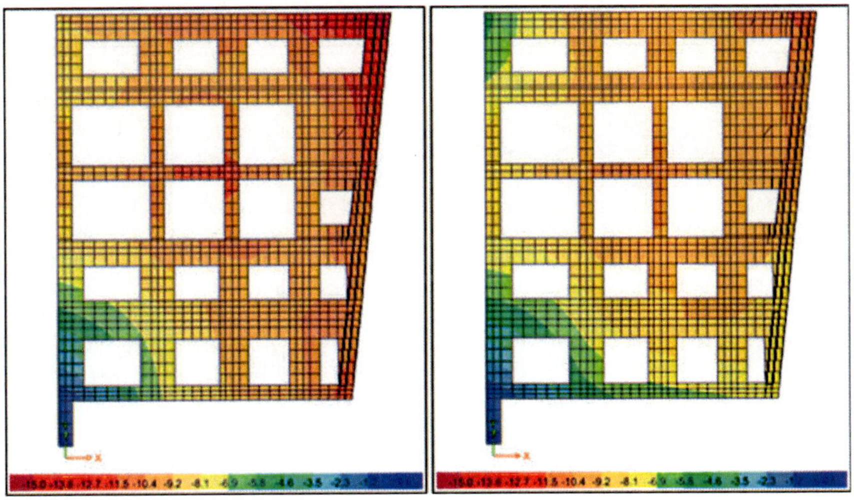

Fig. 13 Displacement distribution for foundation of the building

to the finite element model of the structure, and analyses were performed on the strengthened state. The obtained results are listed below.

- For the "Life Safety" performance level, the vertical elements in all directions are designated in the "Minimum Damage" or "Significant Damage" regions in x- and y- earthquake directions. There is no vertical structural element in the "Advanced Damage" region.

- All beams are found to be in the "Minimum Damage" region, and there are no beams that can be classified in the "Advanced Damage" region. Accordingly, the beams of the building ensure "Life Safety" performance-level conditions in case the proposed strengthening is applied. Limiting the lateral displacement of the whole structure by strengthening led these beams to get fewer damages in case of seismic action.
- In both x and y directions of the building, there are no brittle damaged elements under the target earthquake which ensures the conditions specified for "Life Safety" performance level in the code.
- Sufficient structural strength is achieved in the strengthened foundation.

As a result, it is determined that the presented strengthening proposal for the examined RC building has provided the "Life Safety" performance level, and can be assumed suitable considering the design earthquake (with a 10% probability of exceedance in 50 years) specified in the code. It is believed that the steps and sequence of the methodology, details of application and analysis procedures presented on a real structure will form an example for the literature and experts.

References

1. B. Yıldızlar, B. Sayın, Betonarme Yapıların Perde Duvar İlavesi ile Güçlendirme Tekniği. Dünya İnşaat Dergisi **26**(5), 99–103 (2010)
2. J. Yüzbaşı, H.R. Yerli, Betonarme Yapıların Deprem Etkisi Altında Performans Analizlerinin Yapılması ve Güçlendirilmesi. Çukurova Üniversitesi, Mühendislik Mimarlık Dergisi **33**(2), 273–286 (2018)
3. Ö Keleşoğlu, H. Çakar, A. Polat, Mevcut Betonarme Bir Yapının 2007 Deprem Yönetmeliğine Göre Performansının Belirlenmesi ve Güçlendirme Önerisi. Int. J. Pure Appl. Sci. **3**(2), 58–67 (2017)
4. K. Arı, H. Elcuman, E. Uncuoğlu, B. Somuncu, F. Altun, H.B. Kara, T. Haktanır, Afyon Depreminde Hasar Görmüş Betonarme Bir Yapının Güçlendirme Çalışması. Afyon Kocatepe Üniversitesi Fen Bilimleri Dergisi **6**(1), 81–96 (2006)
5. M. Severcan, B. Sınanı, Mevcut Betonarme Yapıların Deprem Performansının Analizi. Ömer Halisdemir Üniversitesi Mühendislik Bilimleri Dergisi **8**(2), 936–947 (2019)
6. M. Olbak, S. Naimi, Kentsel Dönüşüm Uygulanmış 5 Katlı İki Yapı Örneğinin Deneysel Verileri Kullanılarak Doğrusal Olmayan Analiz Yöntemleri ile Güçlendirme Sonuçlarının İrdelenmesi. İstanbul Aydın Üniversitesi Dergisi **8**(3), 145–166 (2016)
7. M. Tekin, R.T. Erdem, C. Gündüz, Betonarme Bir Yapının Performansa Dayalı Tasarımı: Deplasmana Dayalı Sismik Yaklaşım ve Yerdeğiştirme Katsayıları Yöntemi. Celal Bayar Üniversitesi Fen Bilimleri Dergisi **7**(2), 1–11 (2011)
8. H. Chaluagain, H. Rodrigues, E. Spacone, H. Varum, Assessment of seismic strengthening solutions for existing low-rise RC buildings in Nepal. Earthq. Struct. **8**(3), 511–539 (2015)
9. H. Yalciner, A.A. Hedayat, Repairing and strengthening of an existing reinforced concrete building: a North Cyprus perspective. Am. J. Eng. Appl. Sci. **3**(1), 109–116 (2010)
10. T. Kap, E. Özgan, M.M. Uzunoğlu, Betonarme Bir Okul Binasının 2007 Deprem Yönetmeliğine Göre Performans Analizi. İleri Teknoloji Bilimleri Dergisi **6**(1), 1–10 (2017)
11. M.B.D. Hueste, J.W. Bai, Seismic retrofit of a reinforced concrete flat-slab structure: part I seismic performance evaluation. Eng. Struct. **29**(6), 1165–1177 (2007)

12. A. Ghobarah, M.El. Attar, N.M. Aly, Evaluation of retrofit strategies for reinforced concrete columns: a case study. Eng. Struct. **22**(5), 490–501 (2000)
13. K.A. Korkmaz, Çelik Çapraz Elemanlarla Güçlendirilen Betonarme Yapıların Deprem Davranışlarının İncelenmesi. Doğuş Üniversitesi Dergisi **8**(2), 191–201 (2007)
14. Y.S. Tama, S. Yılmaz, H. Kaplan, A.T. Görgülü, Mevcut Betonarme Yapıların Dış Çelik Perde ile Güçlendirilmesi: Uygulama Çalışması, in *4th International Advanced Technologies Symposium*, 28–30 Eylül 2005, Konya (2005), pp. 1194–1197
15. A. Inoue, M. Youcef, Seismic evaluation and retrofit plan of existing reinforced concrete buildings of Algiers in Algeria, in *The 14th World Conference on Earthquake Engineering*, 12–17 Ekim 2008, Beijing (2008)
16. A. Bal, A.M. Kılıç, Deprem Etkisindeki Mevcut Betonarme Bir Sanayi Yapısının Güçlendirilmesinde CFRP ve Çelik Ceketleme Yöntemlerinin Uygulanmasının Karşılaştırılması, IV. Uluslararası Deprem Sempozyumu, 25–27 Eylül 2019, Kocaeli, Kocaeli, Küv Yayınları (2019), pp. 207–212. ISBN: 978-605-69403-1-6
17. P. Rocha, P. Delgado, A. Costa, R. Delgado, Seismic retrofit of RC frames. Comput. Struct. **82**(17–19), 1523–1534 (2004)
18. A.H. Mutlu, Mevcut Yapıların Güçlendirilmesi ya da Yıkılmasına Karar Verilmesi Aşamasında Göz Önüne Alınması Gereken Kriterler, 3. Türkiye Deprem Mühendisliği ve Sismoloji Konferansı, 14–16 Ekim 2015, İzmir (2015)
19. C. Cicen, Mevcut Bir Betonarme Kamu Binasının Performans Analizi ve Sismik İzolasyon ile Güçlendirilmesi. M.Sc. thesis, İstanbul Teknik Üniversitesi Fen Bilimleri Enstitüsü (2014)
20. TEC 2007. Turkish Earthquake Code, Specification for buildings to be built in earthquake regions, Ministry of Public Works and Settlement, Government of the Republic of Turkey (2007)
21. TS708, Steel for the reinforcement of concrete–reinforcing steel, TSE (Turkish Standards Institute), Ankara, Turkey (2010)
22. TS 498, The loads and loading cases due to use and occupancy in residential and public buildings, TSE (Turkish Standards Institute), Ankara (1997)
23. TS500, Requirements for design and construction of reinforced concrete structures, TSE (Turkish Standards Institute), Ankara, Turkey (2000)
24. ETABS Software, *Extended 3D Analysis of Building Systems* (Computer and Structures, Inc. CA 94704, Nonlinear Version 15, University Avenue, Berkeley, 2015)

Seismic Performance of Existing RC Framed Buildings Using Pushover Analysis Method

Baris Gunes, Atakan Mangir, Turgay Cosgun, and Baris Sayin

Abstract The aim of this study is to present a methodology on the determination of the seismic performance of existing reinforced concrete (RC) structures subjected to earthquake effects. This paper focuses on the assessment of existing RC structures with a nonlinear analysis (pushover analysis) method. The research is presented with the case study of an existing RC structure. The investigations on this RC structure is done with the field surveys, lab studies and numerical analyses. The first stage of the study includes the assessment of the compliance between the architectural and structural design projects of the related structure. Necessary checks have been made whether the projects of the structure are consistent with its existing state. Secondly, on-site mechanical tests were conducted on the structural members, and laboratory tests were performed on the samples collected from the structure. In the last stage, the seismic performance of the structure was determined based on the three-dimensional finite element model prepared for the existing state of the structure. The analysis results are discussed considering the target performance level parameter. The results have shown the importance of designating the appropriate target performance level in the evaluation of the seismic performance of RC structures.

Keywords Seismic performance · Existing RC structures · Fibre modelling method · Pushover analysis

1 Introduction

Various methods are presented in the seismic codes to provide adequate structural performance for newly constructed reinforced concrete structures. These methods are generally grouped in two parts as linear and nonlinear methods. Although the

B. Gunes · T. Cosgun · B. Sayin (✉)
Department of Civil Engineering, Istanbul University-Cerrahpasa, Istanbul 34320, Turkey
e-mail: barsayin@istanbul.edu.tr

A. Mangir
School of Engineering and Natural Sciences, Istanbul Medipol University, Istanbul 34810, Turkey

S. M. Nigdeli et al. (eds.), *Proceedings of 6th International Conference on Harmony Search, Soft Computing and Applications*, Advances in Intelligent Systems and Computing 1275, https://doi.org/10.1007/978-981-15-8603-3_21

uncertainties regarding the design and construction of new buildings based on the current codes reflecting the level of knowledge required by the era are negligible, there are different parameters in determining the current structural performance of existing reinforced concrete buildings. The seismic performance of similar structures, designed based on previous codes, should be determined considering the current knowledge level and codes in the civil engineering discipline. In general, three different outcomes are obtained as a result of the analyses performed on an old building considering the current codes. The first one is that the examined building has sufficient structural safety in terms of seismic resistance. The second one is the requirement of a strengthening approach to provide necessary structural safety for the structure. The last one is that the examined building should be demolished. In the Turkish seismic code, the seismic performance of the existing buildings is associated with the possible cross-sectional damages in the building under the effect of the applied earthquake and is defined on the basis of four different damage states [1]. Accordingly, the seismic performance level of the building is determined by using linear or nonlinear analysis methods and assessing the damage level of structural members. The required performance levels under a specified earthquake are mainly associated with the purpose of use (type) of the building. The earthquakes are specified considering their return periods (probability of exceedance).

A number of researches have been performed to investigate the seismic performance of existing RC buildings [2–16]. Wahyuni presented the evaluation of the vulnerability of RC buildings against earthquake loads. His work was conducted to find out the behaviour of damage and deviation that occurred in a case study based on numerical analysis. By comparing the testing and numerical analysis of the buildings, the analyses were used to make a recommendation whether the building was still in a state of immediate occupancy, life safety or collapse prevention [4]. Melani et al. performed a seismic evaluation on typical low-rise reinforced concrete frames. Three frames were analysed with capacity design concepts considering shear capacity, flexural capacity and contribution of slab reinforcements to beams. Maximum inter-storey drift ratios obtained from time history analyses are plotted against ground motion intensities [6]. Mosleh et al. presented the application and methodology for the seismic assessment of existing buildings. In this research, two concrete buildings which were designed with older codes were chosen, and pushover and time history analyses were performed in longitudinal and transverse directions. The seismic vulnerability was analysed for the earthquakes with different return periods, and the seismic demands were compared with the limits proposed in international codes. The conclusions are drafted in terms of safety [8]. Ghobarah et al. conducted nonlinear static and dynamic analyses of the reinforced concrete frames. In this study, a probabilistic approach is adopted where a large number of artificially generated ground motion records is used as input motion for the structure. The results of the analysis indicated the probability of various levels of damage to be expected when the existing frame is subjected to different ground motion levels [9]. Cherifi et al. aimed to estimate the seismic vulnerability of the existing buildings. For this purpose, capacity curves are developed for the reinforced concrete buildings using pushover method [11]. Halder and Paul evaluated the seismic vulnerability of a low-rise RC frame building which is designed for gravity loads according to the seismic code. The nonlinear static analysis is performed to find the capacity curve of the building. The results show that the damage of considered building varied from moderate to severe damage state under corresponding different seismic hazard levels [14]. Chaulagain et al. studied the performance of an existing building. To achieve this goal, seismic behaviour of four different three-storey RC residential building structures was considered. The buildings were subjected to static pushover analysis with different load patterns as triangular, uniformly and the first mode distributed loading. Besides, adaptive pushover analysis and nonlinear dynamic time history analysis were also performed [15]. Sobaih et al. specified an essential step for the required strategy to evaluate the seismic vulnerability of existing RC buildings on the national level. A proposed methodology is presented for that purpose by developing qualitative norms for factors that supposed to have a significant effect on the seismic behaviour of the buildings [16].

In the literature, some procedures are also developed for seismic assessment of existing RC structures [17–25]. Yakut presented a preliminary process to assess the seismic performance of existing reinforced concrete buildings rapidly. The method classifies the buildings either as safe, meaning the building might suffer from no severe damage or as unsafe, indicating that the life safety performance level would not be provided [19]. Kreslin and Fajfar presented a practice-oriented procedure for the seismic evaluation of structures. The applied approach proved to be a feasible tool for the seismic assessment of complex structures. However, due to the large randomness and uncertainties which are involved in the determination of both the seismic demand and capacity, only rough estimation of the seismic behaviour of such structures can be obtained [21]. Stylianos et al. suggested simple procedures in which the estimation of seismic demands and seismic capacities are used as a basis for the development of an evaluation flow chart that may be used by site engineers as a guide for field assessment. In the study, practical application examples of seismic assessment and evaluation conducted on existing reinforced concrete buildings that suffered from different levels of damage during past strong earthquakes are included for illustration [22]. El-Betar determined seismic vulnerability of buildings to suggest a suitable procedure for seismic evaluation of existing RC buildings. In the study, the seismic evaluation was applied to the selected two case studies, one represents the buildings designed according to gravity loads without seismic provisions, and the other represents the buildings designed by current codes. Moreover, pushover analysis was conducted to investigate the vulnerability of these buildings [24]. Hosseini et al. studied providing the life safety performance level in RC multi-storey regular buildings with a particular moment framing system. For this purpose, a set of multi-storey buildings up to sixteen storeys were considered in the highest seismic hazard zone and were designed based on the codes. Then, a set of near-source three-component accelerograms were employed and scaled according to the code, and a series of nonlinear time history analyses were conducted for all buildings. Roof displacement, acceleration and base shear values were calculated, and also the formation trend of plastic hinges and their distribution in the structures were investigated for evaluating the seismic performance. Results show that for some earthquakes, the performance of the buildings exceeds life safety performance level, and even in some cases, collapse level is reached [25].

In this study, the seismic performance level of an existing reinforced concrete structure is found by field studies, laboratory tests and numerical analysis. Firstly, the damage assessment of the structure is completed, and properties related to the reinforcements in the structural members are detected by field studies. In the second phase, concrete core specimens are obtained from various parts of the structure, and these samples are tested in the laboratory to find the strength of concrete of the building. The dynamic characteristics and material properties of the soil are found through testing the soil samples obtained by the drilling survey conducted near the building. In the last step, the collected data is evaluated and used in the finite element analyses of the structure, which are performed on the 3D analysis model of the building modelled in ETABS software [26]. The analysis results are evaluated considering the requirements of the code, and the seismic performance of the structure is determined.

2 The Examined Building

The examined building is located in Kartal district, Istanbul. The area of the building is calculated as 644 m^2, and storey heights are measured as approximately 3.2 m. The building has one ground and two standard storeys with a regular plan geometry. The dimensions of the building are 5 and 15 m in the x-direction, and 54 m in the y-direction (Fig. 1). The primary structural system of the building is formed with the reinforced concrete framing supported by shear walls. Some RC plate slabs in the building are supported by columns and walls, directly, and some others are carried by beams. All of the vertical structural members have vertical continuity in the structure.

As a result of on-site observations, no significant structural damage (crack, crush, etc.) was detected in the structural or nonstructural elements in the building.

2.1 Rebar Properties

The determination of rebars in the structural members of the building was made in accordance with the principles specified in the seismic code [1], using two different methods according to the TS708 [27] code. In the first method (undamaged), the location and diameter of rebars are determined by x-ray detection, while in the second method (damaged), rebars are visually detected and measured by stripping the cover concrete of the member.

Undamaged rebar detection procedures are performed on a total of 34 different structural members, including columns and shear walls on each floor [28]. The concrete cover of some of these members is stripped and obtained information about the rebars are verified by visual observations and measurements. A general summary

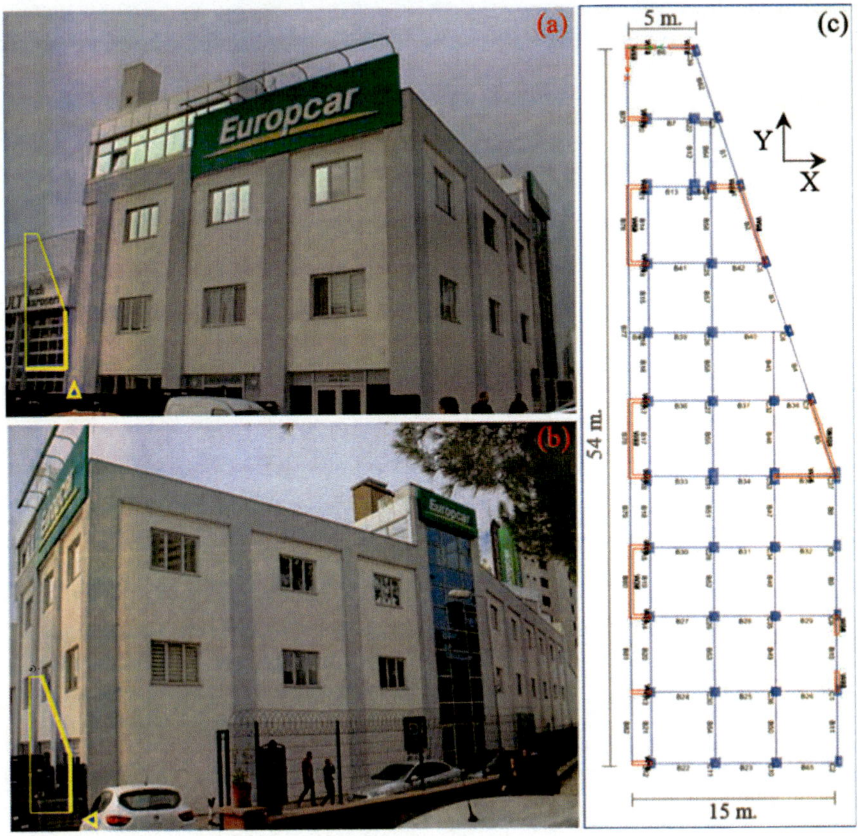

Fig. 1 The examined structure; **a**, **b** Overview, **c** Layout plan

of the rebar properties of some member types determined by these methods is given in Table 1.

Table 1 Rebar properties of columns and beams

Structural member type	Longitudinal rebar quantity/diameter	Cover concrete (mm)	Stirrup diameter/spacing (mm)
Columns	10ϕ16–12ϕ16	40	ϕ10/150
Beams	4ϕ18	40	ϕ10/200

Table 2 The results of compressive test for the specimens

Storey	Core specimen	[a]h/D	[a]A (mm^2)	[a]f_c (MPa)	$f_{c,cube,av.}$ (MPa)
Ground	01	1.0	7850	45.05	41.86
	02			34.99	
	03			42.30	
	04			40.19	
First	05	1.0	7850	30.54	
	06			41.68	
	07			48.83	
	08			47.33	
Second	09	1.0	7850	45.77	

[a]h: height, D: diameter (100 mm), A: cross-section area, f_c: compressive strength

Table 3 The evaluation of concrete compressive strength test results

$f_{c,cube,av}$	σ	$f_{c,cube} = f_{c,cube,av} - \sigma$	$f_{c,cyl} = 0.85 f_{c,cube}$	E_c
41.86	5.60	36.26	30.82	31,042

All units: MPa, σ: standard deviation, E_c: Young's modulus of concrete

2.2 Concrete Strength

In order to find the material strength values to be used within the finite element model of the building, the laboratory studies are carried on the samples gathered from the building. For this purpose, concrete strength tests are performed on 9 cylinder concrete core samples obtained from various parts of the building considering the conditions presented in the code [29–31]. The samples were obtained from all storeys of the building. Details and results related to the performed strength tests are given in Table 2.

The obtained results from the compressive strength tests are evaluated through the formulation presented in the code [32]. Accordingly, the unconfined concrete strength of the building based on these samples is found as 30.82 MPa, and the modulus of elasticity of concrete is found as 31,042 MPa (Table 3). These values are used in the finite element analysis model for the seismic performance analyses of the building.

2.3 Soil Parameters

In order to use specify the data input related to soil parameters in numerical analysis, soil samples were taken from different locations around the building by drilling core

Table 4 Characteristic properties of the soil [33]

Parameter	Soil layer
Soil class[a]	B
Local site class[a]	Z2
Vertical deposition coefficient (t/m^3)	5967
Allowable bearing stress, σ_z (kg/cm^2)	4.97

sampling. And then, characteristic properties of the soil are identified with laboratory test and code specifications. The results of this study are presented in Table 4.

3 Evaluation of the Existing Buildings

The nonlinear analysis method is effectively and widely used in the determination of the seismic performance of structures under earthquake loads. With this approach, the damage that can occur in the structural members under specific levels of ground motions is estimated numerically, and it is checked whether the estimated damage remains below the allowable damage limits. Allowable damage limits are defined compatible with performance targets for the building under the effects of various earthquake levels. Since the predicted seismic damage should be calculated as the cross-sectional damage of members, this situation generally corresponds to nonlinear deformations beyond linear elastic boundaries under severe earthquakes. Thus, the design and assessment approaches according to performance are directly related to nonlinear analysis.

3.1 Damage Limits and Regions Defined for the Structural Members

Three limit conditions have been defined for ductile elements in terms of the cross-sectional damages. These are minimum damage limit (MN), safety limit (SL) and collapsing limit (CL). Minimum damage limit defines the beginning of the behaviour beyond elasticity; safety limit defines the limit of the behaviour beyond elasticity that the section is capable of safely ensuring the strength, and collapsing limit defines the limit of the behaviour before collapsing. This classification does not apply to elements damaged in a brittle condition.

Elements that the damages with critical sections do not reach MN are assumed in the minimum damage region. Those in-between MN and SL are in the marked damage region, those in-between SL and CL are in advanced damage region, and those reached CL are in the collapsing region (Fig. 2).

Damage regions of the cross-sections shall be decided according to the comparison of the internal forces and/or deformation calculated using the linear and nonlinear

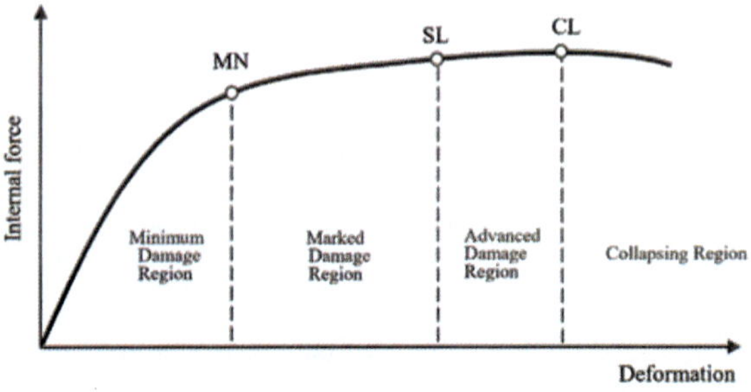

Fig. 2 Cross-sectional damage regions [1]

methods with the numerical values corresponding to the described damage limits. The damage of the structural element shall be decided according to the cross-section of the element with the highest damage.

According to the scope of the data obtained from the examination of the buildings concerning the existing situation, information level for each building type, and consequently, the information level coefficients shall be specified. In the code, the building information levels are classified as limited, medium and comprehensive, respectively. These coefficients are used in the capacity calculations of the structural elements. For the selection of the limited information level, there should be no structural design projects of the structural system. The structural system should be determined with on-site measurements. In the limited information level, in case the architectural projects of the building are present, they can be used in the preparation of measured drawings. If the medium level is selected, such projects should be present, and the compliance of the existing geometry with the project shall be checked. In case there is no detailed information in these projects, then measured drawings shall be prepared by fieldwork. In the comprehensive information level, the projects of the structural system are present. Sufficient measurements are performed for the confirmation of the information presented in the project. The information level coefficients to be applied to element capacities according to the level of information obtained from the examined buildings are given in Table 5. Unless otherwise specified, material strengths are not divided by material coefficients given in the relevant design codes. The determined existing material strength values are used in the calculation of the capacity of structural members. Medium information level is selected in this case study. The compliance of the design project of the building is verified by the on-site measurements.

Table 5 Information level coefficients for buildings [1]

Information level	Coefficient
Limited	0.75
Medium	0.90
Comprehensive	1.00

3.2 The Seismic Performance of Reinforced Concrete Buildings

The seismic performance of the buildings is related to the condition of the predicted damages under the effects of applied earthquakes and is defined based on four different performance levels. The seismic performance is determined by using the linear and nonlinear calculation methods and deciding on the damage regions of the components. There should be no brittly damaged members in the structure, and if any, they should be strengthened in a building to satisfy the performance criteria for the Life Safety in the code. For satisfying the ready for use performance level (RU), at most 10% of the beams exceed the minimum damage limit, and all other structural components remain in the minimum damage region. For satisfying the life safety performance level, the buildings shall satisfy the following conditions. At most 30% of the beams except for the secondary ones (that does not support the main structural system directly) can be in the advanced damage region. The total contribution of the columns in the advanced damage region to the storey shear force in each floor should not exceed 20%. For the top floor, this ratio can be increased to 40%. All other structural components should be in the minimum damage region or marked damage region. However, the total shear forces on the columns which exceed the minimum damage limit both in upper and lower sections in any floor should not be more than 30% of the storey shear. For satisfying the pre-collapse performance level following conditions should be provided. All components that are brittly damaged are in the collapse region. At most, 20% of the beams except for the secondary ones can be in the collapse region. All other structural components are in minimum, marked or advanced damage zone. However, the total shear forces on the columns which exceed the minimum damage limit both in upper and lower sections in any floor should not be more than 30% of the storey shear. This performance level does not provide life safety conditions. The buildings that don't satisfy the pre-collapse performance level are assumed to be in the collapse level.

3.3 Targeted Performance Levels for the Buildings

The general principle of earthquake-resistant design according to the seismic codes is to prevent structural and nonstructural elements of buildings from any damage in low-intensity earthquakes; to limit the damage in structural and nonstructural elements at repairable levels in medium-intensity earthquakes and to prevent the overall or partial collapse of buildings in high-intensity earthquakes in order to avoid the loss of life. The design earthquake considered in the code corresponds to the high-intensity earthquake with a 10% probability of exceedance within a period of 50 years for buildings with building importance factor of $I = 1$. Life safety performance level should be ensured under the effects of the design earthquake for the residential/office buildings.

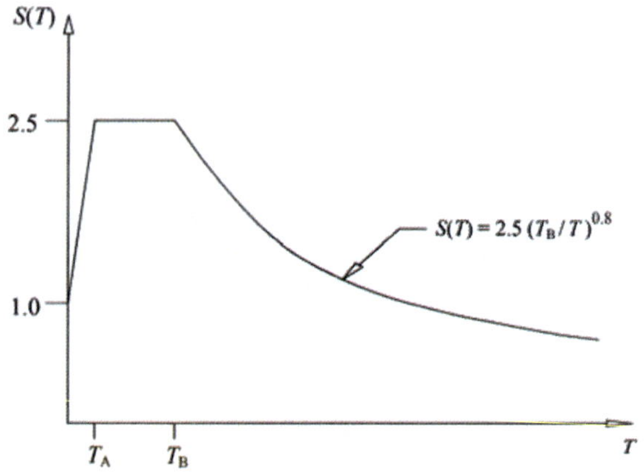

Fig. 3 Acceleration spectrum of the design earthquake [1]

Table 6 Minimum building performance targets anticipated for different earthquake levels [1]

The purpose of use and the type of RC structures	Probability for the earthquake to be exceeded in 50 years		
	50%	10%	2%
The buildings that should be used after earthquakes		RU	LS
The buildings that people stay in for a long time period		RU	LS
The buildings that people visit densely and stay in for a short time period	RU	LS	
Buildings containing hazardous materials		RU	PC
Other buildings		LS	

RU:*Ready for use*; LS: *Life safety*; PC: *Pre-collapse*

The acceleration spectrum defined in Fig. 3 represents the data of the design earthquake. Minimum building performance targets anticipated for different earthquake levels are given in Table 6.

4 Analysis and Results

A 3D finite element analysis model of the examined building is prepared in ETABS software (Fig. 4). The numerical analyses are performed on this model. The structure is examined by using nonlinear static pushover analysis method. The calculations related to earthquake loads and all assumptions and analysis parameters are determined in accordance with the Turkish seismic code (Table 7). Existing material strengths obtained by testing are not reduced with material coefficients in the analysis.

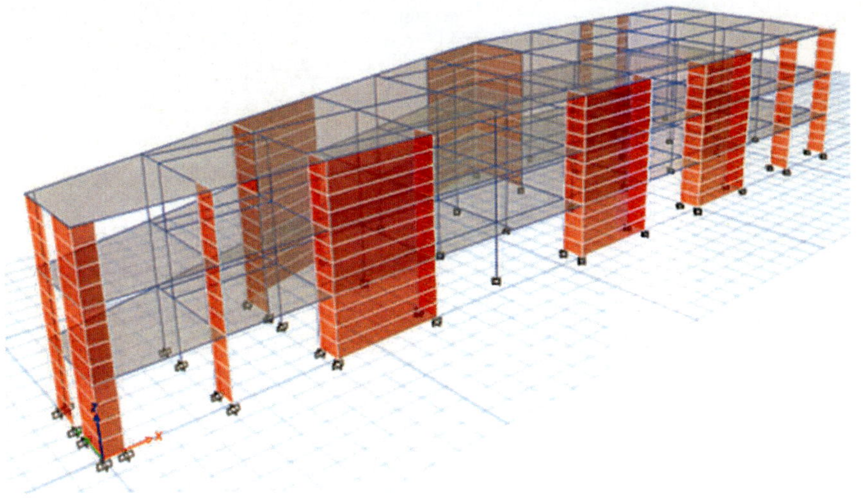

Fig. 4 3D Finite element model of the examined structure

Table 7 Analysis parameters

Parameter	Definition	Value
A_o [a]	Effective ground acceleration coefficient	0.40 g
I	Building importance factor	1.0
–	Building information coefficient	0.90
$T_{x,y}$ [b]	Natural vibration periods of the building in x-and y-directions	$T_x = 0.23$ s, $T_y = 0.25$ s
T_A, T_B [c]	Spectrum characteristic periods	$T_A = 0.15$ s, $T_B = 0.40$ s
$S(T_1)$	Spectrum coefficient	2.5
$R_a(T_1)$	Seismic load reduction factor	2.0
R	Structural system behaviour factor	1.0

[a]For seismic zone I, [b]Analysis result, [c]For Z2 local site class

Dead and live loads acting on the structure are determined considering the provisions given in TS498 code [34]. Building information level is assumed as "medium", and the building information coefficient is taken as 0.9. The seismic performance of the structure is investigated for the design earthquake which has 10% probability of exceedance in 50 years. Necessary checks are made for the "Life Safety" criteria presented in the code.

Demand spectrum, capacity curves and performance lines were plotted on spectral acceleration—spectral displacement chart for the "Life Safety" performance level considering the design earthquake in both directions (Fig. 5).

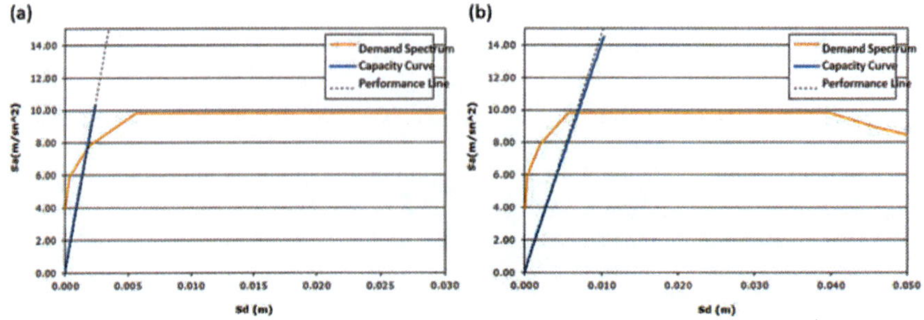

Fig. 5 Demand spectrum—capacity curves in **a** x- and **b** y-directions

5 Conclusions

This research is presented with the case study of an existing reinforced concrete structure. The investigations on this RC structure consists of the field surveys, lab studies and numerical analyses. The nonlinear static pushover analysis method is applied for the determination of the seismic performance level of the building. Analysis results have shown that the building does not ensure "Life Safety" provisions under the effects of the design earthquake, which has 10% probability of exceedance in 50 years. Thus, the life safety performance level is not achieved for the building. In general, the structural members are in minimum damage region, but there are some shear walls which were subjected to brittle failure in the analyses. Accordingly, the building could not present nonlinear behaviour entirely because the shear walls dominate the structural behaviour completely under seismic actions. The capacity curves also show this linear trend of the structure until the failure where the analysis was stopped. It is believed that, if the shear strength of these walls is increased, a more ductile behaviour will be achieved and the life safety conditions will be provided for this building. It is concluded that the building should not be demolished and a strengthening application is required for the shear walls of the building in order to continue its service.

References

1. Turkish Earthquake Code, *Specification for Buildings to be Built in Earthquake Regions*. (Ministry of Public Works and Settlement, Government of the Republic of Turkey, 2007)
2. F. Jalayer, I. Iervolino, G. Manfredi, Structural modeling uncertainties and their influence on seismic assessment of existing RC structures. Struct. Saf. **32**, 220–238 (2010)
3. F. Mazza, Modelling and nonlinear static analysis of reinforced concrete framed buildings irregular in plan. Eng. Struct. **80**, 98–108 (2014)
4. E. Wahyuni, Seismic vulnerability evaluation of existing R.C. buildings. Proc. Earth Planet. Sci. **14**, 76–82 (2015)
5. A. Kocak, B. Zengin, F. Kadioğlu, Performance assessment of irregular RC buildings with shear walls after Earthquake. Eng. Fail. Anal. **55**, 157–168 (2015)

6. A. Melani, R.K. Khare, R.P. Dhakal, J.B. Mander, Seismic risk assessment of low rise RC frame structure. Structures **5**, 13–22 (2016)
7. D. Combescure, F. Wang, Assessment of existing RC structures under severe dynamic loading using nonlinear modelling. INIS Repostory Search **39**, 23 (2007)
8. A. Mosleh, H. Rodrigues, H. Varum, A. Costa, A. Arede, Seismic behavior of RC building structures designed according to current codes. Structures **7**, 1–13 (2016)
9. A. Ghobarah, N.M. Aly, M. El-Attar, Seismic reliability assessment of existing reinforced concrete buildings. J. Earthq. Eng. **2**(4), 569–592 (1998)
10. C. Repapis, E. Vintzileou, C. Zeris, Evaluation of the seismic performance of existing RC buildings: I. Suggested Methodology. J. Earthq. Eng. **10**(2), 265 (2006)
11. C. Repapis, C. Zeris, E. Vintzileou, Evaluation of the seismic performance of existing RC buildings:II a case study for regular and irregular buildings. J. Earthq. Eng. **10**(3), 429–452 (2006)
12. F. Cherifi, M.-N. Farsi, S. Kaci, O. Belaidi, F. Taouchei-Kheloui, Seismic vulnerability of reinforced concrete structures in Tizi-Ouzou City (Algeria). Proc. Eng. **114**, 838–845 (2015)
13. T.Y. Qian, J.L. Fengge, Displacement-based seismic behavior evaluation for existing RC frame structures. Build. Struct. 2001–2007 (2001)
14. L. Halder, S. Paul, Seismic damage evaluation of gravity load designed low rise RC building using nonlinear static method. Proc. Eng. **144**, 1373–1380 (2016)
15. H. Chaulagain, H. Rodrigues, J. Jara, E. Spacone, H. Varum, Seismic response of current RC buildings in Nepal: a comparative analysis of different design/construction. Eng. Struct. **49**, 284–294 (2013)
16. M.E. Sobaih, M.A. Nazif, A proposed methodology for seismic risk evaluation of existing reinforced school buildings. HBRC J. **8**(3), 204–211 (2012)
17. U. Albayrak, M. Canbaz, G. Albayrak, A rapid seismic risk assessment method for existing building stock in urban areas. Proc. Eng. **118**, 1242–1249 (2015)
18. D. Perrone, M.A. Aiello, M. Pecce, F. Rossi, Rapid visual screening for seismic evaluation of RC hospital buildings. Structures **3**, 57–70 (2015)
19. A. Yakut, Preliminary seismic performance assessment procedure for existing RC buildings. Eng. Struct. **26**, 1467–1481 (2004)
20. Z. Sen, Rapid visual earthquake hazard evaluation of existing buildings by fuzzy logic modeling. Expert Syst. Appl. **37**, 5653–5660 (2010)
21. M. Kreslin, P. Fajfar, Seismic evaluation of an existing complex RC building. Bull. Earthq. Eng. **8**, 363–385 (2010)
22. Stylianos, I., Pardalopoulos, S.I., Pantazopoulou, S.J. Lekidis, V.A.: Simplified method for rapid seismic assessment of older R.C. buildings, Engineering Structures, 154, 10–22 (2018).
23. F.G. Gulay, K. Kaptan, E.I. Bal, S.S. Tezcan, P25-scoring method for the collapse vulnerability assessment of R/C buildings. Proc. Eng. **14**, 1219–1228 (2011)
24. S.A. El-Betar, Seismic vulnerability evaluation of existing R.C. buildings. HBRC J. (2016)
25. M. Hosseini, B. Hashemi, Z. Safi, Seismic design evaluation of reinforced concrete buildings for near-source earthquakes by using nonlinear time history analyses. Proc. Eng. **199**, 176–181 (2017)
26. ETABS Extended 3D Analysis of Building Systems, Computer and Structures, Inc. CA 94704, Nonlinear Version 15. (University Avenue, Berkeley, 2015)
27. TS708 Steel for the reinforcement of concrete–Reinforcing steel, *TSE*. (Turkish Standards Institute, Ankara, Turkey, Turkish, 2010)
28. Technical Report "Rebar test", Hedef test laboratory, Report no. KR (2016)
29. TS EN 12504-1 Testing concrete in structures–Part 1: Cored specimens—Taking, examining, and testing in compression, *TSE*. (Turkish Standards Institute, Ankara, Turkey, 2010)
30. TS EN 13791 Assessment of in-situ compressive strength in structures and precast concrete components, TSE (Turkish Standards Institute), Ankara, Turkey (2010).
31. Technical Report "Concrete compressive strength test", Hedef test laboratory, Report no. KR 1420 (2016)

32. TS500 Requirements for design and construction of reinforced concrete structures. *TSE.* (Turkish Standards Institute, Ankara, Turkey, 2000)
33. Soil report, *The Report of Soil Characteristics*. (Enfa Zemin Mühendislik Co., Turkish, 2016)
34. TS 498, The loads and loading cases due to use and occupancy in residential and public buildings. *TSE.* (Turkish Standards Institute, Ankara, Turkish, 1997)

Computation of Axial Symmetric Cylindrical Reinforced Concrete Walls with Domes

Aylin Ece Kayabekir

Abstract This paper presents a computation methodology for optimum design of axially symmetric cylindrical reinforced concrete walls. These walls are made of thin shells. In programming, the flexibility theory is used to analyses axial symmetric cylindrical walls. Reinforced concrete (RC) design is done according to ACI 318-Building code requirements for structural concrete. The generated computer code was performed for different numerical cases of the RC walls including a dome. The investigated cases are for various dome heights. The generated computer code is suitable for the future implementation of metaheuristic-based optimization methods.

Keywords Shell structures · Flexibility theory · Axially symmetric cylindrical wall · Dome

1 Introduction

The shells are curved plates, the thickness of which is small compared to other dimensions and curvature radius. Shells are frequently used in a varying structural system such as places of worship, dams, water tanks, etc.

When equilibrium equations are written on a small element of the cylindrical shell walls, it is seen that the solution of the obtained differential equation is an equation containing four integral constants determined depending on the support conditions of the cylindrical wall. Several studies have been carried out so far to solve the differential equation. In some of these studies, the approach of beams on elastic foundation is used. The reason for this is the similarity of the differential equation of the cylindrical wall and the beams on elastic foundation. This similarity makes a significant contribution and acceleration to the cylindrical wall studies. In this section, some studies on both the beams on elastic foundation and the cylindrical wall are presented.

A. E. Kayabekir (✉)
Department of Civil Engineering, Istanbul University-Cerrahpaşa, 34320, Avcılar Istanbul, Turkey
e-mail: ecekayabekir@gmail.com

S. M. Nigdeli et al. (eds.), *Proceedings of 6th International Conference on Harmony Search, Soft Computing and Applications*, Advances in Intelligent Systems and Computing 1275, https://doi.org/10.1007/978-981-15-8603-3_22

The soil model used in the analysis of beams on the elastic foundation was first introduced by Winkler [1]. According to this model, the floor consists of an infinite number of springs independent from each other. Reaction forces on the soil are proportional to the displacement of the beams at that point. This proportion is equal to the spring coefficient of elastic springs. In addition, the model assumes that the elements that change shape under the applied load will return to their original state by lifting the load [2, 3]. Winkler used this soil model for stress analysis on railways, modeling the rails as beams and obtained the expression of the displacement from the solution of the differential equation of the model.

Although it is not technically very difficult to determine the four integration constants in the displacement equation of the beam that sits on the elastic ground, difficulties arise in the calculations in the case of the uniformly distributed load or the presence of discontinuous loads and point loads such as single load or moment. The reason for the difficulty in the calculations is that the integral coefficients change along with the discontinuities resulting from loads. There are some methods in the literature regarding the solution to this problem. Among these, the methods commonly used in the studies are the initial parameter method developed by Umansky [4] and the superposition method developed by Hetenyi [5].

There are several books written for the analysis of beams on the elastic foundation [6–9]. The book written by Hetenyi in 1946 is regarded as one of the main sources of many solutions related to the subject. After the publication of this book of Hetenyi, many methods have been developed in order to analyze the beams on elastic foundation [10–17].

Bilington [18] used the similarity of the beam on the elastic foundation in the axial symmetrical cylindrical wall analysis. In the reference book, analysis is done using the flexibility method for different support conditions of the cylindrical walls and structural members attached to the wall. Some other studies related subject can be found in Ref.[19–21].

In this study, the computer code developed to carry out the analysis and design of the axial symmetrical cylindrical wall with dome is introduced. The analysis of the system is done according to the flexibility theory, and for the design requirement of the ACI 318-regulation are considered. In numerical examples, the results of the analysis of many conditions, consist of different support, dome feature and load conditions, are presented. In addition to contributing to the scientific studies to be carried out on such special structural systems, it is important in terms of enabling various additions to the program such as optimization and machine learning in the future.

2 The Analyses of Axial Symmetric Cylindrical Reinforced Concrete Walls via Flexibility Theory

Axial symmetrical cylindrical shell structures, i.e., cylindrical walls, are usually constructed with other shell structural members. For example, shell wall and dome, shell wall and foundation plate, etc. In this section, firstly, the application of the force method on the wall for fixed support conditions will be explained, and then the calculation of wall and dome combination will be introduced.

In Fig. 1, fixed based wall (second-degree hyperstatic system) with liquid loads and its equivalent isostatic system with redundant forces are given. In order to find these lateral (X1) and rotational (X2) redundant forces, firstly, two compatibility equations for the system must be written. Considering the fixed support conditions, these compatibility equations defined as follows:

$$D_{10} + F_{11}X_1 + F_{12}X_2 = 0 \text{ and } D_{20} + F_{22}X_2 + F_{21}X_2 = 0 \tag{1}$$

where D_{10} shows the displacement that occurs in the isostatic system at the point corresponding to X1 force and in the same direction as X1. Similarly, D_{20} represent the rotational displacement at the point corresponding to X2 force. F_{11}, F_{12}, F_{21}, and F_{22} are flexibility coefficients. These coefficients can be determined with F_{ij} that means displacement in freedom i in case of unit force applied to freedom j.

The flexibility coefficients for the wall can be written as follows:

$$F_{11} = \frac{1}{2\beta^3 D}, F_{22} = \frac{1}{\beta D} \text{ and } F_{12} = F_{21} = -\frac{1}{2\beta^2 D} \tag{2}$$

and displacement terms can be calculated as follows:

$$D_{10} = -\frac{\gamma r^2 H}{Eh} \text{ and } D_{20} = \frac{\gamma r^2}{Eh} \tag{3}$$

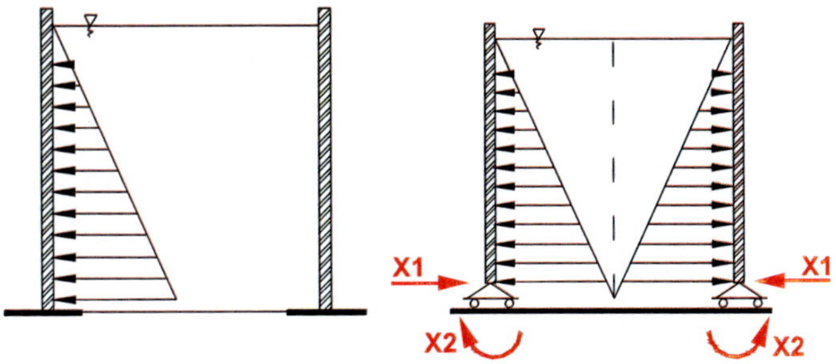

Fig. 1 Fixed based wall and its equivalent isostatic system

In the Eqs. (2, 3) β, D, γ, r, and H are a parameter given in Eq. (4), flexural rigidity (Eq. 5), density of liquid, radius, and height of the wall, respectively.

$$\beta = \sqrt[4]{\frac{Eh}{4r^2D}} = \sqrt[4]{\frac{3(1-\nu^2)}{r^2h^2}} \tag{4}$$

$$D_w = \frac{Eh^3}{12(1-\nu^2)} \tag{5}$$

If these equations are substituted in Eq. (1), redundant forces can be found as follows:

$$\begin{bmatrix} X_1 \\ X_2 \end{bmatrix} = F^{-1}\begin{bmatrix} -\frac{\gamma \cdot r^2 \cdot H}{E \cdot h} \\ \frac{\gamma \cdot r^2}{E \cdot h} \end{bmatrix} \tag{6}$$

In the system where the dome and the wall are together, the flexibility and displacement terms of both structural members are mutually superposed. The flexibility and displacement of the dome corresponding to the terms in the wall are given in Eqs. 7 and 8.

$$D_{10d} = \frac{r_d^2 q}{E_d h_d}\left(\frac{1+\nu_d}{1+\cos\alpha} - \cos\alpha\right)\sin\alpha \text{ and } D_{20d} = \frac{r_d q}{E_d h_d}(2+\nu_d)\sin\alpha \tag{7}$$

$$F_{11d} = \frac{2r_d\lambda_d\sin^2\alpha}{E_d h_d},\ F_{12d} = F_{21d} = \frac{2\lambda_d^2\sin\alpha}{E_d h_d} \text{ and } F_{22d} = \frac{4\lambda_d^3}{E_d h_d r_d} \tag{8}$$

In these equations, r_d, ν_d, α, E_d, and h_d are radius, Poisson ratio, starting angle, elasticity modules, and thickness of the dome, respectively. λ_d is a constant parameter that consists of rigidity, geometrical properties, and q is a distributed load of the dome. By superposing related flexibility and displacement terms of wall and dome, the unknown forces (X_3, X_4) at the top of the wall can be calculated as follows:

$$\begin{bmatrix} X_3 \\ X_4 \end{bmatrix} = F_s^{-1}\begin{bmatrix} D_{10s} \\ D_{20s} \end{bmatrix} \tag{9}$$

Further detailed information can be found in the reference book of Billington [18]. After the determination of internal forces on the wall, the reinforced concrete (RC) design process is applied. For the RC design, the requirement of ACI 318 is considered. These design constraints are summarized in Table 1.

All of this process is summarized in the flowchart given in Fig. 2.

Table 1 Design constraints

Description	Constraints
Flexural strength capacity, M_d	$M_d \geq M_u$
Shear strength capacity, V_d	$V_d \geq V_u$
Minimum steel ratio, $\rho_{\min}$	$A_s \geq A_{s\min}$
Maximum steel bars spacing, $S_{\max}$	$S \leq S_{\max}$
Minimum steel bars spacing, $S_{\min}$	$S \geq S_{\min}$
Minimum concrete cover, $c_{c\min}$	$c_{c\min} \geq 40\ mm$

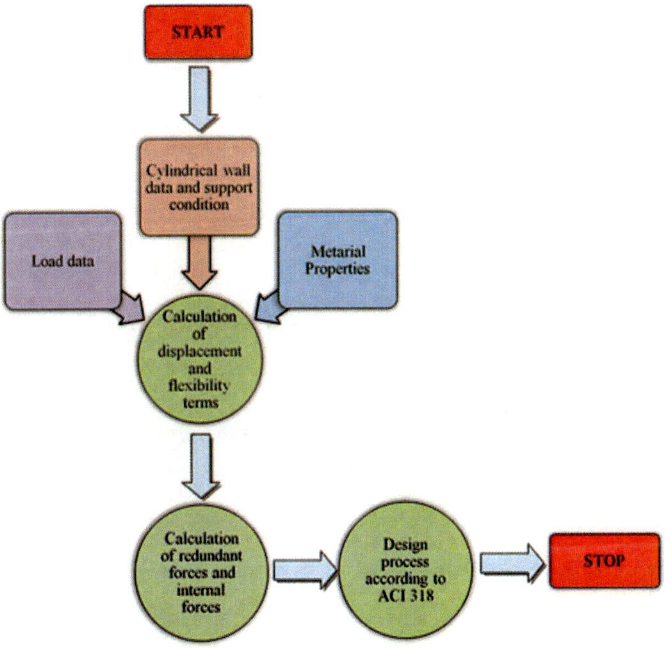

Fig. 2 Flowchart of the process

3 The Numerical Investigation

In this section, the results of the analysis for fixed and simple support condition at the base of the wall, several dome thickness ($h_d = 0.2$ m, 0.3 m) and height ($H_d =$ 2.5 m,4.5 m) features, and several load conditions due to stored liquid density ($\gamma = 7$ kN/m^3, 10 kN/m^3, 13kN/m^3) are presented. Analyses were done for 6-, 8-, and 10-m wall heights.

The radius of the wall, yield strength, concrete cover of steel, compressive strength of concrete, Poisson ratio of concrete, and minimum reinforcement ratio is 20 m, 420 MPa, 50 mm, 30 MPa, 0.2, and 0.008, respectively. The modulus of elasticity is calculated by using the expression defined in ACI 318 4700$(f'_c)^{1/2}$. In the equation,

f'_c is the compressive strength of concrete. Material unit costs used in calculating the cost of the wall are 200 TL/m^3 for concrete and 5000 TL/ton for steel.

The analysis results are presented in Tables 2 and 3 for simple and fixed support walls, respectively. In order to obtain economical results, the solution giving the lowest thickness (h_w) that provides design constraints is found by 0.01 m increments from 0.2 to 1 m. In the tables, the critical values of positive and negative moments, shear force, and hoop tension along the wall are shown with M_p, M_n, V, and N, respectively. It has been found in some cases that design constraints are not provided within the limits determined for wall thickness. These are shown in Table 2 as (–).

4 Conclusions and Future Works

In this study, a computer program developed to design and analyze the axial symmetric cylindrical wall with a dome is presented. The analyses were carried out according to the flexibility theory and the ACI 318-regulation of the design. Numerical research results are presented for a total of 72 different combinations consisting of different wall height, dome thickness and height, and load conditions.

The study has three important results that should be emphasized. The first one is that the developed program can be used both in engineers' designs and in scientific studies.

The second is that with the incremental thickness approach, the lowest thickness is determined for the wall that provides design constraints. In addition, it is easily seen whether the structural system meets the design constraints within the limits determined for the wall thickness.

The third is that, with the developed method, it is quite fast (under 0.1 s) with its structural design time. This is important in terms of optimization and artificial intelligence studies planned to be carried out in the future.

Table 2 The analysis results for simple support cases

H (m)	H_d (m)	h_d (m)	γ (kN/m^3)	h_w (m)	Cost (TL)	M_p (kN.m)	M_n (kN.m)	V (kN)	N (kN)
6.00	2.50	0.20	7.00	0.47	114,724.65	0.00	−41.38	−48.30	360.35
8.00	2.50	0.20	7.00	0.69	230,642.50	0.00	−78.56	−79.12	515.24
10.00	2.50	0.20	7.00	0.95	397,539.11	0.00	−130.61	−116.31	681.37
6.00	4.50	0.20	7.00	0.47	114,411.54	0.00	−38.71	−48.82	355.36
8.00	4.50	0.20	7.00	0.70	231,018.59	0.00	−75.63	−79.85	514.55
10.00	4.50	0.20	7.00	0.96	398,644.83	0.00	−128.47	−117.12	683.55
6.00	2.50	0.30	7.00	0.47	114,860.37	0.00	−42.30	−48.13	362.26
8.00	2.50	0.30	7.00	0.69	230,486.27	0.00	−80.12	−78.75	515.81
10.00	2.50	0.30	7.00	0.95	396,517.57	0.00	−132.32	−115.70	678.75
6.00	4.50	0.30	7.00	0.47	114,426.76	0.00	−39.11	−48.73	355.90
8.00	4.50	0.30	7.00	0.70	230,736.81	0.00	−76.22	−79.62	513.71
10.00	4.50	0.30	7.00	0.95	398,123.46	0.00	−128.96	−116.84	682.11
6.00	2.50	0.20	10.00	0.64	158,613.22	0.00	−81.84	−80.37	441.34
8.00	2.50	0.20	10.00	0.97	324,200.68	0.00	−158.30	−132.86	632.76
10.00	2.50	0.20	10.00	–	–	–	–	–	–
6.00	4.50	0.20	10.00	0.63	156,196.89	0.00	−75.11	−81.08	426.14
8.00	4.50	0.20	10.00	0.96	321,422.18	0.00	−150.17	−133.89	623.29
10.00	4.50	0.20	10.00	–	–	–	–	–	–

(continued)

Table 2 (continued)

H (m)	H_d (m)	h_d (m)	γ (kN/m^3)	h_w (m)	Cost (TL)	M_p (kN.m)	M_n (kN.m)	V (kN)	N (kN)
6.00	2.50	0.30	10.00	0.65	160,415.36	0.00	−85.92	−80.13	450.73
8.00	2.50	0.30	10.00	0.97	324,995.84	0.00	−164.89	−132.13	640.84
10.00	2.50	0.30	10.00	–	–	–	–	–	–
6.00	4.50	0.30	10.00	0.63	156,302.31	0.00	−76.32	−80.84	428.07
8.00	4.50	0.30	10.00	0.96	321,482.71	0.00	−152.34	−133.48	625.37
10.00	4.50	0.30	10.00	–	–	–	–	–	–
6.00	2.50	0.20	13.00	0.80	200,250.09	0.00	−132.86	−117.03	493.19
8.00	2.50	0.20	13.00	–	–	–	–	–	–
10.00	2.50	0.20	13.00	–	–	–	–	–	–
6.00	4.50	0.20	13.00	0.79	196,134.25	0.00	−121.90	−117.64	470.32
8.00	4.50	0.20	13.00	–	–	–	–	–	–
10.00	4.50	0.20	13.00	–	–	–	–	–	–
6.00	2.50	0.30	13.00	0.81	203,082.82	0.00	−140.57	−116.68	508.13
8.00	2.50	0.30	13.00	–	–	–	–	–	–
10.00	2.50	0.30	13.00	–	–	–	–	–	–
6.00	4.50	0.30	13.00	0.79	196,573.92	0.00	−124.71	−117.30	473.89
8.00	4.50	0.30	13.00	–	–	–	–	–	–
10.00	4.50	0.30	13.00	–	–	–	–	–	–

Table 3 The analysis results for simple support cases

H (m)	H_d (m)	h_d (m)	γ (kN/m3)	h_w (m)	Cost (TL)	M_p (kN.m)	M_n (kN.m)	V (kN)	N (kN)
6.00	2.50	0.20	7.00	0.49	122,692.03	71.32	−33.90	−79.33	244.72
8.00	2.50	0.20	7.00	0.74	247,210.78	153.83	−64.58	−133.97	345.25
10.00	2.50	0.20	7.00	–	–	–	–	–	–
6.00	4.50	0.20	7.00	0.47	120,214.01	70.55	−28.13	−79.14	226.67
8.00	4.50	0.20	7.00	0.73	244,118.25	152.90	−56.84	−133.99	332.19
10.00	4.50	0.20	7.00	–	–	–	–	–	–
6.00	2.50	0.30	7.00	0.50	123,861.03	71.94	−36.55	−79.62	257.23
8.00	2.50	0.30	7.00	0.75	249,756.12	154.09	−69.63	−133.79	357.59
10.00	2.50	0.30	7.00	–	–	–	–	–	–
6.00	4.50	0.30	7.00	0.47	120,326.54	70.48	−28.89	−79.04	228.56
8.00	4.50	0.30	7.00	0.73	244,241.29	152.90	−58.65	−133.85	333.58
10.00	4.50	0.30	7.00	–	–	–	–	–	–
6.00	2.50	0.20	10.00	0.66	163,881.96	121.71	−65.15	−126.37	271.76
8.00	2.50	0.20	10.00	–	–	–	–	–	–
10.00	2.50	0.20	10.00	–	–	–	–	–	–
6.00	4.50	0.20	10.00	0.64	160,078.26	121.11	−55.12	−125.96	244.40
8.00	4.50	0.20	10.00	–	–	–	–	–	–
10.00	4.50	0.20	10.00	–	–	–	–	–	–
6.00	2.50	0.30	10.00	0.68	168,802.70	122.25	−71.63	−126.79	293.85
8.00	2.50	0.30	10.00	–	–	–	–	–	–
10.00	2.50	0.30	10.00	–	–	–	–	–	–
6.00	4.50	0.30	10.00	0.64	160,680.99	121.25	−57.43	−126.00	247.96

(continued)

Table 3 (continued)

H (m)	H_d (m)	h_d (m)	γ (kN/m3)	h_w (m)	Cost (TL)	M_p (kN.m)	M_n (kN.m)	V (kN)	N (kN)
8.00	4.50	0.30	10.00	–	–	–	–	–	–
10.00	4.50	0.30	10.00	–	–	–	–	–	–
6.00	2.50	0.20	13.00	0.83	207,793.16	176.25	−104.19	−177.32	272.35
8.00	2.50	0.20	13.00	–	–	–	–	–	–
10.00	2.50	0.20	13.00	–	–	–	–	–	–
6.00	4.50	0.20	13.00	0.80	200,088.44	177.42	−90.36	−177.16	239.39
8.00	4.50	0.20	13.00	–	–	–	–	–	–
10.00	4.50	0.20	13.00	–	–	–	–	–	–
6.00	2.50	0.30	13.00	0.85	212,656.52	175.18	−114.52	−177.36	301.15
8.00	2.50	0.30	13.00	–	–	–	–	–	–
10.00	2.50	0.30	13.00	–	–	–	–	–	–
6.00	4.50	0.30	13.00	0.81	201,952.16	176.95	−94.29	−177.00	245.54
8.00	4.50	0.30	13.00	0.37	–	–	–	–	–
10.00	4.50	0.30	13.00	0.41	–	–	–	–	–

References

1. E. Winkler, Die Lehre von der Elasticitaet und Festigkeit—mit besonderer Rücksicht auf ihre Anwendung in der Technik, für polytechnische Schulen, Bauakademien, Ingenieure, Maschinenbauer, Architekten, etc., Verlag H. Dominicus Prag. (1867)
2. Y.H. Wang, L.G. Tham, Y.K. Cheung, Beams and plates on elastic foundations: a review. Prog. Struct. Mat. Eng. **7**, 174–182 (2005)
3. H.A. Karaşin, P. Gülkan, Elastik zeminlere oturan plakların sonlu ızgara yöntemi ile yaklaşık çözümü. İMO Teknik Dergi, Yazı **293**, 4445–4454 (2008)
4. A.A. Umansky, *Analysis of Beams on Elastic Foundation* (Central Research Institute of Auto-Transportation, Leningrad, 1933)
5. M. Hetenyi, *Analysis of Bars on Elastic Foundaiton*. Final Report of the second International Congress for Bridge and structural Engineering (English Edition), (Berlin, Munich, Verlag von Wilhelm Ernst and Son, 1936), pp 846–851
6. M. Hetenyi, *Beams on Elastic Foundation* (The University of Michigan Pres, Ann Arbor, 1946)
7. F.D.C. Henry, *The Design and Construction of Engineering Foundations* (Mc-Graw Hill, New York, 1956)
8. G. Jones, *Analysis of Beams on Elastic Foundations Using Finite Difference Theory* (Thomas Telford Publishing, New York, 1997)
9. E.D. Melerski, *Design Analysis of Beams, Circular Plates and Cylindrical Tanks on Elastic Foundations* (Taylor & Francis Group, London, 2000)
10. H. Cross, Analysis of continuous frames by distributing fixed-end moments. Proc. ASCE **57**, 919–928 (1930)
11. Z. Levinton, Elastic foundation analysed by the method of redundant reactions. Proc. ASCE J. Struct. Div. **73**, 1529–1541 (also in Transactions ASCE, 114, 40–52) (1947)
12. J. Penzien, Discontinuity stresses in beams on elastic foundation. Proc. ASCE J. Struct. Div. **86**, 67–97 (1960)
13. C. Miranda, K. Nair, Finite beams on elastic foundaiton. Proc. ASCE J. Struct. Div. **92**, 131–142 (1966)
14. F.W. Beaufait, Numerical analysis of beams on elastic foundaitons. Proc. ASCE J. Mechan. Div. **103**, 205–209 (1977)
15. B.Y. Ting, Finite beams on elastic foundaiton with restraints. Proc. ASCE J. Struct. Div. **108**, 611–621 (1982)
16. D.Z. Yankelevsky, M. Eisenberger, M.A. Adin, Analysis of beams on nonlinear winkler foundaiton. Comput. Struct. **31**, 287–292 (1989)
17. L. Vianello, Graphische untersuchungen der knickfestigkeit gerader stabe. Z. Ver. Dtsch. Ing. **42**, 1436–1443 (1898)
18. D.P. Billington, *Thin Shell Structures* (McGraw-Hill, New York, 1965)
19. G. Bekdaş, Harmony search algorithm approach for optimum design of post-tensioned axially symmetric cylindrical reinforced concrete walls. J. Optim. Theory Appl. **164**(1), 342–358 (2015)
20. G. Bekdaş, New improved metaheuristic approaches for optimum design of posttensioned axially symmetric cylindrical reinforced concrete walls. Struct. Des. Tall Special Build. **27**(7), e1461 (2018)
21. G. Bekdaş, Optimum design of post-tensioned axially symmetric cylindrical walls using novel hybrid metaheuristic methods. Struct. Des. Tall Special Build. **28**(1), e1550 (2019)

A Review on the Mechanical Properties of Natural Fiber-Reinforced Composites

Celal Cakiroglu and Gebrail Bekdaş

Abstract Natural fibers have drawn attention from the researchers and engineers in recent years due to their mechanical properties comparable to the conventional synthetic fibers and due to their low cost, eco-friendliness, and bio-degradability. Therefore, natural fibers such as kenaf or flux are being considered as a viable replacement for glass, aramid, or carbon. Extensive experimental studies are being carried out in order to determine the mechanical behavior of different natural fiber types such as the elastic modulus tensile strength, flexural strength, and the Poisson's ratio. This paper is a review of the various experimental studies in this field and it summarizes the findings of the researchers about the elastic properties of the major types of natural fiber-reinforced composites.

Keywords Composite materials · Natural fibers · Mechanical properties

1 Introduction

Composite materials have been playing a crucial role in the construction of lightweight structures in the last decades. The increasing use of the composites as a structural material is largely due to their superior mechanical properties compared to the conventional structural materials such as steel and aluminum. The higher stiffness and lighter weight of these materials reduce the cost of production, while at the same time, increasing the product quality. In today's industry the composite materials that are most widely used consist of fibers made of carbon, boron, or glass embedded in a polymer matrix such as epoxy. Even though these materials are capable of delivering a satisfactory structural performance, their recyclability and biodegradability

C. Cakiroglu (✉)
Department of Civil Engineering, Turkish-German University, Sahinkaya Cad 86, Istanbul 34820, Turkey
e-mail: cakiroglu@tau.edu.tr; bekdas@istanbul.edu.tr

G. Bekdaş
Department of Civil Engineering, Istanbul University, Cerrahpaşa, Istanbul 34310, Turkey

S. M. Nigdeli et al. (eds.), *Proceedings of 6th International Conference on Harmony Search, Soft Computing and Applications*, Advances in Intelligent Systems and Computing 1275, https://doi.org/10.1007/978-981-15-8603-3_23

is a major concern. In order to overcome these issues, the usage of natural fibers in composite materials has been proposed in the literature [1, 2]. Using natural fibers instead of synthetic fibers has certain advantages including their worldwide availability and the ease of manufacturing since these materials are mostly available in nature in stringy forms. A wide range of natural materials has been reported to have the capability to serve as fiber reinforcement in composites. Among these materials, some of the most widely mentioned ones are hemp, jute, kenaf, sisal, and banana leaf. Particularly, kenaf differentiates itself in this group of materials due to its abundant availability, low water usage, and high absorption of carbon dioxide.

2 Mechanical Properties of Natural Fiber-Reinforced Composites

Akil et al. [3] presented an overview of the general properties and the application of kenaf fibers in composites. In this work, kenaf was presented as a potential candidate to replace glass fibers due to the high energy consumption associated with the glass fiber production. It was reported that the energy needed to produce 1 kg of kenaf is 15 MJ, whereas the energy required to produce 1 kg of glass fiber is 54 MJ [4]. Furthermore, it was observed that within three months after sowing the seeds kenaf is able to grow to a height of more than 3 m and a base diameter of 3–5 cm under a variety of weather conditions [5]. Due to these conditions, the usage of kenaf as the fiber material has clear economic and ecological advantages over glass fibers. The mechanical properties of kenaf-reinforced composites have been investigated by various researchers. Among the factors that affect the mechanical properties such as ultimate tensile strength or flexural modulus of the fiber-reinforced composites, the type and orientation of the fibers (unilateral or randomly oriented), as well as the fiber volume fraction and the type of resin used, can be mentioned. Ochi [6] investigated the mechanical properties of biodegradable composites with unidirectional fibers made of kenaf and a resin made of PLA (polylactic acid). The kenaf used in the fabrication of the biodegradable composites was cultivated in two different ambient temperatures (22 and 30 °C) and the effect of the ambient temperature on the tensile strength of the fibers was investigated. In addition to the effect of the ambient temperature also, the effect of the location of the fiber on the plant, i.e., its distance from the root was investigated by dividing the kenaf into 500-mm sections and testing each section separately. Besides the effect of the ambient temperature on the mechanical properties of kenaf, its effects on the growth of kenaf were also examined. It was found that at the end of 168 days, the height of kenaf grown in a 30 °C environment was on average 1650 mm taller than kenaf grown in a 22 °C environment. Furthermore, the kenaf fibers taken from the longer rods grown at 30 °C exhibited greater tensile strength and elasticity modulus compared to the fibers taken from shorter rods. A comparison of the average tensile strength of the fibers taken from the top 500 mm and bottom 500 mm of the kenaf rods revealed that the tensile strength of the top fibers is 80%

of the tensile strength of the bottom fibers. Since the fiber tensile strength was found to be increasing with ambient temperature and inversely proportional to the distance from the roots, only fibers taken from the bottom sections of the kenaf rods cultivated at 30 °C were used in the fabrication of kenaf fiber-reinforced composite specimens that were tested in [6]. The kenaf/PLA composites were found to have tensile and flexural strengths of 223 MPa and 254 MPa, respectively. These values are in the same order of magnitude with glass fiber-reinforced composites (GFRP) [7]. Nishino et al. [8] experimentally investigated Young's modulus of kenaf/PLA (polylactic acid) composites and found it to be comparable to Young's modulus of synthetic fiber-reinforced composites. The average Young's modulus (in the fiber direction) of the kenaf-reinforced composites in these experiments was reported as 6300 MPa, while Young's modulus in the direction perpendicular to the fibers was on average 1500 MPa. Andre et al. [9] investigated the effect of kenaf fiber volume fraction on the tensile properties and the Poisson's ratio of the composites. It was reported that the highest tensile properties and Poisson's ratio occur at a volume fraction of $V_f = 0.42$. According to the research data collected by multiple researchers in the field [9, 10], a volume fraction around 0.4 appears to be a threshold value after which the tensile strength of the composite material begins to decrease due to insufficient filling by the matrix materials [11, 12]. In the study of Andre et al. [9], the modulus of elasticity(E_C) and Poisson's ratio(ν_C) of the kenaf fiber-reinforced composites were predicted using the equations developed by Tsai–Pagano, Manera, and Cox–Krenchel. The predicted values as well as the experimental outcome are demonstrated in Fig. 2.

Similarly, the predicted values of the Poisson's ratio were compared to the experimental values as demonstrated in Fig. 3.

Fig. 1 Kenaf plants [3]

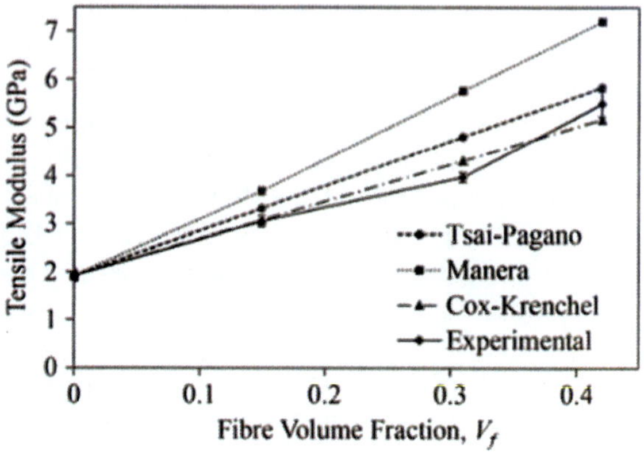

Fig. 2 Experimental and predicted tensile moduli for kenaf-reinforced composite [9]

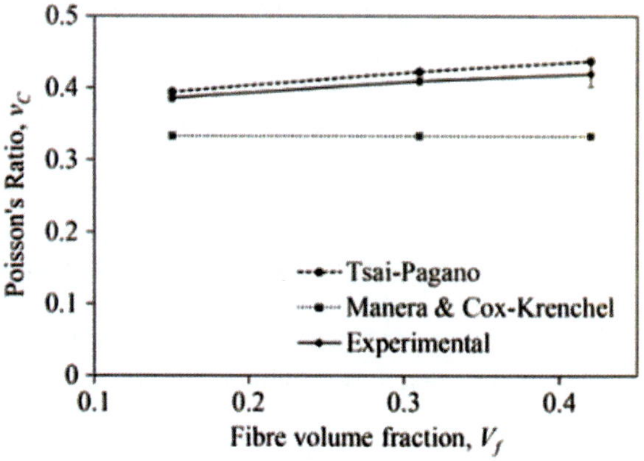

Fig. 3 Experimental and predicted Poisson's ratios for kenaf-reinforced composite [9]

Another type of environmentally friendly fiber that can be used in composites is flax fiber. Oksman et al. [13] investigated the mechanical properties of flax/PLA composites. In this study, it was shown that as a matrix material for composites with flax fibers, PLA performs 50% better than polypropylene (PP). The modulus of elasticity and the tensile strength reported in their experiments were 8300 MPa and 53 MPa, respectively. Andersons and Spa [14] carried out tensile experiments with composites which consist of flax fiber mats embedded in polypropylene matrix in order to determine the mechanical properties. The experiments were carried out with rectangular specimens (250 mm long and 25 mm wide) cut out of plates with 3 mm thickness manufactured for these experiments. An average elasticity modulus of 8700 MPa was reported in this study. Pozzi and Sepe [15] also carried out tensile tests with flax fiber-reinforced composites in order to determine their elasticity modulus, tensile strength, and Poisson's ratio. In these experiments, a tensile modulus of 7882 MPa and a tensile strength of 103 MPa were reported. The Poisson's ratio was reported to vary in a range between 0.24 and 0.51 depending on the orientation of the flax fibers (Fig. 4).

Fig. 4 Flax fibers [13]

Another natural fiber type that was investigated in the literature is hemp fiber. Sawpan et al. [11] carried out tensile tests with hemp fiber-reinforced UPE (unsaturated polyester) composites which is a type of natural composite that drew attention due to its low density and good stiffness. The result of these tensile tests is shown in Fig. 5.

Figure 5 shows the stress–strain response of hemp-reinforced composites at four different levels of fiber volume fraction in addition to the stress–strain behavior of

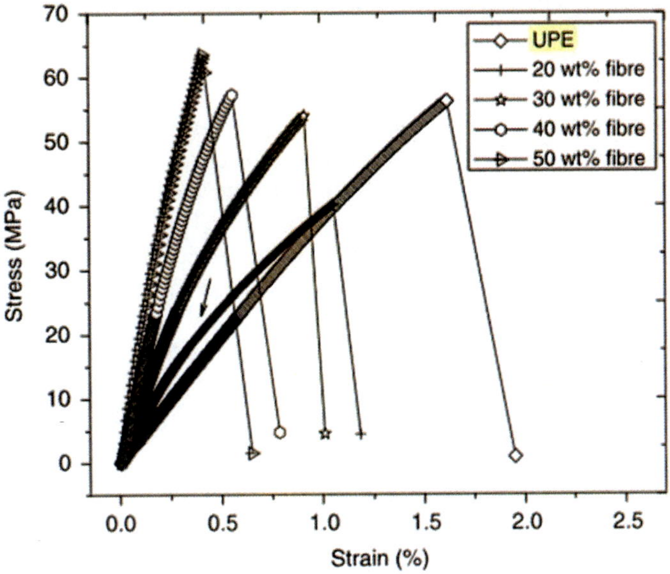

Fig. 5 Stress–strain response of hemp-reinforced UPE (unsaturated polyester) composites at various fiber volume fractions [11]

a matrix only specimen. It can be observed that the composites with 20% of fiber volume fraction have less tensile strength than the matrix only specimen, while the tensile strength tends to increase in proportion to fiber content. The decrease in the tensile strength due to the addition of the fibers into the matrix has been investigated in the literature. Piggott [16] introduced an equation that predicts the minimum fiber volume fraction ($V_{f,\min}$) that leads to increased tensile properties that is given below:

$$V_{f,\min} = \frac{(\sigma_m - \sigma_m^*)}{(K_1 K_2 \sigma_f - \sigma_m^*)} \tag{1}$$

In this equation, σ_m is the tensile strength of the matrix, σ_m^* is the strength of the matrix at the failure strain of the fiber, and σ_f is the strength of the fiber. The values recommended in the literature for the fiber orientation coefficient K_1 and the stress transfer factor K_2 are 0.2 and 0.96, respectively [17, 18].

Lu et al. [19] tested hemp fiber-reinforced composites with high-density polyethylene matrix at fiber volume fractions ranging from 20 to 40%. In these tests, a maximum tensile strength of 60 MPa was recorded, while the best mechanical properties were demonstrated by the composites with 40% of fiber volume fraction. Furthermore, an increase in stiffness was observed in proportion to the fiber volume fraction. A maximum elastic modulus of 2574 MPa was recorded for composites with 40% of fiber volume fraction.

3 Conclusions

The low cost, worldwide availability, and eco-friendliness of natural fiber-reinforced composites lead to an increase in the research activities in this field in recent years. Particularly, the high CO_2 consumption by the natural fiber-producing plants and their easy disposability makes natural fibers an attractive material to be used in structural applications. However, in order to use natural fiber-reinforced composites in the industrial applications in large scale, their mechanical properties need to be assiduously investigated. As one of the challenges in manufacturing with natural composites, it has to be mentioned that the mechanical properties of natural fibers are prone to exhibit wide variability. This is also confirmed by the large differences in the values of mechanical properties reported in the literature [20]. Furthermore, the material used as the matrix of the composite as well as the percentage fiber volume content can lead to variations in the mechanical properties.

A summary of the mechanical properties of kenaf, flax, and hemp fibers reported in the literature is presented in Table 1. Besides the aforementioned advantages of using natural fibers in composites, the wide variability in the documented mechanical properties of these materials makes numerical modeling and simulation of structures made of natural composites a challenging task.

Table 1 Ranges of the mechanical properties of kenaf, flax, and hemp fiber composites

Fiber	Tensile strength [MPa]	Young's modulus [GPA]
Kenaf	60–223	1.5–5.4
Flax	53–103	8.3–8.7
Hemp	60–65	2.6–16.3

References

1. M.R. Sanjay, G.R. Arpitha, B. Yogesha, Study on mechanical properties of natural—glass fibre reinforced polymer hybrid composites: a review. Mater. Today **2**(4–5), 2959–2967 (2015)
2. J. Holbery, D. Houston, Natural-fiber-reinforced polymer composites in automotive applications. JOM **58**(11), 80–86 (2006)
3. H.M. Akil, M.F. Omar, A.A.M. Mazuki, S. Safiee, Z.A.M. Ishak, A. Abu Bakar, Kenaf fiber reinforced composites: A review. Mater. Des. **32**(8–9), 4107–4121 (2011)
4. T. Nishino, Natural fibre sources. Green Composites 49–80 (2004)
5. S.H. Aziz, M.P. Ansell, S.J. Clarke, S.R. Panteny, Modified polyester resins for natural fibre composites. Compos. Sci. Technol. **65**(3–4), 525–535 (2005)
6. S. Ochi, Mechanical properties of kenaf fibers and kenaf/PLA composites. Mech. Mater. **40**(4–5), 446–452 (2008)
7. K. Naresh, K. Shankar, R. Velmurugan, N.K. Gupta, Statistical analysis of the tensile strength of GFRP, CFRP and hybrid composites. Thin-Walled Structures, **126**, 150–161 (2018)
8. T. Nishino, K. Hirao, M. Kotera, K. Nakamae, H. Inagaki, Kenaf reinforced biodegradable composite. Compos. Sci. Technol. **63**(9), 1281–1286 (2003)
9. N.G. Andre, D. Ariawan, Z.A. Mohd Ishak, Elastic anisotropy of kenaf fibre and micromechanical modeling of nonwoven kenaf fibre/epoxy composites. J. Reinf. Plast. Compos. **35**(19), 1424–1433 (2016)
10. H. Ku, H. Wang, N. Pattarachaiyakoop, M. Trada, A review on the tensile properties of natural fiber reinforced polymer composites. Compos. B Eng. **42**(4), 856–873 (2011)
11. M.A. Sawpan, K.L. Pickering, A. Fernyhough, Analysis of mechanical properties of hemp fibre reinforced unsaturated polyester composites. J. Compos. Mater. **47**(12), 1513–1525 (2013)
12. A.C.E. Dent, C.R. Bowen, R. Stevens, M.G. Cain, M. Stewart, Tensile strength of active fibre composites—prediction and measurement. Ferroelectrics, **368**(1 part 2), 209–215 (2008)
13. K. Oksman, M. Skrifvars, J.F. Selin, Natural fibres as reinforcement in polylactic acid (PLA) composites. Compos. Sci. Technol. **63**(9), 1317–1324 (2003)
14. J. Andersons, E. Spa, Stiffness and strength of flax fiber/polymer matrix composites. Polym. Compos. **37**(1), 915–924 (2016)
15. A. Pozzi, R. Sepe, *Mechanical Properties of Woven Natural Fiber Reinforced Composites*. Proceedings of the 15th European Conference on Composite Materials, ECCM 2012—composites at venice (2012)
16. M.R. Piggott, Short fibre polymer composites: a fracture-based theory of fibre reinforcement. J. Compos. Mater. **28**(7), 588–606 (1994)
17. H.L. Bos, J. Müssig, M.J.A. van den Oever, Mechanical properties of short-flax-fibre reinforced compounds. Compos. A Appl. Sci. Manuf. **37**(10), 1591–1604 (2006)
18. G. Kalaprasad, K. Joseph, S. Thomas, C. Pavithran, Theoretical modelling of tensile properties of short sisal fibre-reinforced low-density polyethylene composites. J. Mater. Sci. **32**(16), 4261–4267 (1997)
19. N. Lu, R.H. Swan, I. Ferguson, Composition, structure, and mechanical properties of hemp fiber reinforced composite with recycled high-density polyethylene matrix. J. Compos. Mater. **46**(16), 1915–1924 (2012)
20. S.M. Yukseloglu, H. Yoney, *The Mechanical Properties of Flax Fibre Reinforced Composites*, vol. 12. (RILEM Bookseries, 2016)

Choosing the Appropriate Branch for Participation Banks Through Machine Learning

Bulut Karadağ and Eda Arıkan

Abstract The size of a bank is directly proportional to the number of branches it has. In the traditional banking system, banks try to reach more customers by opening new branches. Each new branch to be opened carries a cost for the bank. Therefore, a feasibility study is required for the new branch planned to be opened. While conducting feasibility studies, demographic information is generally used and it is aimed to open the branch in a place that will provide maximum efficiency. For this purpose, in the first phase of this study, the branch to be opened by a newly established participation bank was found through machine learning. Location data of other participation bank branches that have been in service for many years were used as datasets. With the OPTICS algorithm, a clustering structure was created with at least one kilometer and at least three branches. In the second phase of the study, the priority of the branches decided to open was found through analytical work. In the analytical study, development indices and levels of the cities were used.

Keywords Location selection · Optics algorithm · New branch selection · Participation bank

1 Introduction

With the global developments in the world, companies compete with each other. In these competitive environments, the choice of establishment locations is important. In particular, the choice of branch location of banks is of strategic importance. The branch to be selected should have the qualifications that the bank will achieve its goals in the long run and provide low costs and high profits. Therefore, many factors need to be analyzed in site selection. The place to be chosen should be qualified to

B. Karadağ (✉) · E. Arıkan
Vakıf Participation Bank, Istanbul, Turkey
e-mail: bulut.karadag@vakifkatilim.com

E. Arıkan
e-mail: eda.arikan@vakifkatilim.com

S. M. Nigdeli et al. (eds.), *Proceedings of 6th International Conference on Harmony Search, Soft Computing and Applications*, Advances in Intelligent Systems and Computing 1275, https://doi.org/10.1007/978-981-15-8603-3_24

meet the needs of the enterprise and expert experiences should be used during the installation phase. These studies should be taken into consideration not only in the installation but also in the changes that will occur over time [1].

Banks have started to invest in alternative service channels instead of opening branches with the developments in technology. In particular, transactions made over ATM, telephone, internet, and mobile applications have increased significantly in recent years. In this way, banks also provide services to their customers through the branch through alternative channels. With the active use of these services, especially by the X and Y generations, they concentrate their investments on technology in opening the branch [2]. On the other hand, branches are an important option for the development of the relationship between the bank and the customer. Branches are the first access point for most customers who prefer face-to-face communication. Although technological developments affect the balances in banking, reasons such as people's habits, lack of trust in digital channels, considering that transactions in branches are wider, big money transfers, opening new bank accounts, and big problems in bank accounts still lead people to use branches.

The number of banks operating in the banking sector as of the end of 2018; 32 of them are deposits, 13 of them are development and investment, and 5 of them are participation banks. The volume of participation banking in Turkey is quite low compared to conventional banking. Contributions received from the banking assets of the entire banking market share to 5.3% as of year-end 2018 operating and, Turkey, five contributions of domestic and foreign banks in the total number of branches is 1,122. The number of branches of participation banks accounts for approximately 9.7% of the banking sector branch network [3].

According to Birkin and his friends [4], strategic planning and decision-making in the financial sector are similar to those in other sectors. The location of the branches is an important part of the bank strategy that requires taking into account all the driving forces of branch growth, such as the spatial distribution of customers, distribution of its own and competitor branches, distribution of inputs, labor, and technology development of the target region. When opening new branches in the banking sector, local values such as the economic development of that region and human capital are important. The existence of human capital directly affects the revenues obtained through banking. While banks are branching, they tend to cluster to benefit from qualified workforce, potential customers, and accumulation shared within a certain region. Considering these factors, placing the branch to be opened in a central location increases the number of households and businesses, which is an important parameter of the branch use function.

Clustering of bank branches provides some advantages such as unsaturated markets, potential customers, and reduction of bank customer acquisition costs. Access to banking services for customers is convenient in time and way. There are also advantages such as the low cost of maximizing the strategic market share due to the development of the market by competitors [5]. For these reasons, it is important to consider the locations of the rival branches when determining the location to open the new branch.

Competition in the banking sector is an important and critical issue. People who have the power to decide must take action quickly. In this respect, it is important to predict the locations of new branches that the bank can open [6].

In this study, a method has been developed that will be used to determine the most appropriate locations of the newly opened bank branches. Branches that can be opened by looking at the locations of other participating banks were determined through OPTICS [7] (Ordering Points To Identify Clustering Structure), a density-based clustering algorithm. After the possible locations were calculated, an analytical study was carried out by considering the development levels of the cities and a priority order was prepared between the locations.

2 Related Works

Various studies have been conducted on the selection of suitable locations for planning the strategic locations of the bank branches. Although various studies have been made on the problem of location selection, studies in the field of banking are somewhat more limited. Çınar [8] has determined 5 main items (demographic, socio-economic, sectoral employment, banking, trade potential) and 21 sub-items for the new branch that a selected bank will open in the Southeastern Anatolia Region. Fuzzy AHS (Analytical Hierarchy Process) was used to determine the best place among the alternatives, to determine the main and sub-item weights, and the TOPSIS method to rank between cities considering the weights of these criteria. Similarly, Rahgan and Mirzazadeh [9] identified 5 main items (cost, demography, banking, geographical conditions, and accessibility) and 14 sub-criteria in his study. Fuzzy AHS was used for the weighting of the criteria and the evidential conclusion method was used to rank between cities. The AHS and Electre method for strategic bank location selection has also been examined in the study by Soba [10].

Başar [11] has developed a tabu search approach for determining the appropriate location selection for bank branches by determining different sub-criteria. In the study conducted by Fırat [6], 36 sub-criteria were determined and suitable branch locations were tried to be selected by using multivariate statistical analysis techniques such as reliability analysis of data, data reduction, and discriminant analysis.

These studies conducted in the literature are studies in which more than one sub-criterion is taken into consideration and these criteria vary. It has been tried to obtain suitable positions for banks with various statistical methods.

The OPTICS algorithm used to find existing branches of other participation banks was developed by Hans-Peter Kriegel and his team from the University of Munich [7]. OPTICS algorithm is an improved version of the DBSCAN algorithm. In researches on OPTICS, generally, synthetic data were compared with other density-based clustering methods. Bilgin and Çamurcu [12] evaluated DBSCAN (Density-Based Spatial Clustering), OPTICS, and K-Means algorithms for cluster finding performance. It has been concluded that DBSCAN and OPTICS algorithms are much more successful than K-means algorithm in finding non-spherical clusters.

Similarly, the difference between DBSCAN and OPTICS has been covered in the study by Dubey and Rajavat [13]. It was emphasized that the OPTICS algorithm is more successful in obtaining clusters of different density. In the study conducted by Chauhan and her friends [14], DBSCAN, OPTICS, and DENCLUE (Density-Based Clustering) algorithms were compared and their advantages and disadvantages were mentioned. In Berkhin's [15] comprehensive study on clustering techniques, the basis of DENCLUE has been explained and DBSCAN and OPTICS algorithms have been compared. It is stated that the OPTICS algorithm is better at finding different densities.

3 DBSCAN and OPTICS Algorithms

In DENCLUE algorithms, the natural distribution of objects is determined by a density function and regions that exceed a threshold density are defined as clusters. In particular, it is used in clusters that have irregular or undetermined shapes. DENCLUE algorithms are successful in finding clusters that are not properly shaped and have the advantages of being unaffected by noise and achieving a single scan result. DBSCAN, OPTICS, and DENCLUE are the most used density-based algorithms [16].

The DBSCAN algorithm was introduced in 1996 by Ester and his friends [17]. This algorithm performs clustering by calculating the distance of objects with their neighbors by grouping areas with more objects than a predetermined threshold value in a given region [12]. OPTICS was first presented by Ankerst, Breunig, Kriegel, and Sander at the SIGMOD'99 conference [7]. DBSCAN is an algorithm that is sensitive to input parameters. The OPTICS algorithm has been developed to reduce the dependence on the parameters of the DBSCAN algorithm. While other clustering algorithms divide existing objects into clusters, OPTICS aims to sort them in a particular clustering order. The OPTICS algorithm offers the most flexibility in fine-tuning of detected sets [18].

The DBSCAN algorithm uses a specified search distance to separate clusters from noise data. It is suitable if it has a clear search distance and works well for all potential clusters. This requires that all significant clusters have similar densities for the DBSCAN algorithm [18].

The OPTICS algorithm creates a cluster sorting that reveals the DENCLUE structure of objects. This order is a sequence with core distance and reachability distance values [19]. It is not possible to separate A, B, C1, C2, and C3 clusters with a single general density parameter. A general parameter can contain clusters A, B, and C. Or there are clusters C1, C2, and C3, and A and B clusters as noise. In order to overcome this problem, an algorithm should be run that will find clusters of all density levels by sorting objects in the database in a special order as in the example shown in Fig. 1 [7]. If the data has variable density, it can give better results than other DENCLUE algorithms.

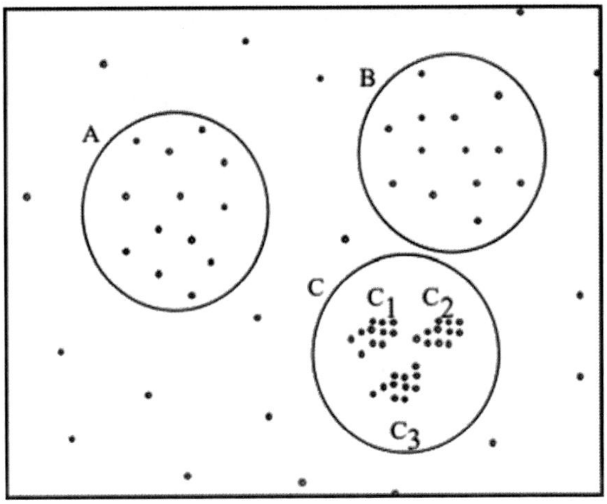

Fig. 1 Clusters with different densities

4 Implementation

Branch location data (latitude–longitude information of branches) and Eps value of three different participation banks were converted into radians when given as input to DBSCAN and OPTICS algorithms.

Eps value is selected as 1 km, MinPts value is determined as 3. Depending on the globalness of the world, the Haversine formula was used as the distance metric to calculate the distance between locations correctly. Thus, clusters containing at least three branches have been formed, at most 1 km between the two branches. Location data not included in a cluster has been ignored. Experiments have been made with OPTICS and DBSCAN algorithms with similar parameters. Since OPTICS algorithm produces more qualified outputs, after obtaining 173 different sets, a polygon was created with the location data in each set, and the center of gravity points of these polygons was found. The center of gravity is obtained by lining the points in the cluster clockwise and creating a polygon. These center of gravity points represent possible locations where branches can be opened.

The location data in the center of gravity of the obtained polygons were queried through the API, and the approximate location information of the new branch that could be opened was obtained. The location of branches detected, opening the State Planning Organization to determine the priority (SPO), Turkey, Participation Banks (TKBB), the Human Development Index (IG-I), the Banking Regulation and Supervision Agency (BRSA) belonging to cities and towns of development using data, an analytical ranking was made according to their levels. Branch positions that can be opened are prioritized by assigning a score in the range of 0–100. Thus, location data and analytical data are related to each other.

4.1 Data Gathering

Three different participating bank branch location located in Turkey, in terms of latitude and longitude are obtained through the API. Obtained branch data was verified through TKBB. Similarly, the location data of the points in the center of gravity of the clusters determined by the OPTICS algorithm were also queried through the API and the city, district, neighborhood, and street information of that point was obtained.

Analytical study phase, the State Planning Organization of the website in social-economic development of the research of the county [20], branch location information is on the website of the Turkish Participation Banks Association [21] Human Development Foundation published the report of the development index [22] and the FINTURK Geographical Distribution [23] on the website of the Banking Regulation and Supervision Agency were used. In the research of SPO, the data of "Socio-Economic Development Index" and "Development Group" of each district were obtained. Using the location information of TPBA, the number of participation branches in each district was reached. More accurate results were obtained by comparing the development indices of the districts included in the report of the Human Development Foundation, the "Economic Development Index" and "Development Group" data included in the DPT's research. Through the FINTURK Geographical Distribution report of the BRSA, "Number of Domestic Branches," "Per Capita Cash Loan," and "Total Deposit per Capita" for each province were included in the calculation.

4.2 Modeling

Branch data of three participation banks were clustered through DBSCAN and OPTICS. Especially in big cities where branch density varies, such as Istanbul and Ankara, more meaningful results were obtained with OPTICS. The Haversine formula was used as the distance metric.

In the study, cluster elements obtained with DBSCAN and OPTICS are represented by different colors and elements not included in any cluster are colored in light gray.

In the results obtained with DBSCAN, Eps value is 1 km, MinPts value is determined as 3 and sample cluster outputs are obtained as in Fig. 2. When the outputs obtained with DBSCAN are evaluated, each cluster is formed to cover a wide area. 146 different clusters were obtained.

Sample cluster outputs obtained in the study performed by reducing the Eps value to 0.5 with DBSCAN are shown in Fig. 3. Similarly, 147 different clusters were obtained, but the number of branches covered by the clusters decreased.

When we decreased the Eps value in DBSCAN outputs, the number of branches covered by the clusters decreased; however, no detailed results were obtained in regions such as Fatih district of Istanbul, where the branches were dense.

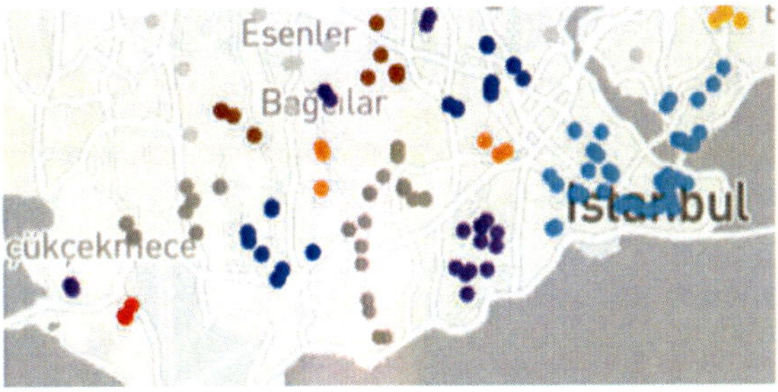

Fig. 2 DBSCAN—Eps: 1, MinPts:3

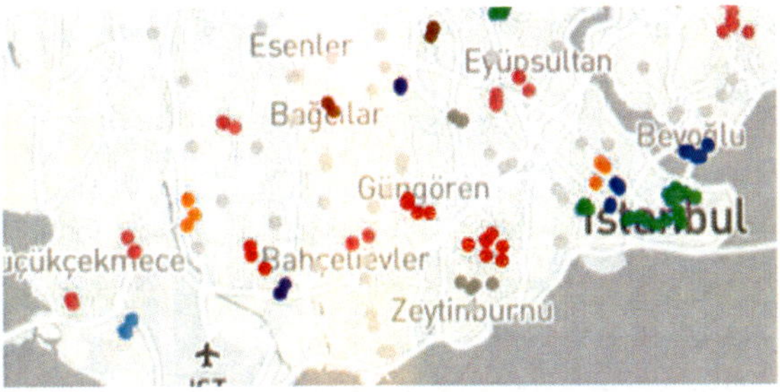

Fig. 3 DBSCAN—Eps: 0.5, MinPts:3

In our studies with OPTICS, MinPts value was selected as 3, with Eps value as 1 km. OPTICS has carried out different meaning clustering and density for branch offices located in every region of Turkey. Sample clusters for the Istanbul obtained are shown in Fig. 4.

When Fig. 4 is examined, it is seen that the clusters formed with OPTICS are more meaningful both in terms of covering the branches and in terms of density. OPTICS algorithm with Turkey achieved across the 173 different clusters. A center of gravity was determined for each cluster obtained. Detailed location information of possible branch locations that can be opened by querying the coordinates of the center of gravity points has been provided. This information was used to evaluate the development levels of provinces and districts. Figure 5 shows the center of gravity of the sample clusters.

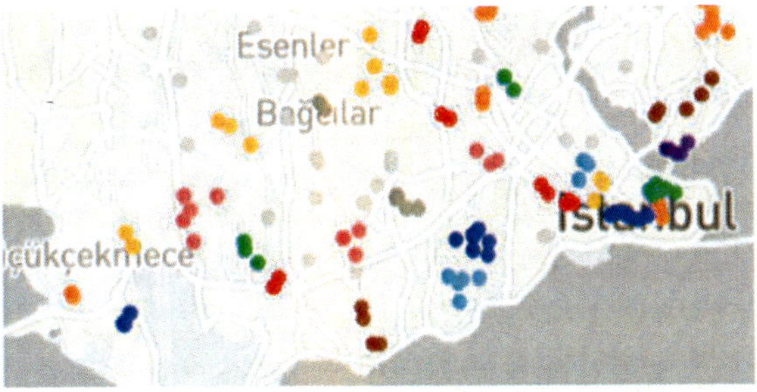

Fig. 4 OPTICS—Eps: 1, MinPts:3

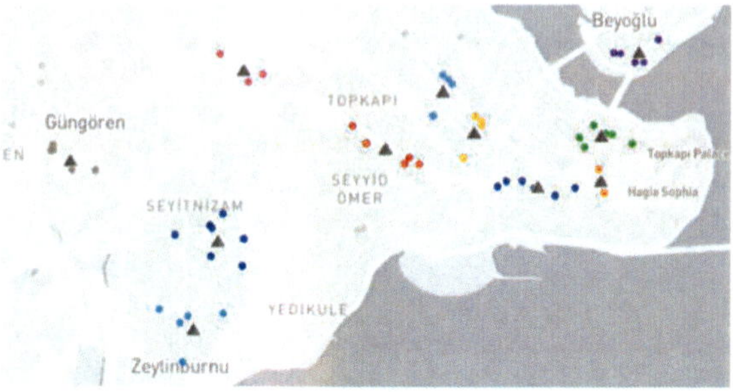

Fig. 5 Weight centers of sample sets

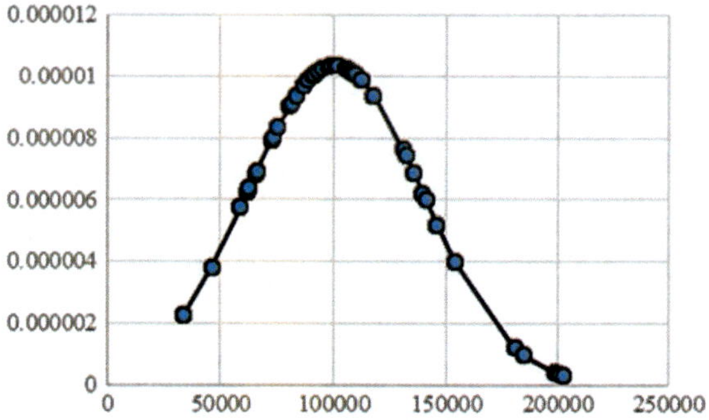

Fig. 6 Population per branch normal distribution

In analytical modeling, the priority of 173 branches that emerged after the OPTICS algorithm was determined. The information needed for this is as follows;

In this way, analytical work was carried out over a total of five categories. Each category can get a maximum of 20 points and the maximum score of a total branch can be 100. The score calculation of each category was calculated with normal distribution as shown in Eq. (1). For example, the average was first calculated to find the number of population per branch. Then standard deviation value was found. With these two values, the normal distribution value of each district was found. Sample dataset of 20 districts with normal distribution value was shared in Table 1. According to the normal distribution, the score that a county can get was shown between 0 and 20. The same was done in the category that required four other analytical calculations. Analytical calculation of the location where each branch will be opened was made according to the total score of the five categories. A potentially openable branch location can get a maximum of 100 points. Table 2 shows the points received by the sample branches. The highest rated branches are the branches that should be opened with priority (Fig. 6 and Table 3).

$$y = \frac{1}{\sigma\sqrt{2\pi}} e^{-\frac{(x-u)^2}{2\sigma^2}} \tag{1}$$

$u = Mean$

Table 1 Sample dataset for the population per branch

City	County	Population per branch	Normal distribution
Kastamonu	Abana	93,093	0.0000101573027
Nevşehir	Acıgöl	73,091	0.0000080132501
Denizli	Acıpayam	101,873	0.0000103513756
İstanbul	Adalar	33,773	0.0000022913916
Kahramanmaraş	Afşin	140,952	0.0000059964889
Aksaray	Ağaçören	80,480	0.0000090257857
Elazığ	Ağın	83,381	0.0000093627931
Kastamonu	Ağlı	93,093	0.0000101573027
Manisa	Ahmetli	117,753	0.0000093877187
Trabzon	Akçaabat	65,527	0.0000068290051
Malatya	Akçadağ	112,382	0.0000098894761
Şanlıurfa	Akçakale	180,523	0.0000012088745
Düzce	Akçakoca	75,522	0.0000083670832
Yozgat	Akdağmadeni	139,550	0.0000062249105
Manisa	Akhisar	117,753	0.0000093877187
Sivas	Akıncılar	88,757	0.0000098718137
Kayseri	Akkışla	72,459	0.0000079185722

Table 2 Mean and standard deviation values of sample data

Mean	100,681.7799
Standard deviation	38,521.60

Table 3 Score of sample branches

City	County	District	Possible branch location	Score
Sakarya	Adapazarı	Semerciler Mah	373 Nolu Sok	65
Mersin	Akdeniz	Cami Şerif Mah	Mücahitler Cad	75
Ankara	Altındağ	Ulubey Mah	Kopça Cad	100
Ankara	Altındağ	Hacı Bayram Mah	Anafartalar Cad	100
Hatay	Antakya	Haraparası Mah	Yavuz Sultan Selim Cad	55
İstanbul	Arnavutköy	Merkez Mah	Ayşıl Sok	85
Mardin	Artuklu	Yenişehir Mah	Vali Ozan Blv	35
İstanbul	Ataşehir	İçerenköy Mah	Of Sok	100
İstanbul	Ataşehir	Örnek Mah	Şehit Cahar Dudayev Cad	100

$\sigma = Standart Deviation$

$\pi \approx 3.14159$

$e \approx 2.71828$

$Mean = \frac{Sum of All Data Points}{Number of Data Points}$

$$SD = \sqrt{\frac{\sum |x - \overline{x}|^2}{n}}$$

5 Result and Discussion

With this study, new branch locations that a participation bank can open in the banking sector, where fast decision-making is important, were found by looking at the branch densities of competing banks and the development levels of the provinces and districts where those densities are located.

The desired results were obtained with OPTICS algorithm both in big cities such as Istanbul, Ankara, Izmir where the branches are concentrated, and in small cities. The obtained location data was supported by analytical work and points were assigned. This study shows that bank branches tend to cluster. Possible locations of a newly opened bank branch were obtained with machine learning and analytical work supporting each other.

The DBSCAN algorithm was not sufficient for this study to distinguish various densities due to its background. OPTICS better identify branches of different density. In light of the data obtained with OPTICS, it is possible to find the approximate location data of new branches that can be opened. When opening a branch, the evaluations made by the authorized people by looking at these locations are also important. In the analytical study conducted in the light of location data, major cities are at the top of the list in the ranking.

In the analytical phase of the study, the districts where the branches will be opened using a normal distribution were given between 0 and 100 points. While making this calculation, firstly, the average value and standard deviation values of the data were found. Using these values, normal distribution values were calculated for each district. Regarding the normal distribution values, the development indices of the districts, the development groups of the districts, the number of participation banks in each district, the per capita deposit amount in each district and the risk amount per person in each district were calculated for each potential branch. The branch with the highest score is expressed as the branch that should be opened with priority.

This study is open to implementation in different sectors such as hotel and restaurant where competition and clustering are observed.

References

1. İ. Barutçugil, Üretim Sistemi ve Yönetim Teknikleri, Uludağ Üniv. Yayınları 73 (1988)
2. TUIK, Turkey in Statistics. Ankara (2015)
3. TKBB, Katılım Bankaları 2018 Verileri, (2019). Available at https://www.tkbb.org.tr/Documents/Yonetmelikler/Katilim-Bankalari-2018-.pdf. Accessed 24 July 2019.
4. M. Birkin, G. Clarke, Using spatial models to solve difficult retail location problems. in *Retail Decision Support* (Nottingham, 2002)
5. K. Polina, *Measuring the Influence of Agglomeration Effects on the U.S. Banking Industry Efficiency* (MA in Economic Analysis, Kyiv School of Economics, 2015)
6. S.Ü. Fırat, U. Kapucu, Bankacılıkta Yeni Şube Lokasyonlarının Çok Değişkenli İstatistik Teknikleri İle Belirlenmesi. in *XI. Üretim Araştırmaları Sempozyumu: "Toplumsal Kalkınmada Üretimin Artan Rolü* (İstanbul, 2011)
7. M. Ankerst, M.M. Breunig, H.P. Kriegel, J. Sander, Optics: Ordering Points to Identify the Clustering Structure. in *Proceedings ACM SIGMOD'99 International Conference on Management of Data* (Philadelphia, PA, 1999)
8. N. Çınar, A decision support model for bank branch location selection. World Acad. Sci. Eng. Technol. Int. J. Math. Comput. Sci. **3** (2009)
9. S. Rahgan, A. Mirzazadeh, A new method in the location problem using fuzzy evidential reasoning. Eng. Technol. **4**, 4636–4645 (2012)
10. M. Soba, Banka Yeri Seçiminin Analitik Hiyerarşi Süreci Ve Electre Metodu İle Belirlenmesi: Uşak İlçeleri Örneği. Mustafa Kemal Üniversitesi Sosyal Bilimler Enstitüsü Dergisi **11**(25), 459–473 (2014)
11. A. Başar, Ö. Kabak, İ. Topçu, Banka Şubeleri İçin Uygun Yer Seçiminin Belirlenmesine Yönelik Tabu Arama Yaklaşımı: Bir Türk Bankası Uygulaması. Endüstri Mühendisliği Dergisi 2–22 (2015)
12. T.T. Bilgin, Y. Çamurcu, DBSCAN, Optics ve K-Means Kümeleme Algoritmalarının Uygulamalı Karşılaştırılması. Politeknik Dergisi **2**, 139–145 (2005)
13. P. Dubey, A. Rajavat, Comparative study between density based clustering—DBSCAN and OPTICS. Int. J. Adv. Comput. Eng. Netw. 4(12) (2016)
14. R. Chauhan, P. Batra, S. Chaudhary, A survey of density based clustering algorithms. IJCST **5**(2) (2014)
15. P. Berkhin, Survey of Clustering Data Mining Techniques (2002)
16. C. Gülce, *Veri Madenciliğinde Apriori Algoritması Ve Apriori Algoritmasının Farklı Veri Kümelerinde Uygulanması, Trakya Üniversitesi, Fen Bilimleri Enstitüsü, Bilgisayar Mühendisliği Anabilim Dalı* (Yüksek Lisans Tezi, Edirne, 2010)
17. M. Ester, H.P. Kriegel, J. Sander, X. Xu, A Density-Based Algorithm for Discovering Clusters in Large Spatial Databases with Noise. in *2nd International Conference on Knowledge Discovery and Data Mining* (1996)
18. How Density-based Clustering works, ArcGISPro. Available at https://pro.arcgis.com/en/pro-app/tool-reference/spatial-statistics/how-density-based-clustering-works.htm. Accessed 27 June 2020
19. S. Brecheisen, H.P. Kriegel, P. Kröger, M. Pfeifle, Visually Mining Through Cluster Hierarchies. *in Proceedings 4th Sıam International Conference On Data Mining* (Lake Buena Vista, Florida, 2004)
20. Sosyo-Ekonomik Gelişmişlik Araştırmaları, D.P. Teşkilatı (2004). Available at https://www3.kalkinma.gov.tr/PortalDesign/PortalControls/WebContentGosterim.aspx?Enc=51C9D1B02086EAFB13C239EE2D6723BE. Accessed 19 Aug 2019
21. Sektör Bilgileri, TKKB. Available at https://www.tkbb.org.tr/banka-genel-bilgileri#1703. Accessed 19 August 2019
22. M. Şeker, Ç Bakış, B. Dizeci, *İnsani Gelişme Endeksi—İlçeler* (Çizge Tanıtım & Matbaacılık Ltd., Şti, İstanbul, 2018)
23. FinTürk-Coğrafi Dağılımı, BDDK. Available at https://www.bddk.org.tr/bultenfinturk. Accessed 27 June 2019

Prediction of Soil Plasticity Index with the Use of Regression Analysis and Artificial Neural Networks: A Specific Case for Bakırköy District

Zülal Akbay Arama, Melda Yücel, Muhammed Selahaddin Akın, Said Enes Nuray, and Oğuzhan Alten

Abstract Consistency limits are simple and precious soil parameters that are directly representing the mineralogical and physical properties and indirectly used to estimate the strength and rigidity characteristics of fine-grained soils. The consistency tests are assumed as the easiest and the basic tests of geotechnical engineering and the application details of the tests are depending on the experience of the operator and the environmental conditions. Therefore, the reliability and correctness of the tests are still in the concern of geotechnical studies. Besides, the development of information and computer technologies also present alternative solution perspectives for the attainment of geotechnical properties. In this connection, the regression analysis and artificial neural network applications become the most preferred techniques to estimate the consistency limits with the use of only a variable. In this study, it is aimed to obtain the consistency properties of high plastic clays that are procured from the site investigation reports conducted for Bakırköy district (Istanbul, Turkey) depending on only the evaluation of liquid limit test results. A database is arranged with the use of approximately 450 consistency limit tests. Linear and multi-linear regression analyses and artificial neural network applications were conducted with the use of different variants such as depth and fine content to determine the plastic limit and the plasticity index value. As a result, comparisons of the applied techniques are done to query the appropriateness of the methods and to find the approximation rate to the actual value. Consequently, the success of the applied techniques was tried to be shown with the developed equation set.

Keywords Plasticity index · Regression analysis · Artificial neural network · Consistency limits

Z. Akbay Arama (✉) · M. Yücel · S. E. Nuray · O. Alten
Istanbul University-Cerrahpaşa, Avcilar, 34320 Istanbul, Turkey
e-mail: zakbay@istanbul.edu.tr

M. S. Akın
Kartal Municipality, Kartal, 34860 Istanbul, Turkey

S. M. Nigdeli et al. (eds.), *Proceedings of 6th International Conference on Harmony Search, Soft Computing and Applications*, Advances in Intelligent Systems and Computing 1275, https://doi.org/10.1007/978-981-15-8603-3_25

1 Introduction

An integral section of site investigation is to procure knowledge as to the stratification, thickness, content, and characterization, as well as the engineering properties of the sub-soil strata [1]. During the sub-surface exploration, field and laboratory tests are tried to be conducted if the conditions of the site are proper for investigation and the undisturbed soil samples are available to identify the geotechnical properties of dominant soil formation of the project site. But sometimes, depending on the lack of project income or the complication of the working conditions of the soil profile or the absence of the undisturbed soil specimens, these envisaged exploration works cannot be conducted. In such a case, due to the indefinite nature of the soil formations, regression analysis and artificial intelligence techniques can also be used as an alternative solution technique to identify the characterization. These techniques associate the empirical information and/or pattern matching techniques are ideally suited to help both the researchers and the engineers to remean knowledge from the previously conducted simple site and laboratory investigations. In this regard, the most frequently used and the most common characterization experiment is the determination of water content to identify the existing physical condition of soil site. But only the natural water content of the soil medium is not adequate to interpret the holistic strength and rigidity properties of the site. Therefore, the logic of consistency limits which are representing the boundary water content values, especially to define the consistency of fine-grained soils, had been developed. The most known method that is used to characterize the consistency in relation to the water content percentage is developed by Atterberg [2, 3]. Atterberg limits include the definition of liquid limit and plastic limit that are representing the boundary water content values between the liquid—plastic and the plastic—semi-plastic states, respectively. The comparison of the natural water content of the soil specimen with the boundary consistencies ensures to obtain specific behavior characteristics and the determination of plasticity index which can be calculated by the difference of liquid limit and plastic limit value can give the chance to classify the soil formation that is existed in the project site. Therefore, it is important to acquire these limit values correctly as possible. The advantage of the attainment process of liquid limit and plastic limit values by the tests concern about the capability of the usage of the disturbed soil samples for the tests. But in other respects, the disadvantage of the tests is the dependency of the tests to the operator's experience and attention. This situation is especially significant for the plastic limit test that the application method only involves hand working to remold and roll. Therefore, several researches have been conducted to make the calculation of plasticity index independent of the test challenges of the plastic limit [4–10]. In a similar vein, this paper explores the prediction capabilities of regression analysis and artificial neural network applications with the use of liquid limit test results to determine the plasticity index value directly. For this purpose, a special database is prepared including a huge number of consistency test results procured from the site investigation reports performed for Bakırköy (Istanbul, Turkey). Linear regression analysis (*LR*), multi-linear regression analyses (*MLR*) and artificial neural network

(*ANN*) applications were conducted with the use of different arrangements of the parameters such as liquid limit (*LL*)—depth (*H*), liquid limit (*LL*)—fine content (*FC*), liquid limit (*LL*)—depth (*H*)—fine content (*FC*) to determine the plastic limit and the plasticity index value. Consequently, comparisons of the *LR*, *MLR*, and *ANN* techniques are made to query the appropriateness of the methods and to find the approximation rate to the actual value.

2 Material and Method

Within the context of this study, linear (*LR*), multi-linear regression (*MLR*) analysis and artificial neural network (*ANN*) applications are conducted to obtain the plasticity index value of clayey soils. An extended database has been prepared with the use of 250 site investigation reports that are supplied with special permission from the Departments of Investment Monitoring and Coordination (DIMCs), Turkey. Totally 423 data's including the values of liquid limit and plasticity index are used to perform the numerical analysis. The coordinates of the data zones are given with the use of a regional map of Bakırköy district (Istanbul, Turkey) in Fig. 1 and the locations of the data points according to the obtained liquid limit and plasticity index values are given on the Casagrande chart is shown in Fig. 2. In Fig. 2, it is shown that all the used soil formations present a high plastic consistency that can be specified by having a liquid limit bigger than 50%. Besides, all the data points are located above the A-Line and that condition represents clayey soils. In Fig. 2, high-plastic clays are given with green dots and named as *CH* according to the Unified Soil Classification System.

The liquid limit tests that are given in the soil investigation reports are all conducted based on the Casagrande cup method [11, 12]. Liquid limit test device, named Casagrande cup, includes a brass cup and a hard rubber base that can be dropped

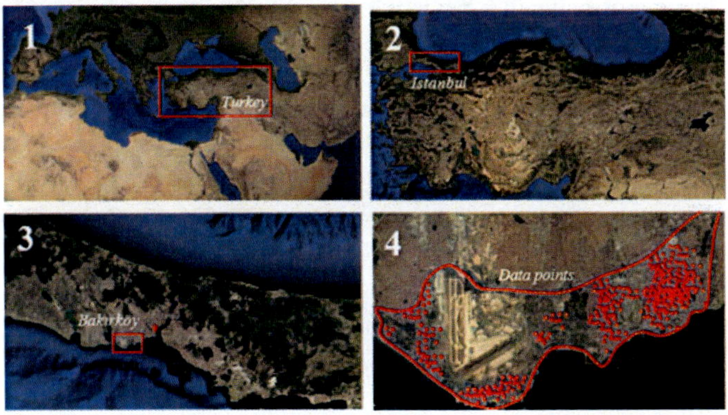

Fig. 1 The coordinates of the data points on the regional map of Bakırköy district

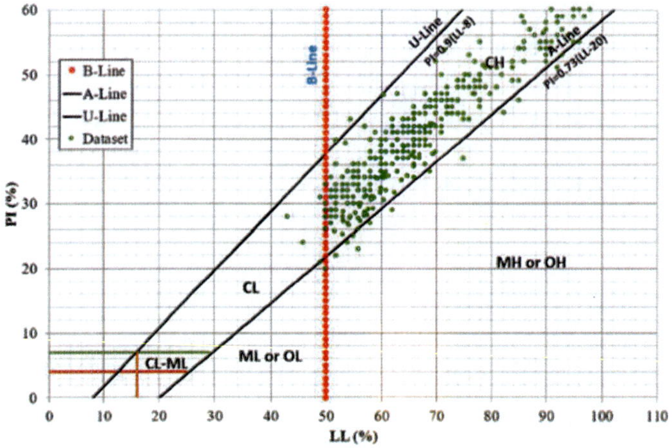

Fig. 2 The location of data points on the Casagrande chart

onto the base by a manually operated by a crank [12]. This situation means that a standardized equipment is used to perform the liquid limit test and thus wise the liquid limit test becomes more independent from the human impact. The procedure of the plastic limit test is defined in ASTM in Test Designation D-4318 [13]. The plastic limit is defined as the boundary water content at which the soil crumbles when rounded into threads of 4.2 mm in diameter. The test is simple but it is conducted by repeated rolling of an ellipsoidal-sized soil mass by manually with the human hand on a glass plate. This process is sensitive and operator depended, therefore it is so hard to obtain accurate results at the first experience. Therefore, there is a need to obtain a direct relationship between the determinations of liquid limit and plasticity index value.

In the context of this study, the analyses to determine the plasticity index value directly from liquid limit tests are performed with the evaluation of two different cases. In the first case, the direct relationship between the liquid limit and the plasticity index is investigated by both regression analysis and artificial neural network applications. In the second case, the attainment process of plasticity index via the liquid limit and the depth is observed and, additionally, the fine content of the soil sample evaluates to the solution sequence as a variable. The second case is conducted with both types of analysis techniques that are foreseen in this study, but the mathematical expression of the solutions is given only in the regression analysis. The outcomes of the analyses are given below respectively in accordance with the cases.

3 Result and Discussion

In this section of this paper, abbreviations are used to define the variables of the analyses. *H* is used to define the depth and *FC* is used to define the fine content percentage.

3.1 Case 1: Prediction of PI Against LL with Both LR and ANN

Prediction of *PI* against *LL* value is applied with both *LR* and *ANN* techniques. The regression analysis is conducted with the use of Python Python programming language with the Spyder 4 application. The Spyder is a powerful scientific idea written in and for Python for the use of engineers, scientists, and also the data analysts. Besides, the artificial neural network applications are performed with Matlab R2018a. The linear regression analyses are performed by the use of two-dimensional graph systems to find the most proper expression of plasticity index. The results of the *LR* analyses are given in Fig. 3. Figure 3a clearly shows that a linear relationship can be available between *PI* and *LL*. Figure 3b represents the obtained linear relationship line. The envisaged mathematical relationship between *PI* and *LL* is defined in Eq. 1. The R-squared (R^2) is the coefficient of regression value that can be defined as the statistical measure of how close the data are in the fitted regression line is determined 0.91 for this linear regression model.

$$PI = 0.696 \times LL - 6.077 \quad (1)$$

The squared sum of the calculated difference between the predicted values of *PI* and the actual data's are calculated 4692.56. Accordingly, the average error per data is determined $\pm$ 3.33.

The results of the conducted *ANN* applications are given in Fig. 4. The input value is the liquid limit and the output value is arranged as the plasticity index value. According to the acquired results, the correlation values of training, validation, and test stages for Case 1 is remarkable regard to can say that *ANN* models are successful in a way to be able to predict the output parameter (Fig. 4a). Besides, the best result occurred in this application via one input parameter by providing 94.808% correlation value. In addition to all these, Microsoft Excel applications have used to obtain the mathematical expression of test results (Fig. 4b). The linear, polynomial, and exponential curve fitting options have been applied to the obtained values and Eqs. 2, 3, and 4 are obtained with 0.9939, 0.9944, and 0.9694, R^2 values, respectively. As a result, the polynomial curve fitting option has shown the most appropriate approximation for predicting the *PI* value. But related to the easiness of the applicability of the obtained relationship, it will be also a proper selection to use a linear relationship that is defining *PI* with a simple expression.

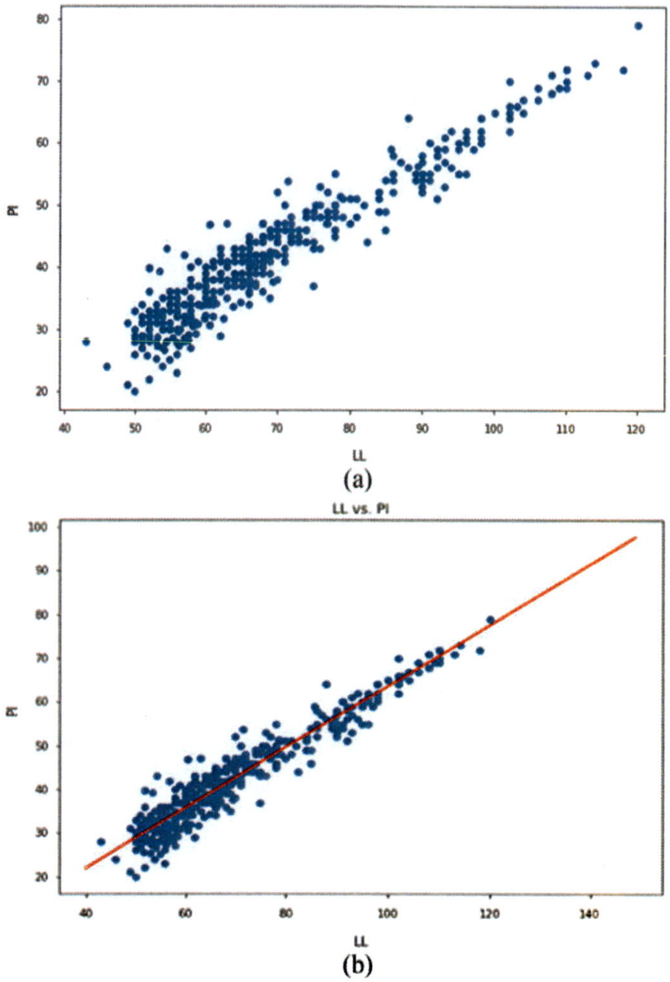

Fig. 3 **a** The distribution of data's *PI* against *LL*, **b** The results of the linear regression analysis between the *PI* and *LL* (input parameter: *LL*)

$$PI = 0.689 \times LL - 5.5168 \tag{2}$$

$$PI = -0.0008 \times LL^2 + 0.8168 \times LL - 10.187 \tag{3}$$

$$PI = 14.25 \times e^{0.0152LL} \tag{4}$$

It can be seen from the comparison of Figs. 3 and 4 that *ANN* applications give the most accurate solution to determine the plasticity index directly from the liquid limit test.

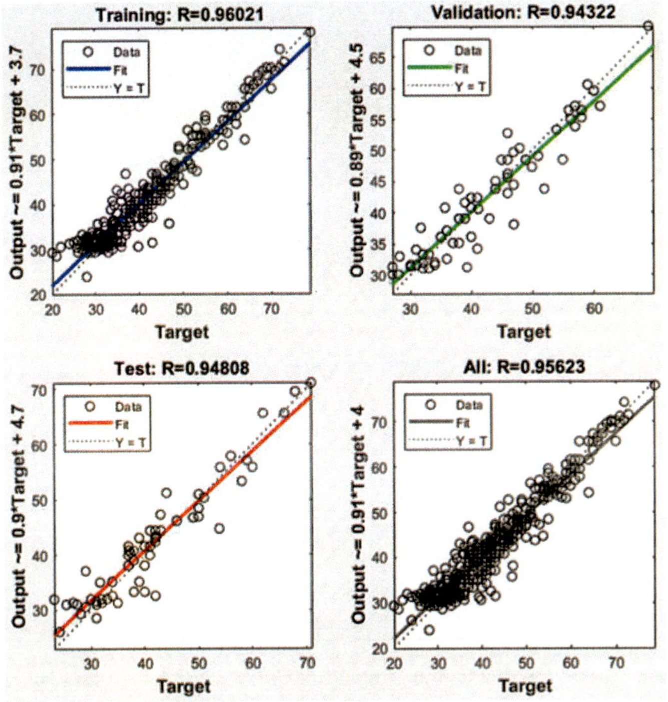

(a)

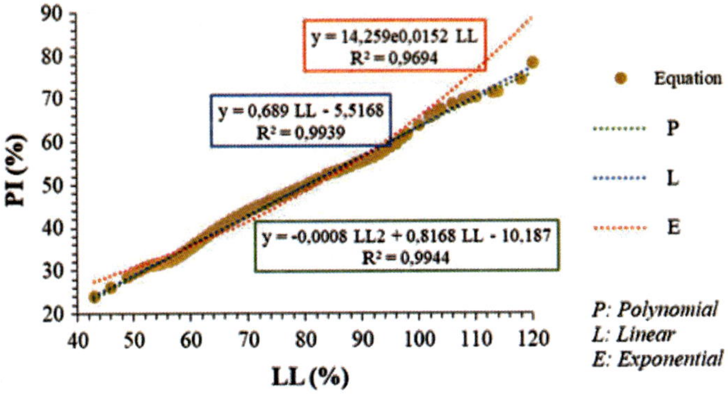

(b)

Fig. 4 The results of the *ANN* analysis for Case 1 (input parameter: *LL*)

3.2 *Case 2: Prediction of PI Against LL, H, and FC with MLR*

In the second case, the attainment process of plasticity index is tried to be conducted by considering the depth and the fine content of the inspected soil, respectively.

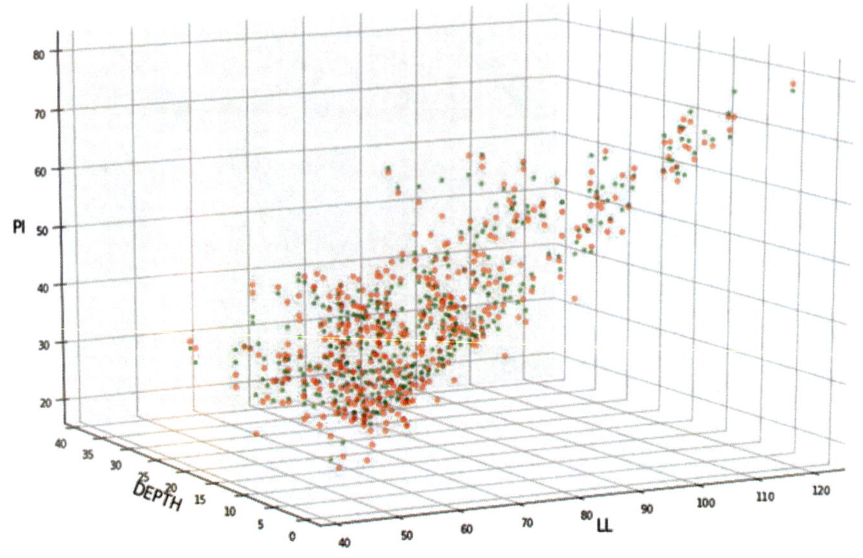

Fig. 5 The results of the *MLR* analysis for Case 2 (input parameters: *LL, H*)

Besides, the interacted effects of all the influencer variables are taken into consideration. These ordered analyses are arranged as the initial class (only regarding the effect of depth and liquid limit—*IC*), second class (only regarding the effect of fine content and liquid limit—*SC*) and third class (regarding the effect of depth, fine content, and liquid limit—*TC*) analysis and the details of the analyses are given below.

The *IC* analyses are performed with both the evaluation of *H* and *LL*. Multi-linear regression analyses are done to determine this envisaged relationship between three variables. Figure 5 presents the real data's with red dots and the results of the regression analyses are given with green dots. Besides, Fig. 6a shows the distribution of the resulting plasticity index values and Fig. 6b gives the resultant plane of *PI*. The acquired relationship is given by Eq. 5.

$$\text{PI} = [0.6965 \times LL] + [-0.0413 \times H] - 5.60 \tag{5}$$

The R^2 is determined approximately 0.91 for this multi-linear regression model. The squared sum of the calculated difference between the predicted values of *PI* and the actual data's are calculated 4650. Accordingly, the average error per data is determined $\pm$ 3.31. The comparison of the *LR* analyses and the *MLR* analyses are shown that the depth variant is not as effective as the value of liquid limit value. The error rate has been reduced to only 0.01 for the mentioned change of the variants.

The *SC* analyses are performed with both the evaluation of *LL* and *FC* of the inspected soil. The distribution of the actual *PI* values and obtained results are given in Fig. 7a. Besides, Fig. 7b. shows the relationship between actual and resultant data's which tends to be linear. The R^2 value is determined approximately 0.927 for

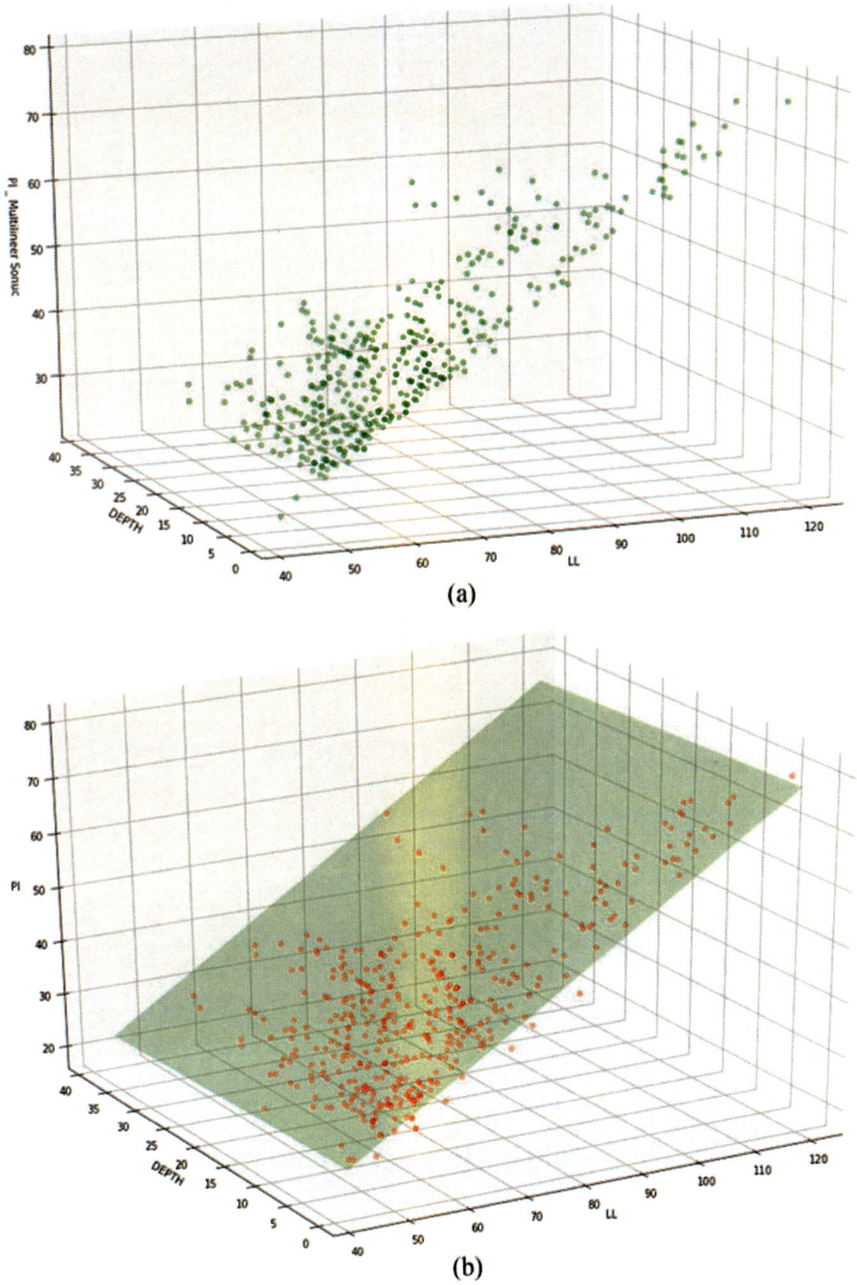

Fig. 6 The results of the *MLR* analysis for Case 2 (input parameters: *LL, H*)

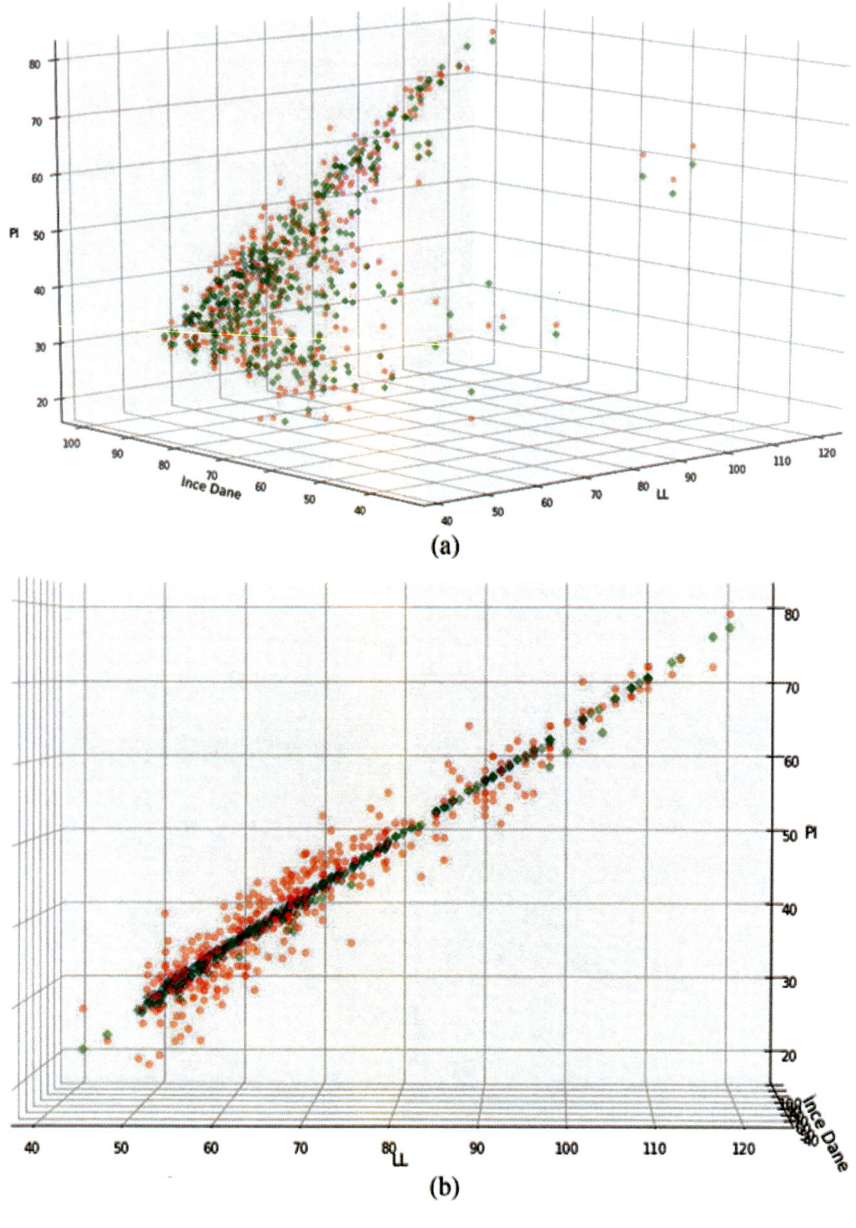

(a)

(b)

Fig. 7 The results of the *MLR* analysis for Case 2 (input parameters: *LL, FC*) this regression model and the obtained mathematical expression is given in Eq. 6. The squared sum of the calculated difference between the predicted values of *PI* and the actual data's are calculated 3669. Accordingly, the average error per data is determined ± 3.063. This regression results show that the addition of *FC* value as a variant increases the prediction accuracy of the analyses.

$$\mathrm{PI} = [0.6881 \times LL] + [0.045 \times FC] - 9.57 \tag{6}$$

The *TC* analyses are performed with both the evaluation of *LL, H,* and *FC* of inspected soil. There cannot be obtained a graphical representation for 4-d *MLR* analyses. The R^2 is determined approximately 0.9274 for this regression model. The squared sum of the calculated difference between the predicted values of *PI* and the actual data's are calculated 3645. Accordingly, the average error per data is determined ± 3.053. The results show that the addition of other variants changes the unit fault per data only ± 0.01.

$$\mathrm{PI} = [0.6871 \times LL] + [0.048 \times FC] + [-0.0322 \times H] - 9.4 \tag{7}$$

In addition to all these *MLR* analyses, *ANN* applications are conducted to search for the relationship of different variants. Figure 8a shows that *TC* analysis that is performed with the usage of the liquid limit, the depth, and the fine content as inputs to the output, namely *PI*. Figure 8b, c shows the *SC* and *IC* analyses, respectively. Here, according to obtained results, correlation values of training, validation, and test stages for all cases are remarkable regards to can say that *ANN* models are successful in a way to be able to predict the output parameter. However, the best result occurred in the *TC* application via three input parameters by providing 96.576% correlation value. In this respect, it can be understood that the usage of these parameters all of one, is the best choice for increasing prediction success. Because, in the *IC* and *SC* applications, correlation showed a little amount decreasement with used the parameters as depth or fine content, separately. Additionally, using *FC* together with *LL* without *H* can be assumed as the second successful prediction case due to that correlation is higher than the other one.

4 Conclusion

In this paper, regression analyses and artificial neural network techniques have been conducted to find a proper relationship to obtain the plasticity index value directly without performing a plasticity limit test for the special clays obtained from Bakırköy district (Istanbul, Turkey). Linear regression analysis and artificial neural network applications have been performed for the bivariate problem of the calculation of plasticity index by only the use of liquid limit. Multi-linear regression analyses and artificial neural networks have been used to analyze the multivariate effects of the influencer parameters on geotechnical behavior. The variants of the analyses have

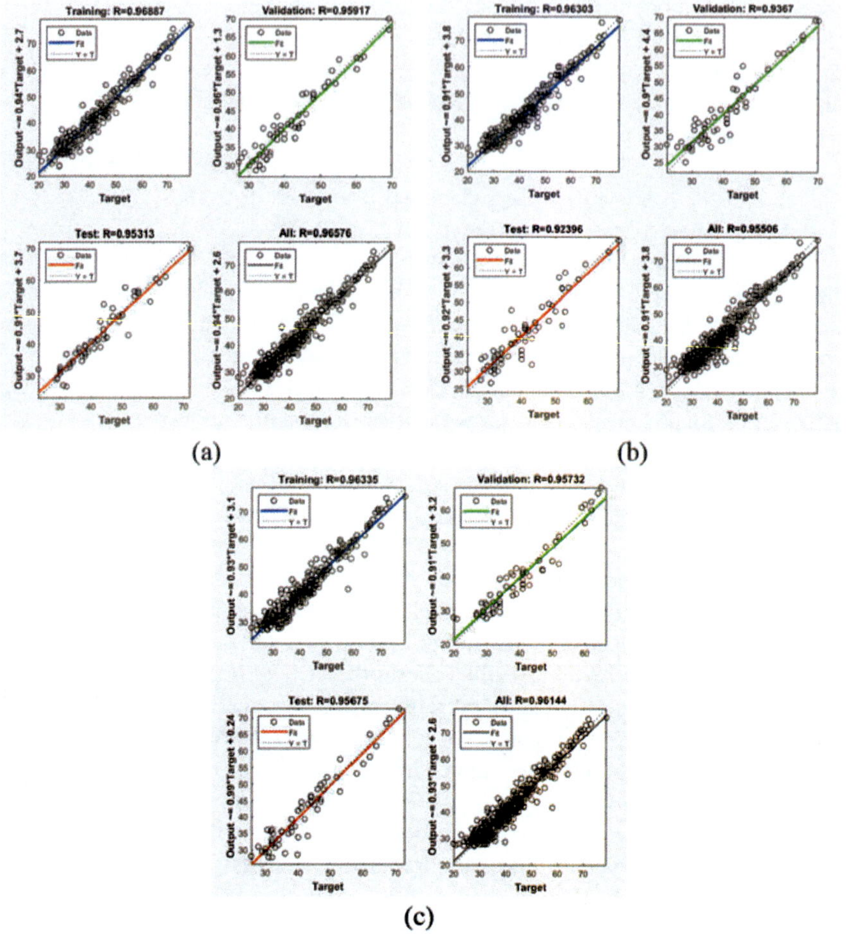

Fig. 8 The results of the *ANN* analysis for Case 2 **a** *TC* analysis, **b** *IC* analysis, **c** *SC* analysis

been selected as the liquid limit, depth, and fine content of the soil specimen. The individual and integrated effects of the variants have also been both investigated. The results of the analyses show that the increase of the variant number also increases the accuracy of the attained relationships. This increase in ratio is more affected by the change of the fine content of the soil specimen for all the analyses. Besides, the application of artificial neural network techniques gives more accurate approximations rather than the regression models, although all the models show an applicable linear relationship. Consequently, the obtained relationships will be beneficial for both researchers and engineers to predict simple geotechnical properties of the investigated sites, especially located within the northwest of Istanbul.

References

1. A.T.C. Goh, Soil laboratory data interpretation using generalized regression neural network. Civil Eng. Env. Sys. **16**, 175–195 (1999)
2. A. Atterberg, Die Plastizität der Tone. Int. Mitt für Bodenk. **1**, 10–43 (1911)
3. A. Atterberg, Die Konsistenz und die Bindigkeit der Böden. Int. Mitt. Für Bodenk. **2**, 149–189 (1912)
4. D. Fall, A numerical model for repaid determination of plasticity of fine-grained soils. Ground Eng. (2000)
5. N. Naveena, S.J. Sanjay, N.S. Chandanshree, Establishing relationship between plasticity index and liquid limit by simple linear regression analysis. Int. J. Res. Appl. Sci. Eng. Technol. **6**(VI), 1975–1978 (2018)
6. C.A. Seybold, M.A. Elrashidi, R.J. Engel, Linear regression models to estimate soil liquid limit and plasticity index from basic soil properties. Soil Sci. **173**(1), 25–34 (2008)
7. B. Sharma, P.K. Bora, A study on correlation between liquid limit, plastic limit and consolidation properties of soils. Ind. Geotechn. J. **45**, 225–230 (2014). https://doi.org/10.1007/s40098-014-0128-0
8. A.T.C. Goh, Modeling soil correlations using neural networks. J. Comput. Civil Eng. **9**(4), 275–278 (1995)
9. C.P. Wroth, D.M. Wood, The correlation of index properties with some basic engineering properties of soils. Can. Geotech. J. **15**, 137–145 (1978)
10. C.O. Modesto, A.M. Bernardini, Determination of clay plasticity: indentation method versus Pfefferkorn method. Appl. Clay Sci. **40**(1–4), 15–19 (2008)
11. A. Casagrande, Research of atterberg limits of soils. Public Roads **13**(8), 121–136 (1932)
12. A. Casagrande, Notes on the design of the liquid limit device. Géotechnique **8**(2), 84–91 (1958)
13. ASTM D4318–17e1, *Standard Test Methods for Liquid Limit, Plastic Limit, and Plasticity Index of Soils* (ASTM International, West Conshohocken, PA, 2017)

The Applicability of Regression Analysis and Artificial Neural Networks to the Prediction Process of Consistency and Compaction Properties of High Plastic Clays

Zülal Akbay Arama, Hazal Berrak Gençdal, Said Enes Nuray, and Melda Yücel

Abstract In all kinds of site investigation reports prepared to acquire the current situation of the project site, it is a common fact to perform the consistency tests which are specialized as Atterberg limit tests. Consistency can be defined as an important term, especially for fine-grained soils, to appoint the current state of the water content of soil formation in the field. Based on the ease and cost-effectiveness of the Atterberg tests, it has become a traditional solution to determine the fundamental design properties such as the rigidity and strength of the soil formation with the use of empirical approaches that are developed according to them. In this context, "compaction" can be an interesting term to investigate the appropriateness of determination of special characteristics of the phenomenon such as the optimum water content and the maximum dry unit weight with the development of a new perspective based on a simplest experimental process formed with only the evaluation of water content. Because it is a complicated and time-consuming process to apply the compaction test beginning of the sample preparation step to the ultimate evaluation step. Hence, in this paper, an integrated study is performed for highly plastic clays to acquire the consistency and the compaction properties together with a direct relationship. A huge database was prepared according to the data's given in the well-accepted literature sources by the transmission of liquid limit and plastic limit test results conducted for only the high plastic clays. Besides, simple equations are tried to be obtained to calculate the plasticity index and approximations are proposed to find the maximum dry unit weight and the optimum water content of the soil, respectively. As a result, the applicability of both the regression analysis and the artificial neural network studies to the attainment process of both consistency characteristics and compaction problem were compared with each other to procure a reliable determination process.

Z. Akbay Arama (✉) · S. E. Nuray · M. Yücel
Istanbul University-Cerrahpaşa, 34320 Avcilar/Istanbul, Turkey
e-mail: zakbay@istanbul.edu.tr

H. B. Gençdal
Istanbul Kültür University, 34158 Bakirkoy/Istanbul, Turkey

S. M. Nigdeli et al. (eds.), *Proceedings of 6th International Conference on Harmony Search, Soft Computing and Applications*, Advances in Intelligent Systems and Computing 1275, https://doi.org/10.1007/978-981-15-8603-3_26

Keywords Regression · ANN · Prediction · Consistency · Compaction · High plastic clay

1 Introduction

This study explores the prediction capabilities of both regression analysis and artificial neural network applications by the interpretation of the liquid limit test data to obtain consistency and compaction characteristics of specific types of clayey soils. The consistency properties of soil formations represent the interacted relationship of soil with water considering the water content percentages to define the basic physical properties of especially fine-grained soils. The complicated nature of the behavior of clayey soils is initially observed by Atterberg [1] by the definition of a conservative technique to identify the consistency with different water contents [1]. The connection between identified water contents is specialized with the term consistency limits and divided to compare the current conditions of soil specimens based on four different basic states such as solid, semi-solid, plastic, and liquid for the classification of the physical states of investigated soils [2]. The water content value that is formed a simple consistency boundary between the liquid and plastic states of the soil is defined as the liquid limit (*LL*) and the consistency boundary that is occurred between the plastic and semi-solid state of the soil is identified as the plastic limit (*PL*). The difference between the liquid limit value and the plastic limit value is defined as the plasticity index (*PI*). The plasticity index value represents a range defining the plastic behavior of the investigated soil specimen and used to classify the soil. Two standardized techniques, such as Casagrande cup method and cone penetrometer method, are suggested to determine the liquid limit value and the application of both the tests are conducted with special equipment. This condition has reduced the margin of error that can be caused by the human effect. But unlike the application of liquid limit test, the identification of the plastic limit value only performed by a method that is mentioned in ASTM D-4318 (2008) and GTM-7 (2007) [3, 4]. The application of the test is mostly dependent on the hand skill and experience of the operator and the environmental conditions such as the temperature and moisture in the environment also influence the attainment of plastic limit value. Therefore, the indirect empirical relationships are preferred to obtain the relationship between the plasticity index and liquid limit tests by researchers and engineers to eliminate the dependency of the tests to human labor. Although the consistency limits are mentioned as simple engineering properties, they are directly used to classify soil formations and also used indirectly to estimate the other characteristic features such as the strength and rigidity of the investigated soil profile when the conditions cannot be appropriate to make a detailed sub-soil exploration. In this context, compaction is the most common geotechnical application that is always applied if an earthwork is existed. Compaction is a process of implementation of mechanical energy to the soil to reduce the void ratio in order to ensure reconfiguration of the soil particles. In such a case, the requirement of the identification of optimum moisture content and maximum dry unit weight of the compressed soil strata is arising. The specification of the compaction characteristics includes a complicated and time-consuming process. Therefore, in recent studies, a tendency to estimate the compaction characteristics by the use of simply conducted tests like Atterberg limit tests is preferred to be used. In this context, regression analysis [5, 6, 7, 8, 9, 10, 11, 12, 13] and the application of artificial neural networks [5, 11, 14, 15] leads the methods which are favored to obtain the complicated loop between consistency and compaction. In this regard, within the context of this study, 357 consistency limit tests and 61 standard proctor tests are collected from the well-known sources in the literature that is only considering the applications conducted for highly plastic inorganic clays (*CH*) [16, 17, 18, 19, 20]. The study is divided into two different stages in relation with the aim to obtain the plasticity index value and the compaction characteristics. In the first stage of the study, it is purposed that to reach the plasticity index value directly by the use of 357 liquid limit tests and control them with accurate results of the actual tests. In the second stage of the study, it is aimed to check if it is possible to acquire the optimum water content and maximum dry unit weight values determined for highly plastic clays with the use of both regression analysis and artificial neural network applications. Comparisons have been conducted between the solution techniques to reach the most appropriate determination way of both the consistency and compaction behavior.

2 Material and Method

The applicability of regression analysis (*Ra*) and artificial neural network (*ANNa*) logic to the prediction process of both consistency and compaction characteristics of highly clayey soils have been controlled with the use of Microsoft Excel and Matlab 2018a software. The database used in the prediction process of the geotechnical properties of the soil specimens is prepared according to a collection of related studies performed on the similar characteristic clay types. Totally, 357 data's including the values of liquid limit and plasticity index are used to perform the first stage numerical analysis and 61 data's including the values of natural water content, optimum water content, and maximum dry unit weight are used to conduct the second-stage analysis. The upper and lower boundaries of the liquid limit percentage have ranged between 50 and 93%. Besides the plasticity index value is changed between 23.3 and 59%. The optimum water content values are range between 17.9 and 40% and the maximum dry unit weight of the highly plastic clayey soils are ranging between 1.315 and 1.82 t/m^3. The locations of the collected data are given in Fig. 1 with the use of Casagrande plasticity chart.

The Casagrande cup method was used all the collected liquid limit tests and ASTM D698 was used to perform the standard proctor tests to obtain the maximum dry unit weight and optimum water content of the investigated soil specimens for all the considered studies [21]. The collected values of optimum water content and maximum dry unit weight is given in Table 1.

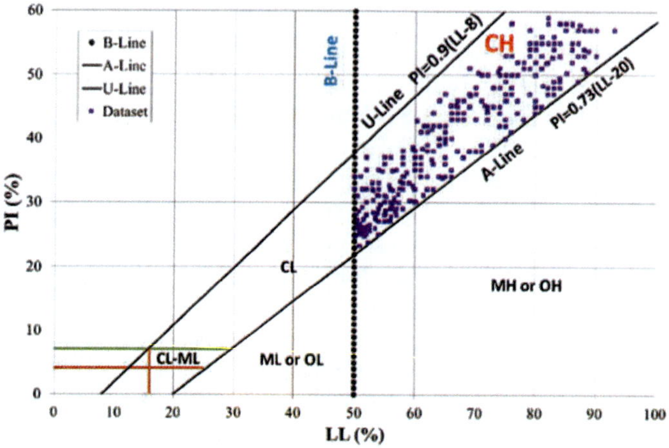

Fig. 1 The locations of the data points on the plasticity chart

Table 1 The collected values of optimum water content and maximum dry unit weight

w_{opt} (%)	γ_{dry} (t/m3)	w_{opt} (%)	γ_{dry} (t/m3)	w_{opt} (%)	γ_{dry} (t/m3)	w_{opt} (%)	γ_{dry} (t/m3)
22.0	1.42	25.8	1.49	35.0	1.31	19.3	1.64
23.0	1.50	20.5	1.66	40.0	1.40	19.0	1.67
21.5	1.55	21.5	1.66	23.0	1.60	24.0	1.60
24.6	1.53	16.1	1.83	29.4	1.47	22.0	1.68
23.5	1.55	23.4	1.54	24.0	1.57	20.0	1.70
24.0	1.52	22.4	1.57	16.0	1.82	31.0	1.33
21.4	1.54	19.9	1.67	21.5	1.66	23.0	1.48
25.0	1.82	27.4	1.48	25.0	1.52	19.0	1.66
28.0	1.40	26.8	1.49	19.6	1.62	19.0	1.68
25.0	1.52	25.0	1.52	24.0	1.46	26.0	1.52
30.0	1.40	19.6	1.62	22.0	1.61	35.0	1.29
35.3	1.46	24.1	1.57	19.0	1.65	25.0	1.58
35.3	1.53	31.6	1.41	19.0	1.63	24.0	1.59
28.0	1.34	21.4	1.59	26.8	1.49	24.0	1.54
19.0	1.63	21.0	1.68	17.9	1.71	22.0	1.64
						22.0	1.68

3 Results and Discussion

The discovery process of the sufficiency of the prediction methods to determine soil geotechnical properties are conducted in two stages. The first stage consists of the estimation of plastic limit and plasticity index value directly from the results of the liquid limit tests and the second stage includes the analyses for the prediction of compaction properties. The outcomes of the analyses are given below, respectively.

3.1 Stage 1: Prediction of Clay Consistency

The prediction of clay consistency has been both analyzed with regression and artificial neural network applications. Besides, the regression analyses have been performed with both Microsoft Excel and Matlab 2018a to obtain the most accurate mathematical expression. The linear, logarithmic, polynomial, and exponential curve fitting options have been also applied to the solution. The application of regression analysis gave similar results in both of the analysis. The results of the curve fitting option are illustrated in Fig. 2. The acquired mathematical expressions of the solution are given in Eqs. 1, 2, 3, and 4, respectively, for linear, logarithmic, polynomial, and exponential curve fitting.

$$PI = 0.7551 \times LL - 9.7996 \tag{1}$$

$$PI = 50.663 \times \ln(LL) - 171.48 \tag{2}$$

$$PI = -0.005 \times LL^2 + 1.4369 \times LL - 32.41 \tag{3}$$

$$PI = 0.1911 \times LL^{1.2733} \tag{4}$$

The proper expression can be seen from Fig. 2 that polynomial curve fitting gives the most accurate results to predict PI. The R^2 is determined at approximately 0.838

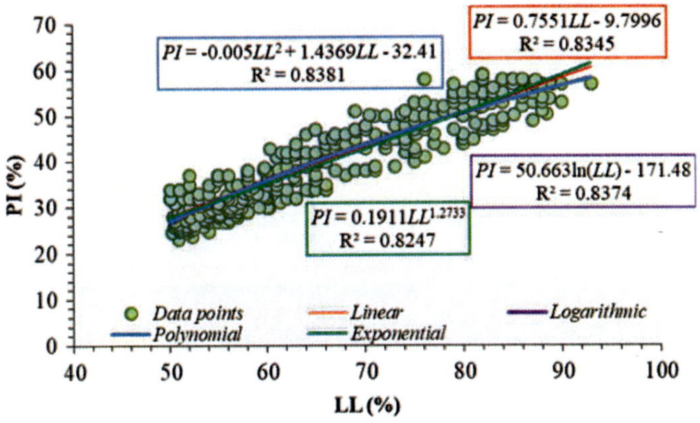

Fig. 2 The change of *PI* against *LL* with *Ra*

for the regression model. The squared sum of the calculated difference between the predicted values of *PI* and the actual data are calculated 5288. Accordingly, the average error per data is determined ±3.84.

The results of the conducted *ANNa* are given in Fig. 3. In order to obtain the mathematical expression of test results, Microsoft Excel applications have used (Fig. 3b). The input value is the liquid limit and the output value is arranged as the plasticity index value. According to the acquired results, the correlation values of training, validation, and test stages for Stage 1 is remarkable regard to can say that *ANNa* models are successful in a way to be able to predict the output parameter. Besides, the best result has occurred in this application via one input parameter by providing 90.451% correlation value (Fig. 3a). The obtained mathematical expressions are given by Eqs. 5, 6, and 7, respectively, for linear, polynomial, and exponential curve fitting.

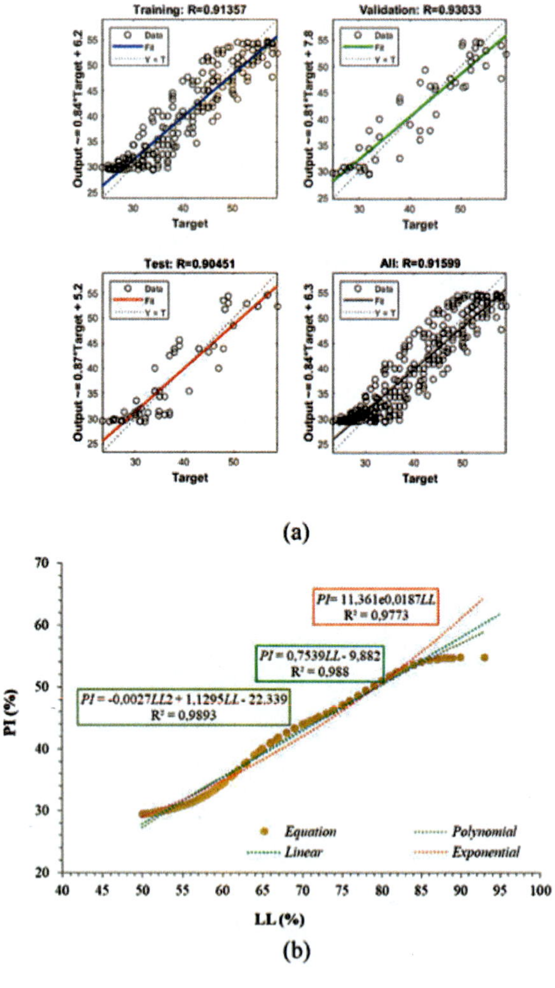

Fig. 3 The change of *PI* against *LL* **a** the results of *ANNa*, **b** curve fitting application

$$PI = 0.7539 \times LL - 9.882 \quad (5)$$

$$PI = -0.0027 \times LL^2 + 1.1295 \times LL - 22.339 \quad (6)$$

$$PI = 11.361\, e^{0.0187LL} \quad (7)$$

It can be seen from the comparison of Figs. 2 and 3 that *ANN*a gives the most accurate solution to determine the plasticity index directly from the liquid limit test. Besides, the proper fitting option is seen to be polynomial and the suggested expression to calculate the plasticity index value is to be Eq. 6.

3.2 Stage 2: Prediction of Compaction Characteristics

Within the context of the second stage, the prediction of compaction characteristics such as the optimum water content (w_{opt}) and the maximum dry unit weight ($\gamma_{dry\text{-}max}$) has been studied. Totally, 61 data couples including both the maximum dry unit weight and the optimum moisture content values, have been used to search for the appropriateness of the prediction methods. The attainment process of compaction characteristics with only the use of plasticity index value has been discussed individually for each parameter. Besides, in order to obtain the most acceptable expression of w_{opt}*and* $\gamma_{dry\text{-}max}$, all types of curve fitting options have been applied to the regression problem. Some of the results of the regression analysis results are shown in Figs. 4 and 5.

In Fig. 4, only the change of optimum moisture content against the plasticity index and in Fig. 5 only the change of maximum dry unit weight is shown. Figures 4 and 5 shows that it is not possible to procure an appropriate evaluation of compaction characteristics with any of the used fitting options for the selected dataset. This

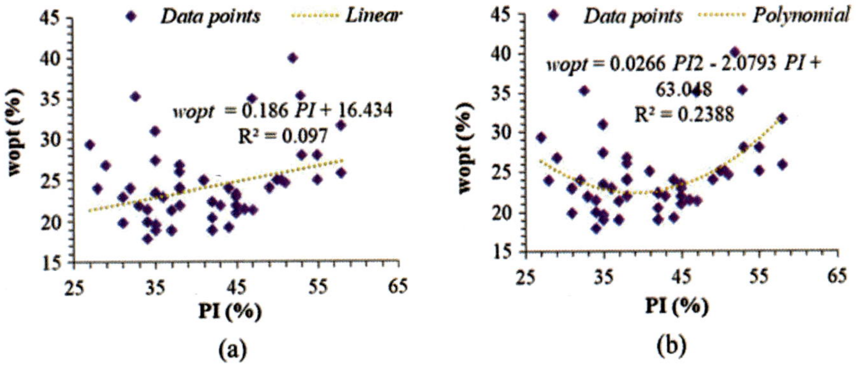

Fig. 4 The change of w_{opt} against *PI* **a** Linear *Ra*, **b** Polynomial *Ra*

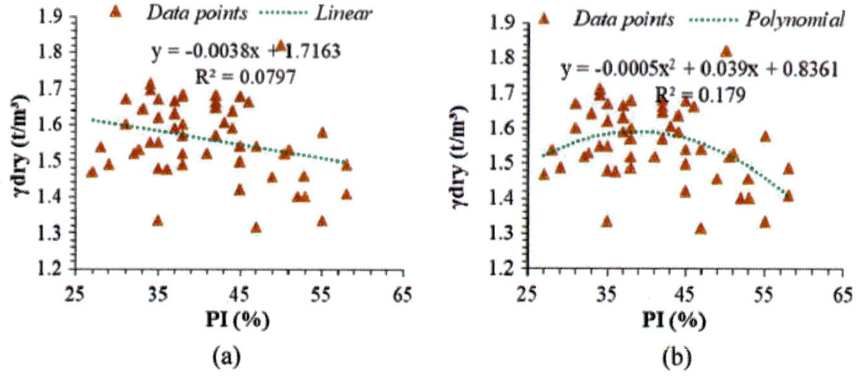

Fig. 5 The change of $\gamma_{\text{dry-max}}$ against *PI* **a** Linear *Ra*, **b** Polynomial *Ra*

situation is the proof of the use of regression techniques leads to form an unsuccessful prediction. The outcomes of the performed *ANNa* are given in Fig. 6. In order to obtain the mathematical expression of test results, Microsoft Excel applications have used (Fig. 6b, c). The input value is the plasticity index and the output value is arranged as the optimum moisture content and maximum dry unit weight value, respectively. According to the obtained results, the correlation values of training, validation, and test stages for Stage 2 is remarkable regard to can say that *ANNa* models are successful in a way to be able to predict the output parameters which are defining the compaction behavior of clayey soils. Besides, the best result has occurred in this application via one input parameter by providing 96.063% correlation value (Fig. 6a). The obtained mathematical expressions for the optimum water content are given by Eqs. 7, 8, and 9, respectively, for linear, polynomial, and exponential curve fitting (Fig. 6b).

The mathematical expressions for the maximum dry unit weight are given by Eqs. 11, 12, and 13, respectively, for linear, polynomial, and exponential curve fitting options (Fig. 6c).

$$w_{\text{opt}} = 0.2637\,PI + 12.713 \tag{8}$$

$$w_{\text{opt}} = 2E - 06\,PI^5 - 0.0006\,PI^4 + 0.0544\,PI^3 - 2.4165\,PI^2 + 50.433\,PI - 370.91 \tag{9}$$

$$w_{\text{opt}} = 15.849\,e^{0.0094\,PI} \tag{10}$$

Consequently, the polynomial fitting option is seemed to be the proper design tool for the estimation of compaction characteristics. Equations 9 and 12 seems to be proper to be the best solution expression for the optimum moisture content and the maximum dry unit weight, respectively.

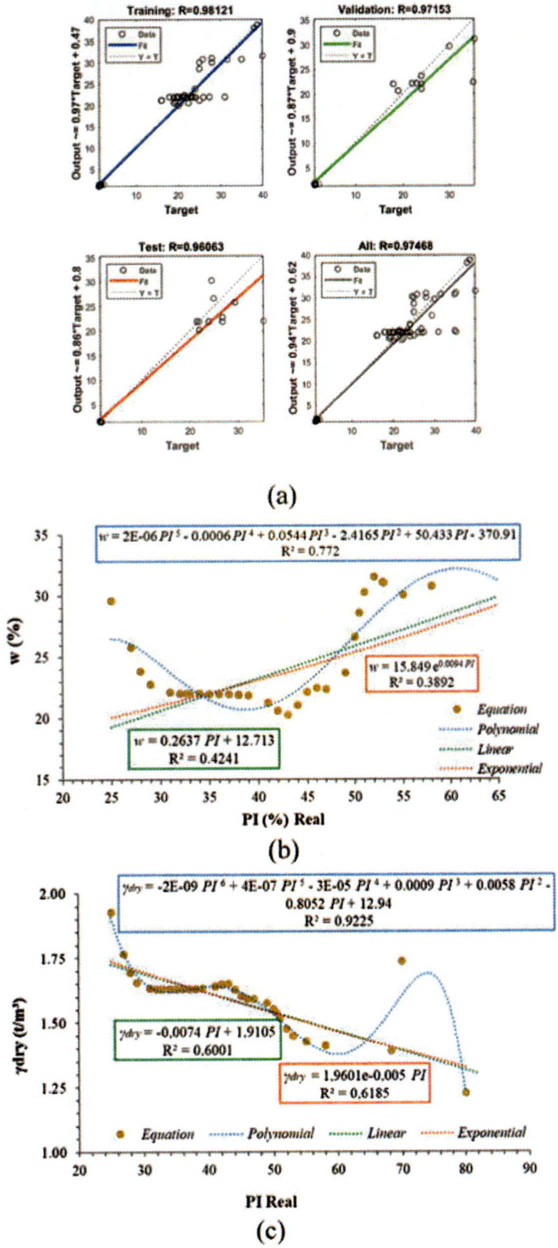

Fig. 6 Application of *ANNa* **a** the results of *ANNa,* **b** the change of w_{opt} against *PI*, **c** the change of $\gamma_{dry\text{-}max}$ against *PI*

$$\gamma_{\text{dry - max}} = -0.0074\,PI + 1.9105 \tag{11}$$

$$\gamma_{\text{dry - max}} = -2E-09\,PI^6 + 4E-07\,PI^5 - 3E-05\,PI^4 + 0.0009\,PI^3 + 0.0058\,PI^2 - 0.8052\,PI + 12.94 \tag{12}$$

$$\gamma_{\text{dry - max}} = 1.9601\,e^{-0.005\,PI} \tag{13}$$

But it is so clear that the suggested expressions are complicated, therefore alternative solutions may be searched with the use of more data to obtain a basic definition.

4 Conclusion

In this study, regression analysis and artificial neural network applications are conducted to procure some of the basic geotechnical characteristics of clayey soils. A sample database was prepared with the use of the studies of the literature done for the purpose of evaluating the compaction behavior of the highly plastic clayey soils. Two-staged analysis has been performed to revive the ordered calculations done, while the evaluation process of a geotechnical report to design a structure. The first stage includes the prediction of plasticity index value that is representing the plasticity characteristics of the soil formation and used in the classification of the soil. The second stage includes the estimation process of the compaction characteristics such as the optimum moisture content and the maximum dry unit weight that are used to model the whole constructions which include earthwork. In addition, the success of the prediction methods has been discussed by comparing the prediction accuracy of the obtained relations. The results of the analyses show that the use of artificial neural network applications gives more accurate approximations to procure the geotechnical properties of clayey soils. Although the results of the regression analyses are satisfying to determine the consistency properties, the success of the estimation cannot be continued for the analyses conducted to calculate the compaction properties. Therefore, it can also be said that this study presents the advantages of using artificial neural network applications in geotechnical engineering practice. Besides, the acquired expressions may guide the researchers and engineers to design the constructions that include compaction works. In addition, the existence of these prediction expressions may extinguish the necessity to conduct laboratory tests.

References

1. A. Atterberg, Über die physikalische Bodenuntersuchung, und über dile Plastizität de Tone. International Mitteilungen Für Bodenkunde **1**, 10–43 (1911)
2. G. Kollaros, Liquid limit values obtained by different testing methods, in *Proceedings of the 14th International Congress of the Geological Society of Greece* (Thessaloniki, 2016), pp. 778–787
3. ASTM D-4318, Standard test methods for liquid limit, plastic limit, and plasticity index of soils. Annual Book of ASTM Standarts (2008)
4. GTM-7, Test method for liquid limit, plastic limit, and plasticity index, geotechnical test method GTM-7, Geotechnical Engineering (2007)
5. O. Gunaydın, Estimation of soil compaction parameters by using statistical analyses and artifical neural networks. Environ. Geol. **57**, 203–215 (2009)
6. M.J. Thompson, D.J. White, Estimating compaction of cohesive soils from machine drive power. ASCE (2008)
7. L.D. Matteo, F. Bigotti, R. Ricco, Best-fit models to estimate modified proctor properties of compacted soil. ASCE (2009)
8. C.E. Canillas, M.V. Saloke, Regression analysis of some factors influencing soil compaction. Soil Tillage Res. **61**, 167–178 (2001)
9. M. Omar, A. Shanabled, A. Basma, S. Barakat, Compaction characteristic of granular soils in United Arab Emirates. Geotech. Geol. Eng. **21**, 283–295 (2003)
10. S. Khuntia, H. Mujtaba, C. Patra, K. Farooq, N. Sivakugan, B.M. Das, Prediction of compaction parameters of coarse grained soil using Multivariate Adaptive Regression Splines (MARS). Int. J. Geotech. Eng. **9**(1), 79–88 (2014)
11. V.K. Viji, K.F. Lissy, C. Sobha, M.A. Benny, Predictions on compaction characteristics of fly ashes using regression analysis and artificial neural network analysis. Int. J. Geotech. Eng. **7**(3), 282–291 (2013)
12. A. Saikia, D. Baruah, K. Das, H.J. Rabha, A. Dutta, A. Saharia, Predicting compaction characteristics of fine-grained soils in terms of atterberg limits. Int. J. Geosynthetic Ground Eng. **3**, 18 (2007)
13. A. Ahangar-Asr, A. Faramarzi, N. Mottaghifard, A.A. Javadi, Modeling of permeability and compaction characteristics of soils using evolutionary polynomial regression. Comput. Geosci. **37**, 1860–1869 (2011)
14. S.K. Sinha, M.C. Wang, Artificial neural network prediction models for soil compaction and permeability. Geotech. Geol. Eng. **26**, 47–64 (2008)
15. A. Ardakani, A. Kordnaeij, Soil compaction parameters prediction using GMDH-type neural network and genetic algorithm. Eur. J. Environ. Civ. Eng. **23**(4), 449–462 (2019)
16. M.M. Dewoolkar, R.J. Huzjak, Drained residual shear strength of some claystones from Front Range, Colorado (2005)
17. B. Mehta, A. Sachan, Effect of mineralogical properties of expansive soil on its mechanical behavior. Geotech. Geol. Eng. **35**, 2923–2934 (2017)
18. C.H. Benson, H. Zhai, X. Wang, Estimating hydraulic conductivity of compacted clay liners. J. Geotech. Eng. **120**(2) (1994)
19. C.H. Benson, D.E. Daniel, G.P. Boutwell, Field performance of compacted clay liners. J. Geotech. Geoenviron. Eng. **125**(5) (1999)
20. S. Singhal, S.L. Houston, W.N. Houston, Swell pressure, matric suction, and matric suction equivalent for undisturbed expansive clays. Can. Geotech. J. **52**, 356–366 (2015)
21. ASTM D698–12e2, Standard test methods for laboratory compaction characteristics of soil using standard effort (12 400 ft-lbf/ft^3 (600 kN-m/m^3)). ASTM International, West Conshohocken, PA, (2012)

A Hybrid Recurrent Model Based on LSTM and Statistical Evaluation to Predict Soil Displacement for Landslide Monitoring

Yilin Li, Xiaojun Pu, Ying Qiao, and Hongan Wang

Abstract Landslides usually cause not only severe economic impacts in industrial and agricultural fields, but also human toll and property losses. Although many efforts have been devoted into landslides research, identifying effective predictors is still a long-term problem. Previous research mainly focused on monitoring landslides by including all available factors to construct models with high prediction accuracy. However, they did not consider model efficiency by selecting effective variables and none of them aimed to detect which particular factors should be focused on in landslides monitoring. This paper proposes a novel hybrid recurrent method based on Long Short-Term Memory (LSTM) model, ANOVA, and stepwise algorithm to identify which meteorological attributes and soil properties affect soil displacement significantly. Due to the difficulty in accessing landslide monitoring data, soil displacement, which can be treated as a warning sign of landslide if it is accumulated to a certain level, is used in this research instead. Starting with a base model to predict soil displacement time series by conducting a LSTM model with revised loss function, each candidate variable is added to the base model separately and the optimal model in this iteration is selected with the lowest mean squared error. Then, ANOVA is used to evaluate whether the current optimal model improves the soil displacement time series prediction significantly, compared to the base model of the previous iteration. Results of our experiments demonstrate that soil moisture content at 10 cm, 20

This work is supported by National Key Research and Development Plan of China (2017YFC0803300).

Y. Li · X. Pu (✉) · Y. Qiao · H. Wang
Institute of Software, Chinese Academy of Sciences, Beijing 100190, China
e-mail: xiaojun2019@iscas.ac.cn

Y. Li
e-mail: liyilin1214@163.com

Y. Qiao
e-mail: qiaoying@iscas.ac.cn

H. Wang
e-mail: hongan@iscas.ac.cn

S. M. Nigdeli et al. (eds.), *Proceedings of 6th International Conference on Harmony Search, Soft Computing and Applications*, Advances in Intelligent Systems and Computing 1275, https://doi.org/10.1007/978-981-15-8603-3_27

cm, and 30 cm depths, and atmospheric pressure are selected as the most significant factors affecting soil displacement prediction.

Keywords Landslide monitoring · Soil displacement · Long short-term memory · Time series · Neural networks

1 Introduction

Landslide is one of the common geological hazards with a high incidence. It usually causes not only severe economic impacts in industrial and agricultural fields, but also human toll and property losses. The key for reducing the impacts of landslide is an effective pre-warning mechanism with early monitoring. Mining the effective predictors of landslides and monitoring them could help evacuate people in advance and reduce economic losses. In the past, landslide prediction was mainly based on expert's experience. Apart from being time-consuming and strenuous, expert's experience also has a limitation leading to lack of accuracy. With the development of statistic models, efforts have been made to model relationships between landslides and triggering factors [1–3]. Deep learning algorithms were employed in several recent studies to predict landslide susceptibility, due to its outstanding ability in exploiting low-level features and high-level representations of information from environmental factors [4–9]. Previous research mainly focused on monitoring landslides by including all available factors to construct models for achieving high prediction accuracy. However, they did not consider model efficiency by selecting effective variables, and none of them aimed to detect which particular factors should be focused in landslides monitoring.

In this paper, we get data from **Insentek** [10], a company which develops a tube-type sensor that can collect soil moisture and temperature at different depths. After conducting LSTM [11] to predict soil displacement, the significance of the feature set used to predict soil displacement is evaluated by Analysis of Variance (ANOVA) [12, 13]. Finally, all significant attributes affecting soil displacement are iteratively selected by stepwise algorithm. When soil displacement reaches a certain level, it may cause landslides. Unlike previous prediction studies, we aim to detect more detailed predictors of soil displacement.

Our work makes the following contributions: (1) We propose a hybrid model based on LSTM and statistical evaluations to explore features related to soil displacement while achieving high prediction accuracy at the same time.

(2) We evaluate our approach on a dataset collected on a certain hillside in Shaanxi Province and discover soil displacement is more relevant to soil moisture content from 10 cm to 30 cm depths and atmospheric pressure.

The rest of the paper is organized as follows: the relevant work is discussed in Sect. 2. Section 3 describes the data source. Section 4 presents the approach that is used to detect the features related to soil displacement. In Sect. 5, we explain experiment design and show the statistical results.

2 Related Work

Statistical regression models were typical methods adopted to infer spatial relationships between landslide occurrence and impact factors [1–3]. Nandi and Shakoor [1] evaluated the multivariate statistical relationship between landslides and various factors including slope angle, proximity to stream, soil erodibility, and soil type based on the logistic regression approach. Yilmaz [3] compared the GIS-based landslide susceptibility mapping methods such as conditional probability, logistic regression, artificial neural networks, and Support Vector Machine (SVM) applied in Koyulhisar (Sivas, Turkey). Several recent studies attempted to predict landslide susceptibility using deep learning algorithm. Mezaal et al. [4] conducted the first study applying Recurrent Neural Network (RNN) on high-resolution airborne laser scanning data in automatic landslide detection, with an accuracy of more than 80%. Huang et al. [5] proposed a novel deep learning-based algorithm, the Fully Connected Spare Autoencoder (FC-SAE), for landslide susceptibility prediction. Results showed that both the prediction rate and the total accuracy of FC-SAE were better than SVM and Back Propagation Neural Network (BPNN). LSTM has recently attracted a wide attention in time series predictions, natural language generation, and so on. Most studies collected relevant geographic data at different moments. Therefore, LSTM has been widely conducted in landslide prediction. Xiao et al. [6] implemented four machine learning algorithms, namely decision tree, SVM, BPNN, and LSTM to predict landslide susceptibility. Xu and Niu [7] predicted the trend component and the total periodic component by using an S-curve estimation and LSTM model. Results obtained from LSTM model were more accurate than those of other statistical models. Xie et al. [8] predicted soil displacement by using LSTM with geological conditions, rainfall intensity, and human activities as inputs. Because it was difficult to access landslide monitoring data, soil displacement, which can be treated as a warning sign of landslide if it was accumulated to a certain level, was used in research instead [14]. Yang et al. [9] proposed a dynamic model to predict soil displacement based on time series analysis and LSTM by analyzing the relationship between landslide deformation and water related factors, such as rainfall and reservoir water level.

Compared to all aforementioned research, data collected for this research has a much higher spatio-temporal resolution. We propose a novel approach to explore how the concrete environmental properties affect soil displacement. According to the conclusion, it is more crucial to monitor certain factors when establishing the pre-warning mechanism.

3 Data

Four categories of triggering data, including soil moisture content [15, 16], soil temperature [17], atmospheric pressure [18], and atmospheric temperature [19, 20], are used to evaluate whether they have significant impacts on soil displacement. Data used in this research were obtained from **Insentek**, a company which develops a tube-type probe that can collect four categories of triggering data and release environmental reports based on scientific ecological data analysis for government and the public. The probe can penetrate deep into the soil and record the recorded time, the moisture content and temperature of soil from 10 to 80 cm depths with 10 cm increment, the angle of the probe offset from geocenter, atmospheric pressure and atmospheric temperature of the site. Data are collected once an hour and sent to a remote server. We use data collected at a station in Shaanxi Province from 00:00:00 January 1, 2019 to 23:00:00, December 31, 2019. Due to the data privacy requirement, this paper keeps the observed location confidential. The sample size N of this dataset is 8760.

X and Y measure the tilt angles of two orthogonal axes on a horizontal plane toward the earth's center. Variation of X and Y can be regarded as measurements of soil displacement. Thus, X and Y are dependent variables to be predicted in the model. We denote the moisture content of soil as M_j^i, the temperature of soil as T_j^i, atmospheric pressure as P_j, and atmospheric temperature as C_j. i means the depth of soil, j represents the j-th time step. For example, soil moisture at the depth of 10 cm at the first moment is represented as M_1^{10}. For simplicity, we describe our algorithm for a single variable and discard subscript j to avoid cluttering the notation in the following sections. The candidate feature set used to predict soil displacement consists of eighteen variables: soil moisture content (M^{10}, M^{20}, ..., M^{20}), soil temperature (T^{10}, T^{20}, ..., T^{80}), atmospheric pressure (P), atmospheric temperature (C).

4 Methodology

In this section, we present a novel approach for discovering the factors affecting soil displacement iteratively.

The approach starts with a base model to predict soil displacement time series by conducting LSTM model, each candidate feature is added separately and the optimal model in this iteration is selected with the lowest Mean Squared Error (MSE). Then, a classical statistical hypothesis testing method, ANOVA [12, 13], is adopted to evaluate whether the added candidate variable in the current optimal model makes a significant contribution compared to the previous model. The combination of selection and evaluation process should be treated as one iteration, and they are integrated and promoted by stepwise algorithm [21], which is a method to identify the optimal model with the efficiently selected feature set. The final feature set consists of factors affecting soil displacement.

LSTM model is introduced in Sect. 4.1. Then, we describe the mechanism of ANOVA in Sect. 4.2. Finally, the integrated model for exploring relationship between soil displacement and candidate features is elaborated in Sect. 4.3.

4.1 LSTM Prediction

LSTM [11] is a kind of artificial neural network with RNN architecture. Yang et al. [9] proved that LSTM can model the dynamic characteristics of landslides more properly because it can make the most of the historic information. The main reason for choosing LSTM is because of its powerful capability in time series prediction, which is suitable for the temporal nature of our dataset.

LSTM model is built for predicting soil displacement X and Y at the j-th moment given the historic information of the previous $j-1$ time steps, relying on soil moisture content (M^{10}, M^{20}, ..., M^{80}), soil temperature (T^{10}, T^{20}, ..., T^{80}), atmospheric pressure (P), and atmospheric temperature (C) as inputs. In each iteration, the candidate attribute variables are added to the base model separately, and the current optimal feature set used to fit LSTM model can be identified by the lowest MSE value. The process of LSTM can be described by Eqs. (1) and (2). Equation (3) displays how to calculate MSE.

$$h_1, \ldots, h_{j-1} = \mathrm{LSTM}(M_1^{10}, M_1^{20}, \ldots, M_1^{80}, P_1, C_1, \ldots, M_{j-1}^{10}, M_{j-1}^{20}, \ldots, M_{j-1}^{80}, P_{j-1}, C_{j-1}, \theta_j - 1) \tag{1}$$

$$\hat{X}_j, \hat{Y}_j = W h_j^{-1} + b \tag{2}$$

$$\mathrm{MSE} = \frac{1}{q} \times \left(\sum_{i=1}^{q} (X^{(i)} - \hat{X}^{(i)} \right)^2 + \sum_{i=1}^{q} (Y^{(i)} - \hat{Y}^{(i)})^2) \tag{3}$$

in which $\hat{X}_j, \hat{Y}_j$ are the predicted soil displacement values at the j-th moment, h_j^{-1} is the hidden layer output of LSTM model at time step $j-1, \theta_j^{-1}$ denotes the parameters of LSTM after training at time step $j-1$, W represents the weight matrix, b denotes the bias vector, and q is the sample size of the training dataset.

4.2 ANOVA

Hypotheses of ANOVA [12, 13] are as follows:

$$H_0 : \text{the full model is not significantly different from the reduced model.}$$
$$H_1 : \text{the full model is significantly different from the reduced model.}$$

We define the optimal model selected from the previous iteration as the reduced model M_{reduced}. After adding a tested variable to the reduced model in the current iteration, the revised model is defined as the temporal full model M_{full} to be tested. RSS_{reduced} and RSS_{full} represent residual sum of squares of the reduced model and the full model, respectively.

$$M_{\text{full}} : y \sim x_1 + x_2 + \ldots + x_k + x_{k+1}$$
$$M_{\text{reduced}} : y \sim x_1 + x_2 + \ldots + x_k$$

$$F = \frac{(RSS_{\text{reduced}} - RSS_{\text{full}})/(K-1)}{RSS_{\text{full}}/(N-K)} \sim F(K-1, N-K) \quad (4)$$

We can calculate F-statistics and look up the corresponding p-values with $K-1$ and $N-K$ degrees of freedom. If the p-value is lower than 0.05, the null hypothesis should be rejected, and researchers can make a conclusion that the full model is significantly different from the reduced model. In other words, the added variable has a significant contribution to soil displacement prediction, and the full model in the current iteration should be chosen as the temporal optimal model. However, the full model is failed to be rejected if the p-value is greater than 0.05, and we choose the reduced model as the optimal model due to the law of parsimony.

4.3 Stepwise Data Mining

The mining process is shown in Algorithm1. S_c is the candidate feature set including features of soil moisture content (M^{10}, M^{20}, ..., M^{80}) and soil temperature (T^{10}, T^{20}, ..., T^{80}) from 10 cm to 80 cm depths, atmospheric temperature (C), and atmospheric pressure (P). S_f denotes the selected feature set affecting soil displacement. R_S represents the MSE of LSTM model with the input feature set S.

In each iteration, if the MSE of the full model is lower than that of the reduced model and meanwhile the p-value of ANOVA test is lower than 0.05, the tested feature should be added into S_f due to its significance. Iterations will stop if one of the three criteria is satisfied (1) the MSE fails to decrease; (2) the p-value is greater than 0.05; (3) S_c reaches empty.

Algorithm 1 Iterative Association Mining

Input: S_c
Output:S_f
Initialize: $S_c \leftarrow M_{10}, M_{20}, \ldots, M_{80}, T_{10}, T_{20}, \ldots, T_{80}, P, C$; $S_f \leftarrow null$
repeat
  find the element $e \in S_c$ minimizing $R_{S_f \cup e}$
  if $R_{S_f \cup e} < R_{S_f}$ and p-value < 0.05 **then**
    $S_f \leftarrow S_f \cup e$
    $S_c \leftarrow S_c \backslash e$
  else
    stop iteration
  end if
until S_c is null

5 Experiment

5.1 *Experimental Design*

LSTM is implemented based on a deep learning library-TensorFlow. The step length of LSTM model is 50. This experiment allocates 80% data to the training set and 20% data to the testing set. Thus, the size of the training set is 6967, while that of the testing set is 1742. Results of the first four effective steps with declining MSE values and significant p-values are displayed in Table 1. We can observe that the MSE decreases gradually when keeping adding the selected variables into the LSTM model. The recurrent process stops until the fifth selected variable does not make a significant contribution in prediction with a p-value greater than 0.05.

5.2 *Results and Analysis*

The final selected feature set S_f includes four attributes: M_{10}, M_{20}, M_{30}, and P. This result is reasonable because soil moisture content and atmospheric pressure are important triggering factors in previous landslides research [15, 16, 18]. Soil moisture content directly affects the slope material, by which the pore water pressure diminishes the soil stability [15, 16, 18]. On the other hand, the shallow soil depth is considered to be more unstable and leading to landslides [22]. Besides proving the combination of these two environmental factors has impacts on soil displacement, this study reveals that only soil moisture contents at the particular range of depths are effective. This work points out the key variables to be concerned for this study area. Experiment results show that observers should pay more attention to the local atmospheric pressure and the soil moisture content from the ground surface to the 30 cm depth when monitoring landslides susceptibility in the study area.

Table 1 Selected variables of the hybrid model

Step	Selected variables	MSE	P-value
1	M_{10}	0.0198	0.00031
2	M_{20}	0.0196	0.00044
3	M_{30}	0.0175	0.00800
4	P	0.0163	0.01900

6 Conclusions

This research proposes a novel hybrid model integrating deep learning approach and classical statistical methods to identify the effective factors for landslide monitoring when obeying the parsimony rule. It detects soil moisture content at 10 cm, 20 cm, and 30 cm depths, and atmospheric pressure as the four significant variables for modeling the pattern of soil displacement at the study location. Unlike the common statistical models that combines ANOVA and stepwise algorithm with linear regression model, this work merges these two methods with LSTM, a non-linear time series prediction model. Our model is a general framework that can be applied to other similar studies for detecting the effective predictors with temporal attributes. For example, this hybrid model can be utilized to detect whether the temporal PM2.5 concentration is affected by atmospheric pressure, temperature, wind direction, wind speed, hours of rainfall, and hours of snow.

This research could be extended from three perspectives. Firstly, precipitation [15, 16] and Normalized Difference Vegetation Index (NDVI) [23, 24] could be incorporated as the potential impact factors to be evaluated. These two factors are not included in this research because no data are available with the matched spatial and temporal scales. Furthermore, ANOVA, which is employed to evaluate whether the added variable is significant, can be replaced by more sophisticated methods, such as Least Absolute Shrinkage and Selection Operator (LASSO). Lastly, one can consider to extend this model from only temporal aspect to spatio-temporal perspectives if data are available.

References

1. A. Nandi, A. Shakoor, A GIS-based landslide susceptibility evaluation using bivariate and multivariate statistical analyses. Eng. Geol. **110**(1–2), 11–20 (2010)
2. S.B. Bai, J. Wang, G.N. Lü, P.G. Zhou, S.S. Hou, S.N. Xu, GIS-based logistic regression for landslide susceptibility mapping of the Zhongxian segment in the Three Gorges area, China. Geomorphology **115**(1–2), 23–31 (2010)
3. I. Yilmaz, Comparison of landslide susceptibility mapping methodologies for Koyulhisar, Turkey: conditional probability, logistic regression, artificial neural networks, and support vector machine. Environ. Earth Sci. **61**(4), 821–836 (2010)

4. M.R. Mezaal, B. Pradhan, M.I. Sameen, H.Z. Mohd Shafri, Z.M. Yusoff, Optimized neural architecture for automatic landslide detection from high-resolution airborne laser scanning data. Appl. Sci. **7**, 730 (2017)
5. F. Huang, J. Zhang, C. Zhou, Y. Wang, J. Huang, L. Zhu, A deep learning algorithm using a fully connected sparse autoencoder neural network for landslide susceptibility prediction. Landslides **17**(1), 217–229 (2020)
6. L. Xiao, Y. Zhang, G. Peng, Landslide susceptibility assessment using integrated deep learning algorithm along the China-Nepal highway. Sensors **18**(12), 4436 (2018)
7. S. Xu, R. Niu, Displacement prediction of Baijiabao landslide based on empirical mode decomposition and long short-term memory neural network in Three Gorges area, China. Comput. Geosci. **111**, 87–96 (2018)
8. F. Huang, J. Zhang, C. Zhou, Y. Wang, J. Huang, L. Zhu, The application of long short-term memory (LSTM) method on displacement prediction of multifactor-induced landslides. IEEE Access **7**, 54305–54311 (2019)
9. B. Yang, K. Yin, S. Lacasse, Z. Liu, Time series analysis and long short-term memory neural network to predict landslide displacement. Landslides **16**(4), 677–694 (2019)
10. Beijing Oriental Ecological Technology Ltd., Co., Beijing, China. https://www.insentek.com/. Accessed 18 March 2020
11. Sepp Hochreiter, Jürgen Schmidhuber, Long short-term memory. Neural Comput. **9**(8), 1735–1780 (1997)
12. R. Henson, W. Penny, *'ANOVAs and SPM', in Technical Report* (Wellcome Department of Imaging Neuroscience, London, 2003)
13. D.M. Bates, D.G. Watts, *Nonlinear Regression Analysis and Its Applications*, (Wiley, New York), pp. 103–104
14. R. Romeo, Seismically induced landslide displacements: a predictive model. Eng. Geol. **58**(3–4), 337–351 (2000)
15. P. Valenzuela, Domínguez-Cuesta, J. María, G. Mora, A. Manuel, M. Jiménez-Sánchez, Rainfall thresholds for the triggering of landslides considering previous soil moisture conditions (Asturias, NW Spain). Landslides **15**, 273–282 (2017). https://doi.org/10.1007/s10346-017-0878-8
16. Marco Bittelli, Roberto Valentino, Fiorenzo Salvatorelli, Paola Rossi Pisa, Monitoring soil-water and displacement conditions leading to landslide occurrence in partially saturated clays. Geomorphology **173–174**, 161–173 (2012)
17. S. Liu, S.X. Zhang, J.F. Li, L.I. Jiangfeng, Investigation of a landslide in the new site of badong county by integrated geophysical survey. Sci. China **044**(005), 426–436 (2001)
18. W.H. Schulz, J.W. Kean, G. Wang, Landslide movement in southwest colorado triggered by atmospheric tides. Nat. Geosci. **2**(12), 863–866 (2009)
19. P. Miščevi, D. Števaníc, N. Štambuk-Cvitanovíc, Slope instability mechanisms in dipping conglomerates over weathered marls: Bol landslide, Croatia. Environ. Geol. **56**(7), 1417–1426 (2009)
20. J.A. Coe, Regional moisture balance control of landslide motion: implications for landslide forecasting in a changing climate. Geology **40**(4), 323–326 (2012)
21. G. Pucciarelli, Stepwise regression: an application in earthquake localization. J. Environ. Sci. Eng. B **7**, 103–110 (2018)
22. H.R. Pourghasemi, A. Gayen, S. Park, W.C. Lee, S. Lee, Assessment of landslide-prone areas and their zonation using logistic regression, logitboost, and navebayes machine-learning algorithms. Sustainability **10**(10), 3697 (2018)
23. A.P. Yunus, X. Fan, X. Tang, D. Jie, Q. Xu, R. Huang, Decadal vegetation succession from MODIS reveals the spatio-temporal evolution of post-seismic landsliding after the 2008 Wenchuan earthquake. Remote Sens. Environ. **236**, 111541 (2020)
24. A.A.J. Deijns, A.R. Bevington, F. van Zadelhoff, S.M. de Jong, M. Geertsema, S. McDougall, Semi-automated detection of landslide timing using harmonic modelling of satellite imagery, Buckinghorse River, Canada. Int. J. Appl. Earth Obs. Geoinf. **84**, 101943 (2020)

Prediction of Optimum 3-Bar Truss Model Parameters with an ANN Model

Melda Yücel, Gebrail Bekdaş, and Sinan Melih Nigdeli

Abstract In this study, a structural engineering benchmark problem from classic literature, which is 3-bars truss model, was handled intended for determination of optimum areas of bars and the minimum volume of the structure, and harmony search (HS) metaheuristic algorithm was utilized to realize this. In here, the main purpose of the study is the prediction of these values directly by using an intelligent and learnt model. For this reason, after optimization process, training of intelligent machine learning algorithm, which is selected as ANN, was performed and following, the prediction operation was carried out with this developed model by considering optimization results. Finally, owing to the development of prediction model and to be successful about accurate and reliable predict, a test model also was generated to evaluate that main model is used replacing of the optimization process and to find the optimum results belonging any structure model. In this regard, the main prediction model has a capability that it can predict design variables as optimal, and minimum volume, directly.

Keywords Optimization · Metaheuristic algorithms · Harmony search · Truss structures · Machine learning · Artificial neural networks · Prediction

1 Introduction

Advanced optimization methods gain importance in view of the optimum (namely the most properly to the specified purpose) design of the various and numerous structure models belonging to the field of structural engineering in a way to provide the desired conditions such as safety, strength, economy, and durability. In this regard,

M. Yücel (✉) · G. Bekdaş · S. M. Nigdeli
Department of Civil Engineering, Istanbul University-Cerrahpaşa, 34,320, Avcılar/Istanbul, Turkey
e-mail: melda.yucel@yahoo.com.tr

S. M. Nigdeli
e-mail: melihnig@istanbul.edu.tr

S. M. Nigdeli et al. (eds.), *Proceedings of 6th International Conference on Harmony Search, Soft Computing and Applications*, Advances in Intelligent Systems and Computing 1275, https://doi.org/10.1007/978-981-15-8603-3_28

it is seen that metaheuristic algorithms are preferred for the mentioned methods. These methods, which are also frequently employed in various science and engineering fields, were developed by benefiting from nature, physical processes, and specific features of human and are commonly preferred about designing of structural engineering models in accordance with desired, too.

From the other respect, the designing of truss systems properly to required constraints such as displacement, compression and tension stresses, and frequency that it is one of the most significant issues in structural engineering. In this respect, various optimization applications were performed by using classical structural analysis or optimization techniques in a way to provide the cost minimization beside of either minimization of structure weight for truss systems or optimally placement of nodal points. Some studies that can be given as examples of these are studies were made by Lipson and Agrawal [1], Cheng [2], Cheng and Guo [3], Wang and Rosen [4]. However, as to come to nowadays, it is seen that the mentioned metaheuristic algorithms are replaced by these methods as convenient to optimum design target. For example, in the study carried out by Miguel and Miguel [5], size and shape optimization were carried out in the direction of minimization of weight for truss systems under multiple natural frequency constraints. To make this, they benefited from the methods including HS and firefly algorithm (FA). On the other hand, Miguel et al. [6] used FA to realize topology optimization of truss systems under displacement, stress and stability constraints besides shape and size as intended for the same purpose, too. Kaveh and Mahdavi [7] tried to solve the truss optimization problem under frequency limitations with a hybrid method, which was developed by combining colliding bodies (CBO) and particle swarm optimization (PSO) methods. Also, Ho-Huu et al. [8] used the improved version of differential evolution (DE) with the aim of determining the optimum layout, which provides to occur of minimum weight for different truss structures. Bekdaş et al. [9] realized an optimization approach by using FPA with the aim of sizing of section areas of bars to minimize the weight for a benchmark model from the literature, which is 25-bar truss structure, without grouping of bars, too. Nevertheless, Talatahari and Goodarzimehr [10] used a hybrid algorithm through combining of teaching–learning based optimization (TLBO) and HS to minimize the weight by providing of shape optimization of space trusses; Salar and Dizangian [11] benefited from ant lion optimizer (ALO) method realized this purpose and studied the design of both space and planar trusses.

Nonetheless, it is recognized that optimization methods are more widely handled coordinately with machine learning techniques thanks to the advanced technology and innovations. Because, process of ensuring of optimum results intended for more than one design combination for any design model takes a lot of time and this case may cause to both time, effort, and also cost lost implicitly. However, the process can be resulted very quickly owing to the mentioned machine learning models, and it is becoming possible that the most convenient information which belongs to the targeted designs are acquired speedily and reliability of them also can be analyzed.

With this aim in the current study, optimal section areas were determined for 3-bar truss structure model by using metaheuristic harmony search (HS) method, which is an optimization algorithm based on memory and musical performances, in a way to minimize the structure volume; and then a machine learning training model is developed via artificial neural networks by using these data and thanks to this, it is made possible that optimal design values and volume of structure are directly determined intended for any different designs.

2 3-Bar Truss Structure Model

In Fig. 1, the structure of 3-bar truss is shown that it is subjected to optimization and machine learning applications [12]. This structure has three bars including same of them as two, due to that they are connected to a common point with a same distance from supports according to the symmetry axis. Hence, two different section areas ($A_1 = A_3$ and A_2) exist, and a P load is applied to the connection node with 45° in the direction of south-west of symmetry axis. In here, section areas are design variables; distance between each support and length of 2nd bar ($l = 100$ cm) and P is design constants. Also, P is handled between 0.1–2.8 kN by increasing with 0.1 kN to generate a training dataset by ensuring of optimum designs belonging to defined load options. All information that belong to truss model and optimization elements were expressed in Table 1.

On the other hand, objective function of the problem is minimization of total truss volume indicated as $f(\boldsymbol{v})$ (Eq. 1). Also, some design limitations are applied related to each bar stress. Bars must not exceed allowable stress value ($\boldsymbol{\sigma} = 2$ kN/cm^2) and

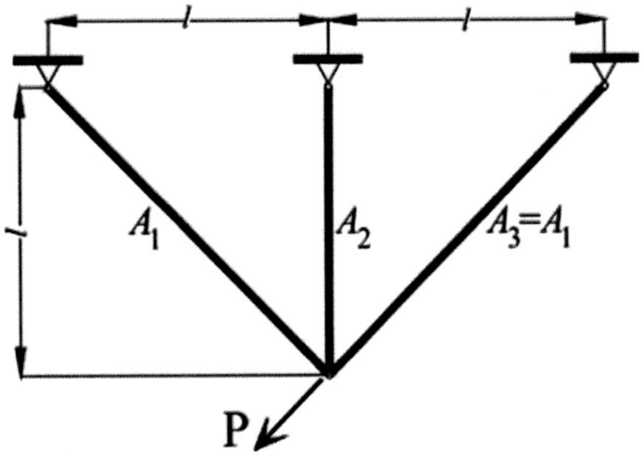

Fig. 1 3-Bar Truss model and design parameters [12]

Table 1 Some expressions defined for optimization process

Element	Notation	Property	Value/limit/range	Unit
Design variables	A	Section area of bars	0–1	cm^2
Design constants	E_s	Elasticity modulus	20,000	kN/cm^2
	P	Applied load	0.1–2.8	kN

are expressed in Eqs. (2)– (4), respectively.

$$\mathrm{Min}_{f(v)} = \left(2\sqrt{2}A_1 + A_2\right)l \quad (1)$$

$$g_1 = \frac{\sqrt{2}A_1 + A_2}{\sqrt{2}(A_1)^2 + 2A_1A_2}P - \sigma \leq 0 \quad (2)$$

$$g_2 = \frac{A_2}{\sqrt{2}(A_1)^2 + 2A_1A_2}P - \sigma \leq 0 \quad (3)$$

$$g_3 = \frac{1}{A_1 + \sqrt{2}A_2}P - \sigma \leq 0 \quad (4)$$

3 Harmony Search Algorithm (HS)

HS is a metaheuristic optimization algorithm, and was developed by Geem et al. [13]. This algorithm has two different stages, which are applied according to state of usage of harmony memory with a harmony memory consideration rate (HMCR) parameter when it is considered that improving musical performance by any musician. According to modification of HS to the problem, the stages used harmony memory (Eq. 5) and randomly generating notes (Eq. 6) by applying HMCR.

$$X_{i,\mathrm{new}} = X_{i,k} + \mathrm{rand}\left(\frac{-1}{2}, \frac{1}{2}\right)\mathrm{PAR}\left(X_{i,\max} - X_{i,\min}\right), \ \mathrm{HMCR} < \mathrm{rand}() \quad (5)$$

$$X_{i,\mathrm{new}} = X_{i,\min} + \mathrm{rand}()\left(X_{i,\max} - X_{i,\min}\right), \ \mathrm{HMCR} > \mathrm{rand}() \quad (6)$$

where pitch adjusting rate (PAR) is a parameter specific to HS. $X_{i,\mathrm{new}}$ is new value of ith design variable corresponding to each candidate vector. $X_{i,\min}$ and $X_{i,\max}$ are minimum and maximum value of the variables, respectively. Also, $X_{i,k}$ is kth candidate vector that is randomly selected from the harmony memory. And $\mathrm{rand}\left(\frac{-1}{2}, \frac{1}{2}\right)$ is a function that provides the generation of a random number ranging in $\frac{-1}{2}$ and $\frac{1}{2}$. rand() is used to generate a random number between 0 and 1.

4 Artificial Neural Networks (ANNs) and Machine Learning Process

Artificial neural networks (ANNs) are a kind of machine learning algorithm that it was developed by inspiring from human nervous system and its working arrangement.

ANNs have ability to make some functions such as estimate, optimize, classify of outputs, and pattern recognition. Also, this algorithm is considered as a calculation method that it can adapt to the environment, work with missing data, decide under uncertainties, and be tolerant of errors [14].

In this respect, a machine learning process was carried out via ANNs to make quick predictions for different 3-bar truss models. In the first stage of this process, a training dataset was generated through the usage of optimum results ensured for defined 28 different design combinations, and ANNs were trained with optimization data to generation of a learning and also prediction model. In here, P load is input, and optimum section area of each bar and minimum truss volume were employed as outputs.

As to the second stage, error rates and correlation coefficients (R) were found for the main ANN prediction model by considering some error metrics to understand reliability and accuracy success of model and validate of performance.

5 Numerical Results and Conclusion

For the main prediction model, mean squared error (MSE) values and R were determined after training of ANN, as seen in Fig. 2. According to the correlations, it was coming out that the model can be used to find optimum design values with volume for any different designs due to small errors and high accuracy.

Following, owing to learnt model, expected optimization data can be acquired directly and rapidly for a test model comprising five designs, which are not encountered and known by ANN model previously. Optimization results belonging to this model were presented in Table 2. Also for each test design, errors (mean absolute error as MAE, mean squared error as MSE, and root mean squared error as RMSE) with optimum section areas ($A_1 = A_3$ and A_2) and minimum volume ($f(v)$) were calculated again to compare predictions reached by ANN model with real optimum outputs (Tables 3, 4, and 5).

Fig. 2 Evaluation of training success of ANN model

	Samples	MSE	R
Training:	20	7.55751e-5	9.99999e-1
Validation:	4	1.84434e-4	9.99999e-1
Testing:	4	7.74328e-4	9.99999e-1

Table 2 Optimization results of test samples

P (kN)	$A_{1=3}(\text{cm}^2)$	$A_2(\text{cm}^2)$	$\textbf{Min}_{f(v)}(\text{cm}^3)$
1.74	0.6866	0.3540	229.5904
2.42	0.9533	0.4969	319.3153
0.85	0.3350	0.1741	112.1577
0.07	0.0277	0.0140	9.2368
2.81	1.0000	0.9626	379.1043

Table 3 Predictions and error calculations of test samples for optimum section area of 1st and 3th bar

Prediction of ANN	Error rates for HS		
$A_1 = A_3$	Error	Absolute error	squared error
0.6880	−0.00138	0.00138	0.0000019
0.9446	0.00872	0.00872	0.0000760
0.3378	−0.00280	0.00280	0.0000079
0.0230	0.00474	0.00474	0.0000224
0.9932	0.00678	0.00678	0.0000460
		MAE	**MSE**
Average		***0.00488***	***0.000031***
RMSE		***0.00555***	

Table 4 Predictions and error calculations of test samples for optimum section area of 2nd bar

Prediction of ANN	Error rates for HS		
A_2	Error	Absolute error	Squared error
0.3497	0.00430	0.00430	0.0000185
0.5214	−0.02445	0.02445	0.0005979
0.1662	0.00797	0.00797	0.0000635
0.0261	−0.01211	0.01211	0.0001465
0.9791	−0.01650	0.01650	0.0002722
Average		***0.01306***	***0.0002197***
RMSE		***0.01482***	

6 Conclusions

When the ANN training and prediction process is evaluated, as it can be seen that the predictions are very close and similar to each other and almost equal to the real optimization results. Reason for this is that the correlation values of main model are very great and calculated different error rates are extremely small between real and prediction values provided for design variables and objective function (namely minimum volume). In this regard, the ANN main prediction model can be considered

Table 5 Predictions and error calculations of test samples for minimum truss volume

Prediction of ANN	Error rates for HS		
Min $_{f(v)}$	Error	Absolute error	Squared error
229.5731	0.01737	0.01737	0.0003015
319.2859	0.02940	0.02940	0.0008646
112.1616	−0.00392	0.00392	0.0000154
9.0810	0.15587	0.15587	0.0242942
378.8521	0.25223	0.25223	0.0636211
Average		***0.09176***	***0.0178194***
RMSE		***0.13349***	

as an effective, successful, and precision tool in the issue of determination of real values for any parameter, with a fair degree of accuracy.

In this direction, according to the error rates, both optimum section areas of truss bars and the best volume could be determined by the ANN model for different test models that they are not seen and known previously by training model and include the designs, which exceed the load limit values used in training. So, generated ANN model can be used for any design belonging 3-bar truss structure as a precision prediction tool intended for detecting of optimum variables and best volume directly and quickly. In future, the optimization problems in structural engineering using the proposed ANN model can be extended.

References

1. S.L. Lipson, K.M. Agrawal, Weight optimization of plane trusses. J. Struct. Div. 100(Proc Paper 10521) (1974)
2. G. Cheng, Some aspects of truss topology optimization. Struct. Optim. **10**(3–4), 173–179 (1995)
3. G.D. Cheng, X. Guo, ε-relaxed approach in structural topology optimization. Struct. Optim. **13**(4), 258–266 (1997)
4. H.V. Wang, D.W. Rosen, Computer-aided design methods for the additive fabrication of truss structure. M.Sc. Thesis, School of Mechanical Engineering, Georgia Institute of Technology
5. L.F.F. Miguel, L.F.F. Miguel, Shape and size optimization of truss structures considering dynamic constraints through modern metaheuristic algorithms. Expert Syst. Appl. **39**(10), 9458–9467 (2012)
6. L.F.F. Miguel, R.H. Lopez, L.F.F. Miguel, Multimodal size, shape, and topology optimisation of truss structures using the Firefly algorithm. Adv. Eng. Softw. **56**, 23–37 (2013)
7. A. Kaveh, V.R. Mahdavi, A hybrid CBO–PSO algorithm for optimal design of truss structures with dynamic constraints. Appl. Soft Comput. **34**, 260–273 (2015)
8. V. Ho-Huu, T. Nguyen-Thoi, M.H. Nguyen-Thoi, L. Le-Anh, An improved constrained differential evolution using discrete variables (D-ICDE) for layout optimization of truss structures. Expert Syst. Appl. **42**(20), 7057–7069 (2015)
9. G. Bekdaş, S.M. Nigdeli, X. Yang, Size optimization of truss structures employing flower pollination algorithm without grouping structural members. Int. J. Theor. Appl. Mech. **1**, 269–273 (2017)

10. S. Talatahari, V. Goodarzimehr, A discrete hybrid teaching-learning-based Optimization algorithm for optimization of space trusses. J. Struct. Eng. Geo-Tech. **9**(1), 0–0 (2019)
11. M. Salar, B. Dizangian, Sizing optimization of truss structures using ant lion optimizer, in *2nd International Conference on Civil Engineering, Architecture and Urban Management in Iran* (Tehran University, 2019)
12. H. Nowcki, Optimization in pre-contract ship design, in *Computer Applications in the Automation of Shipyard Operation and Ship Design*, vol. 2, ed. by Y. Fujita, K. Lind, T.J. Williams (Elsevier, New York, 1974), pp. 327–338
13. Z.W. Geem, J.H. Kim, G.V. Loganathan, A new heuristic optimization algorithm: harmony search. Simulation **76**(2), 60–68 (2001)
14. E. Öztemel, Yapay Sinir Ağlari, Papatya Yayincilik (Istanbul, Türkiye, 2012). ISBN: 9789756797396

Defects Detection in Fruits and Vegetables Using Image Processing and Soft Computing Techniques

V. G. Narendra and Ancilla J. Pinto

Abstract In the science of agriculture, automation helps to improve the country's quality, economic growth, and productivity. The fruit and vegetable variety influences both the export market and quality assessment. The market value of vegetables and fruits is a key sensory feature, which affects consumer preference and choice. Although the process of sorting and grading can be performed manually, it is inaccurate, time-consuming, unreliable, subjective, hard, expensive, and easily influenced by the surroundings. Therefore, intelligent classification technique is necessary for vegetables and fruits, along with the system for defect detection. This research aims to detect external defects in vegetables and fruits-based on morphology, color, and texture. In this proposed work, the various algorithms proposed for quality inspection, including external fruit defects (i.e., RGB to L*a*b* color conversion and defective area calculation methods are used to recognize errors in both apple and orange) and vegetables (i.e., K-means cluster and defective area calculation methods are used to identify defective tomatoes from their color), several image techniques are used. The overall accuracy achieved in quality analysis and defect detection is 87% (apple: 83%; orange: 93%; and tomatoes: 83%) of defective fruits (apple and orange) and vegetables (tomatoes).

Keywords Quality analysis · Fruit and vegetable grading · Defect detection · Image processing · Vegetables · Fruits

1 Introduction

Agriculture is partly described as the art and science of growing fruits and vegetables. This increases the stability of mind and emotions of the individuals and the wealth and health of the community [1]. The agriculture sector also plays a vital

V. G. Narendra (✉) · A. J. Pinto
Department of Computer Science and Engineering, Manipal Institute of Technology, Manipal Academy of Higher Education, Manipal 576104, India
e-mail: narendra.vg@manipal.edu

S. M. Nigdeli et al. (eds.), *Proceedings of 6th International Conference on Harmony Search, Soft Computing and Applications*, Advances in Intelligent Systems and Computing 1275, https://doi.org/10.1007/978-981-15-8603-3_29

role in India's economic development. An increased awareness is needed in the food and health sector. Research needs to be carried out on the production of defined quality and to maintain marketing quality. Therefore, the possibility of evaluating in production process, the quality parameters are integrating. The market value of vegetables and fruits is a key sensory feature, which affects consumer preference and choice. In agriculture science, automation improves the quality, economic growth, and productivity of the country.

Consistency in the size, shape, and other quality parameters of fruits and vegetables is necessary to determine the overall quality of acceptance for customers. The main aspects of fresh inspection are color, size, shape, texture, and number of defects. Defect or damage usually occurs in fruits and vegetables due to various factors such as rotting, bruising, scab, fungal growth, injury, and disease. Proper care should be taken following post-harvest of fruits and vegetables [2]. These defects must be removed to prevent cross-contamination and reduce subsequent processing costs [3]. Labor shortages and a lack of overall consistency in the process resulted in the search for automated solutions [2].

Consider the scenario where the human experts perform quality checks on vegetables and fruits. This kind of manual sorting by visual inspection is labor-intensive and time-consuming. It suffers from the problem of inconsistency in human judgment. Hence, the advent of fast and high-precision technology for automation of the defect detection process is expected to reduce labor costs and improve the sorting process efficiency. The quality of vegetables and fruits can be determined by various image processing algorithms [4, 5].

The input image is provided to the image processing algorithm and the output obtained is in the form of an image feature. Image processing is used for diverse uses, such as pattern recognition, image sharpening, and image retrieval. The main four types of digital images are a binary image, an indexed image, a grayscale image, and a true-color image. Various phases of the processing of the digital image are to be followed to extract the required information from the digital image. The stages of image processing are pre-processing, segmentation, extraction, selection of features, and classification [6, 7].

The primary objective of this work is to design an algorithm that can identify defects and classify fruits and vegetables based on digital image analysis. Section 1 contains a brief introduction to the identification process of defects in fruit and vegetables. Some related work is discussed in Sect. 2. Section 3 discusses the methodology for this research. The experimental result of the proposed work is presented in Sect. 4. Finally, Sect. 5 sets out the conclusion and future scope of this work.

2 Background

In this paper, different systems proposed in previous studies are offered to provide an overview of the location, investigation, and evaluation of apple and orange fruits. Paper [8] presents the late advancement and requisition of the Image Investigation and Machine Vision Framework in the quality assessment of agricultural and food products. The instrument used as part of the image investigation and mechanized sorting and evaluation must be highlighted by the light on the essential ideas, and advances linked to the machine vision framework. In [2] it was discussed that an acoustic estimation could be a valuable device for separating distinctive fruit clumps with a low lapse rate. Starting from the range of the indicator recorded by the mouthpiece after the effect of a small mallet on soil products, 18 key characteristics were recognized and used for the arrangement of fruit fittings in ten separate mixtures. In [9], the mass and external measurements (size and width) of the fruit were measured and the evaluation was centered around and the external measurements were taken. In [10] a new provision was proposed for the evaluation of fruits grown from the ground by machine vision. After precisely dividing impressions, the measurable, textural and geometrical characteristic, factual and linguistic classifications are prepared for two and several kinds of reviews of tree-grown foods. The result is negligible perplexity with stem/calyx zones in multispectral images. The geometric properties of fruit and vegetable skin are difficult to define, as shown in all these studies.

To ensure the quality standard required for fruit and vegetable production lines, companies need to train personnel to inspect fruits and vegetables which are moving on a conveyor belt. These experts categorize fruits and vegetables using visual characteristics into certain categories. However, for fast and accurate quality standards, this approach is not competitive and at times unreliable. So, assumed that [11–16], an automated framework based on smart methods is therefore critical. Several methods based on the acquisition of more specific images have been reported in [17–23]. This research seeks to investigate the use of imaging and artificial intelligence technique for the analysis of external vegetable and fruit defects based on soft computing. In [24], the automated apple defect detection using near-infrared vision system is proposed, and it also records some recent improvement in the image and biometric features [25–29]. Computerized methods have evolved for identifying bruises in apple varieties such as Jonathan, Golden Delicious, and Hermon using the threshold technique. Both brushed and non-brushed apples could be differentiated from the algorithm and did not contribute to online detection. However, neural network models were used to classify apple defects of various types [30].

Many industry procedures such as washing and packing are highly automated, while the world's most important monitoring measures are still carried out manually (e.g., quality inspection and grading). For this purpose, we are proposing to introduce an effective classifier based on the geometric (sizes, shape), color, and skin texture analysis of vegetables and fruits [31]. The following sections contain a new automatic classifier, which is based on geometry, color (HSL), co-occurrence matrix (Texture), and probabilistic neural network. In today's extremely competitive environment, high-quality goods are the source of success. The growing demand for quality guarantees calls for simple and reliable methods of sorting. When using computer vision systems, external defects may be observed. The purpose of this study is to explore the use of an external defect detection method and an automated system for fruit and vegetable classification. For this reason, we assessed the creation of a classifier based on color and textural characteristics of images captured by digital

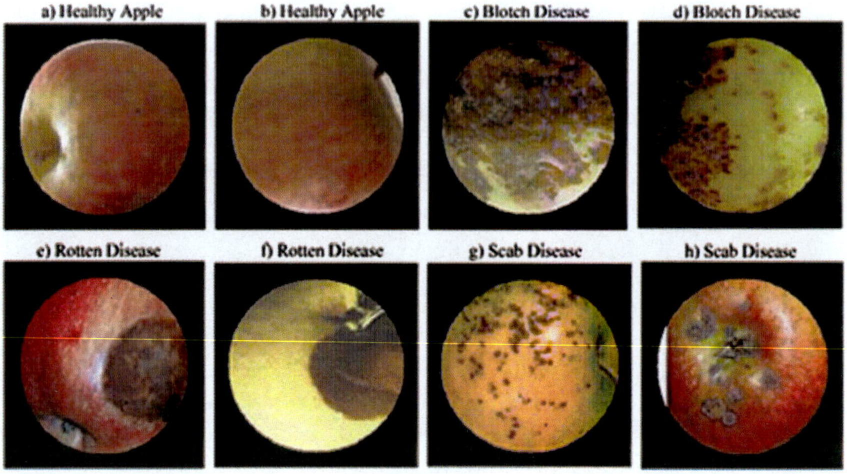

Fig. 1 Apple defects

camera. The method being suggested is very reliable and can distinguish different defects such as surface defects, morphological defects, color defects, black mold or healthy vegetables and fruits.

3 Materials and Methods

3.1 Database Preparation

Experiments were conducted on each of the fifty fruits (apple and orange) and vegetables (tomato) taken from the local market with varying levels of maturity. Samples collected were categorized as standard fruit varieties (Apple-Fig. 1 and Orange-Fig. 2) and vegetables (Tomato-Fig. 3). Samples collected were preserved under normal conditions such as room temperature (22–26 °C) and humidity. Some of the images were also imported from outside to generate a database of some particular type of disease due to the lack of availability of these conditions in the environment.

3.2 Proposed Method

The image of the vegetables and fruits are transformed to the RGB image. In order to enhance image quality, the pre-processing approach has been used. The segmentation process focuses on the desired fruit and vegetable portion. Feature extraction is carried out using the Segmentation-based Fractal Texture Analysis (SFTA)

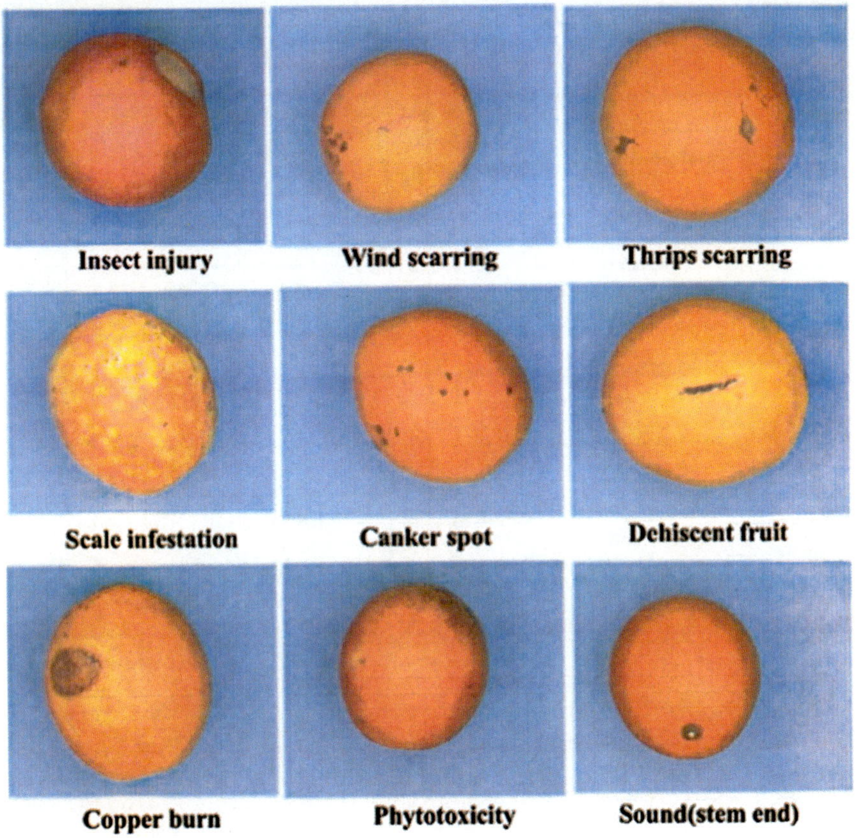

Fig. 2 Orange defects

Fig. 3 Tomato defects

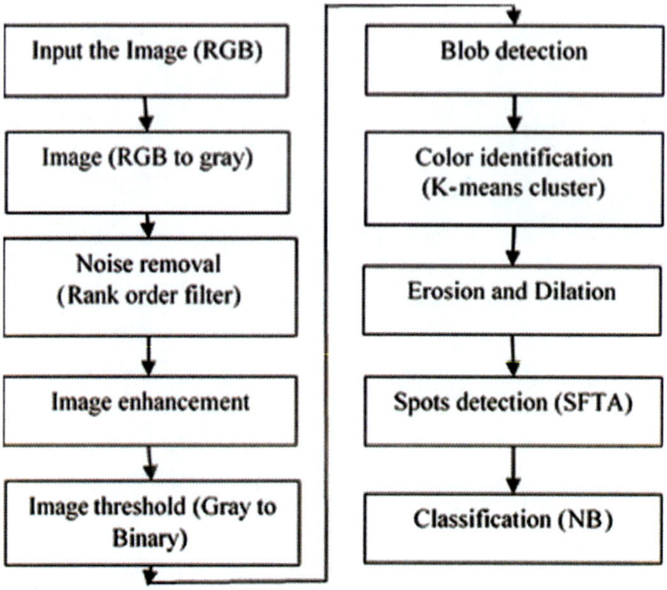

Fig. 4 The flow diagram for the proposed method

[29] method and is categorized according to the data collection. The data collected was trained and tested in the Naïve Bayes classifier. Figure 4 shows the flow diagram of the proposed process.

3.3 *Pre-Processing*

Fruits/vegetables for which a defect is found are captured and considered as input. The input color image is processed into a grayscale image. Pre-processing is performed using a rank order filter. Rank order filters are useful for reducing noise (blurriness) and also for smoothing. The order filtering process is done by selecting a 5 × 5 mask and starting to move it to the whole image. Equation 1 is used to implement the noise removal algorithm.

$$Y(i,j,k) = \sum_{i=3}^{i=h-1} \sum_{j=3}^{j=w-1} \sum_{k=1}^{b} G(i,j,k) * \sum_{p=i-2}^{p=i+2} \sum_{q=j-2}^{q=j+2} \sum_{r=1}^{k} \text{input}(p,q,r) \quad (1)$$

where $G(i, j, k)$ is the kernel, and input (p, q, r) is the input captured image.

The other pre-processing section is the region of interest for segmentation or extraction. The system then captures two following frames, the first without sample of vegetable/fruit at t time and the second one with a $t + 1$ time sample of vegetable/fruit, both subtracting and extracting the image. The segmentation mechanism can be defined as follows using the frame difference approach (Eq. 2). It is a suitable and efficient method for detecting a foreground object with a stationary background:

$$D(x, y, t+1) = \sum_{i=1}^{i=n} \sum_{j=1}^{j=n} \sum_{k=1}^{3} (V(i, j, t+1) - V(i, j, t)) \tag{2}$$

3.4 Blob Detection

The enhanced image is converted into a binary format after completion of the pre-processing phase. A binary image uses the blob detection method. Regions that vary in characteristics, such as size, color, and location, are contrasted and emphasized when detecting blob. The centroid is determined using the coordinate values x and y of the field.

3.5 Color Identification Using K-means Cluster

The K-means clustering algorithm was applied, and the method classifies various clusters using different methods for extracting features (see Figs. 5 and 6). Also algorithm classifies a K number of classes depending on a collection of features. Classification is obtained by minimizing the squares between data objects and their relevant clusters. First, the cluster extraction method transforms the segmented RGB

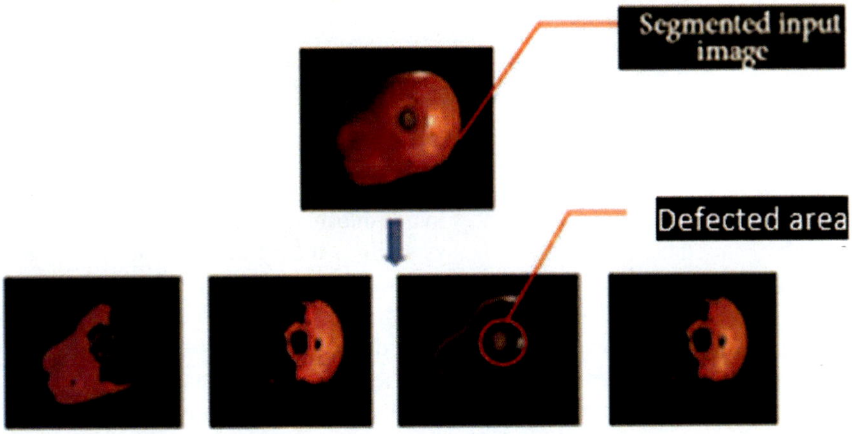

Fig. 5 Extracted clusters using the proposed algorithm for a tomato sample

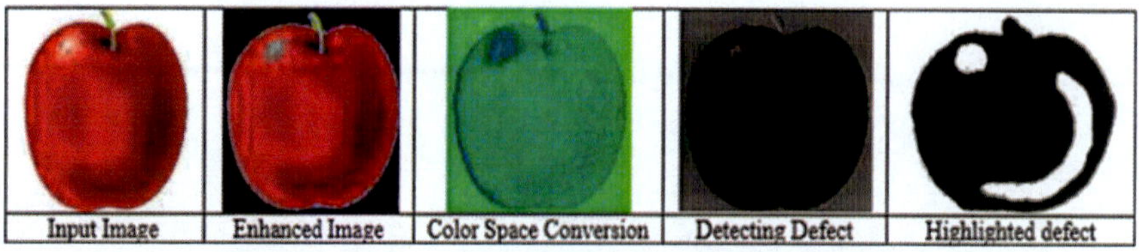

Fig. 6 Extracted clusters using the proposed algorithm for an apple sample

picture into color space L*a*b *. In the 'L*a*b*' space, the system then classifies the different colors. In the segmented image, the system will then execute the labeling phase on the basis of the previous step results. Related label types produce different clusters in an image that is processed.

For each cluster in the processed image, the K-means clustering algorithm selects a random mean pixel and finally extracts the clusters shown in the image on bases of K, which is why K should be intelligently chosen. $K = 4$ has computed the developed system. This means that in every sample of fruit and vegetable the developed system only detects four clusters. $K = 4$ has been selected to be able to extract each cluster present in the samples of fruits and vegetables. After this entire system of processes the mean pixels should be transferred to their respective clusters according to the following measurement of minimum Euclidean distance (Eq. 3):

$$(E.d.)^2 = \min\{\mu_1, \ldots, \mu_k\} \sum_{i=1}^{i=n} \sum_{j=1}^{j=n} \left| X(i, j) - \mu^2 \right|, \mu_{h=} \frac{\sum_{i=1}^{i=n} P(i) * i}{\sum_{i=1}^{i=n} P(i)} \quad (3)$$

3.6 *Watershed Algorithm*

Watershed is a widespread technique for image segmentation. This is a method of converting the grey images, by applying erosion and dilation and to remove unwanted pixels.

Area Calculation

During the transformation of the binary image into an input grayscale image, the area calculation method applies first. The system uses a hard threshold for eliminating unwanted objects in the captured image during the conversion of a binary image. In the transformed binary image, the system estimates the area of the object and the estimated surface location. The calculated area is a scalar quantity that roughly correlates to the total amount of pixels in the image, which may not be the same, because various pixels are weighted differently. In the transformed binary image, the machine determined the region of all pixels by summarizing the areas for each pixel in the image. By looking at its 2 by 2 neighborhood, the location of the individual pixel is calculated. Six patterns, each representing a different area, are accessible. Each pixel belongs to four separate neighborhoods two-by-two.

a. Pixel zero pattern (area = 0).
b. Pixel patterns with one (area = 1/4).
c. Patterns with two adjacent on pixels (area = 1/2).
d. Pixel patterns with two diagonals (area = 3/4).
e. Pixel patterns with three (area = 7/8).
f. Pixel patterns with four (area = 1).

3.7 Feature Extraction

Texture analysis has been performed in this process in order to determine the maximum texture of the fruit/vegetable areas. Intensity distinctions like ruggedness or smoothness are quantified. The strength of the pixel is defined as a variation. The image must be filtered, and the image texture will provide details on the pixel intensity values throughout the texture analysis. The pixel values for the roughness of the texture are minimal and the cost will be greater when the texture is smooth.

For texture removal features in our system, we employed the SFTA technique. Two features are extracted from the fruit/vegetable picture after the segmentation process namely energy and contrast. A threshold value is defined for each feature. A total of 23 characteristics were extracted from every sample of the images (i.e., 6 + 15 + 2 = 23).

i. Morphological characteristics (6)-Size: length, width, area, and perimeter; Shape: elongation and roughness.
ii. Color features (15): RGB to L*a*b*-Mean of L *, a * and b *, standard deviation of L *, a * and b *, range of L *, a * and b *, color distance metric to each L *, a * and b * color components, Chroma, hue, and color score.
iii. Texture features (2): Segmentation-based Fractal Texture Analysis (SFTA) was used to extract the texture properties (energy and contrast).

3.8 Classification

The method used to train and test the fruit/vegetable dataset is the Naïve Bayes Classification [31]. This is based on the training set according to the identified defects to classify fruit/vegetables. The fruit and vegetable defects that are reported in our study are in Table 1.

Table 1 Defects found in fruits/vegetables

Fruit/vegetable	Defects identified in			Accuracy (%)
	Apple	Orange	Tomato	
Apple	25	02	03	83
Orange	01	28	01	93
Tomato	04	01	25	83
Overall accuracy				**87**

4 Results and Discussion

The image size of 640 × 480 pixels was checked in the system. The total image samples obtained for fruit or vegetable are 50 each, with 25 good samples, and 25 samples with different diseases and which are defective. In many of these tests, the exactness of the established method is approximately 87 %. Compared to other vision sensing systems, the method used in this study is very simple and efficient. Developed system predicts (see Fig. 7) various main parameters related to geometry, color, texture, and defect (disease attacks). The system built for the identification of fruit/vegetable samples using vision sensing techniques is one of the most effective and quickest methods. The findings were overwhelmingly acceptable and hence a successful system development is to be considered (see Table 2).

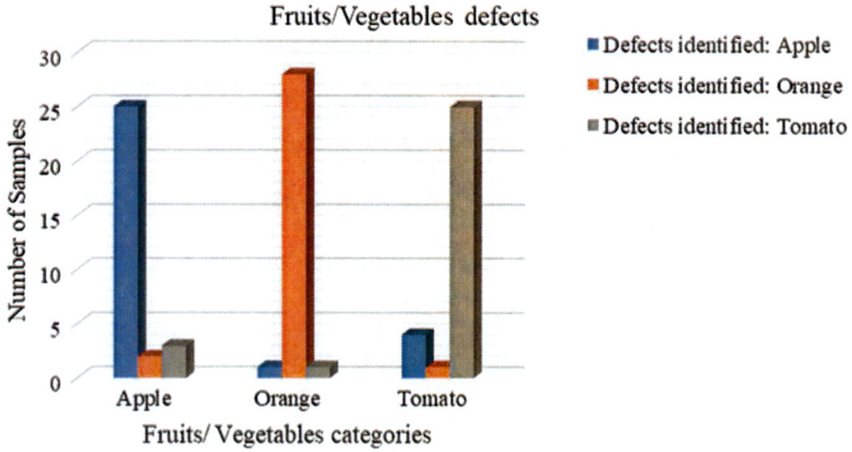

Fig. 7 A clustered columns bar chart compares the values across fruits and vegetable categories

Table 2 Comparative study of existing and proposed method to identify the defect in fruit/vegetable

Sl. no	Fruit/Vegetable	Existing method (%)	Proposed method (%)	
		Features and classification-obtained accuracy	Features and classification-obtained accuracy	
1	Apple	Color (RGB and HSV) and Texture features using Gray-Level Co-Occurrence Matrix and Support Vector Machine [3]–84.02 %	Morphological, Colour ($L^*a^*b^*$), and Texture features along with Area calculation—K-means cluster and Naïve Bayes	83.00
2	Orange			93.00
3	Tomato			83.00

5 Conclusion

This study strongly advises to use the developed system for detecting fruit/vegetable defects. In fruit/vegetable samples the proposed system will detect existing visible defects. Based on the observations made, it can be concluded that the Naïve Bayes classification is an important stage in identifying fruit/vegetable defects. In addition, planning is being made to develop a standardized framework for fruit and vegetable defects identification and classification.

Acknowledgements The author would like to thank Dr. Srikanth Rao, Director of Manipal Institute of Technology, Manipal Academy of Higher Education, Manipal-576104, India for helping us to carry out and publish this research on a project of this nature in Computer Science and Engineering.

References

1. T. Dhanabal, D. Samanta, Computerized spoiled tomato detection. Int. J. Res. Eng. Technol. **2**, 38–41 (2013)
2. S. Arunachalam, H.H. Kshatriya, M. Meena, Identification of defects in fruits using digital image processing. Int. J. Comput. Sci. Eng. **6**(10), 637–640 (2018)
3. D. Sahu, Identification and classification of mango fruits using image processing. Int. J. Sci. Res. Comput. Sci., Eng., Inf. Technol. **2**(2), 203–210 (2016)
4. A. Davenel, T. Labarre, Automatic detection of surface defects on fruit by suing a vision system. J. Agric. Eng. Res. **41**(1), 1–9 (1988)
5. V.H. Pham, B.R. Lee, An image segmentation approach for fruit defect detection using K-means clustering and graph-based algorithm. Vietnam J. Comput. Sci. **2**(1), 25–33 (2015)
6. S.V. Phakade, Automatic fruit defect detection using HSV and RGB colour space model. Int. J. Innov. Res. Comput. Sci. Technol. **2**(3), 67–73 (2014)
7. R.C. Gonzalez, R.E. Woods, *Digital Image Processing* (Prentice-Hall, New Jersey 07548, 2007)
8. V. G. Narendra, K.S. Hareesh, Quality inspection and grading of agricultural and food products by computer vision—a review. Int. J. Comput. Appl. **2**(1) (2010)
9. F. Keshavarzpour, M. Rashidi, Classification of apple size and shape based on mass and outer dimensions. Am.-Eurasian J. Agric. Environ. Sci. **9**(6), 618–621 (2010)
10. D. Unaya, B. Gosselin, O. Kleynen, V. Leemans, Automatic grading of Bi-colored apples by multispectral machine vision. Elsevier (2010)

11. M.S. Pukish, P. Rózycki, B.M. Wilamowski, Polynet: a polynomial-based learning machine for universal approximation. IEEE Trans. Ind. Inform. **11**(3), 708–716 (2015)
12. P.D. Reiner, B.M. Wilamowski, Efficient incremental construction of RBF networks using a quasi-gradient method. Neurocomputing **150**, 349–356 (2015)
13. A. Horzyk, How do generalization and creativity come into being in neural associative systems, and how does it form human-like knowledge? Neurocomputing **144**, 238–257 (2014)
14. M. Wozniak, W.M. Kempa, M. Gabryel, R.K. Nowicki, A finite-buffer queue with a single vacation policy: An analytical study with evolutionary positioning. Appl. Math. Comput. Sci. **24**(4), 887–900 (2014)
15. A. Horzyk, Innovative types and abilities of neural networks based on associative mechanisms and a new associative model of neurons, in *ICAISC 2015, Part I, LNAI 9119* (2015), pp. 26–38
16. M. Wozniak, D. Połap, On some aspects of genetic and evolutionary methods for optimization purposes. Int. J. Electron. Telecommun. **61**(1), 7–16 (2015)
17. C. Napoli, G. Pappalardo, E. Tramontana, Improving files availability for BitTorrent using a diffusion model, in *23rd IEEE International WETICE Conference* (IEEE, 2014), pp. 191–196
18. G. Capizzi, F. Bonanno, C. Napoli. Recurrent neural network-based control strategy for battery energy storage in generation systems with intermittent renewable energy sources, in *IEEE International Conference on Clean Electrical Power (ICCEP)* (IEEE, 2011), pp. 336–340
19. F. Bonanno, G. Capizzi, S. Coco, C. Napoli, A. Laudani, G. Lo Sciuto, Optimal thicknesses determination in a multilayer structure to improve the spp efficiency for photovoltaic devices by hybrid fem–cascade neural network-based approach, in *International Symposium on Power Electronics, Electrical Drives, Automation, and Motion (SPEEDAM)* (IEEE, 2014), pp. 355–362
20. C. Napoli, G. Pappalardo, E. Tramontana, G. Zappalà, A cloud distributed GPU architecture for pattern identification in segmented detectors big-data surveys. The Comput. J. **59**(3), 338–352 (2016)
21. M. Wozniak, Fitness function for evolutionary computation applied in dynamic object simulation and positioning, in *IEEE Symposium on Computational Intelligence in Vehicles and Transportation Systems, Proceedings* (Orlando, Florida, USA, 2014), pp. 108–114
22. S. Fidanova, M. Paprzycki, O. Roeva, Hybrid GA-ACO algorithm for a model parameters identification problem, in *2014 Federated Conference on Computer Science and Information Systems Proceedings* (2014), pp. 413–420
23. I. Lirkov, M. Paprzycki, M. Ganzha, S. Sedukhin, P. Gepner, Performance analysis of scalable algorithms for 3d linear transforms, in *2014 Federated Conference on Computer Science and Information Systems Proceedings* (Warsaw, Poland, 2014), pp. 613–622
24. Z. Wen, Y. Tao, Dual-camera NIR/MIR imaging for stem-end/calyx identification in apple defect sorting. Trans. ASAE **43**(2), 449–452 (2000)
25. M. Wozniak, Z. Marszałek, An idea to apply firefly algorithm in 2D images key-points search, in *Communications in Computer and Information Science (ICIST'2014),* vol. 465 (2014), pp. 312–323
26. C. Napoli, G. Pappalardo, E. Tramontana, Z. Marszałek, D. Połap, M. Wozniak, Simplified firefly algorithm for 2d image key-points search, in *2014 IEEE Symposium on Computational Intelligence for Human-like Intelligence* (IEEE, 2014), pp. 118–125
27. M. Kubanek, D. Smorawa, T. Holotyak, Feature extraction of palm vein patterns based on two-dimensional density function, in *14th International Conference on Artificial Intelligence and Soft Computing (ICAISC 2015) Proceedings- Part II* (Zakopane, Poland, 2015), pp. 101–111
28. M. Kubanek, D. Smorawa, M. Kurkowski, Using facial asymmetry properties and hidden Markov models for biometric authentication in security systems, in *13th International Conference on Artificial Intelligence and Soft Computing (ICAISC 2014) Proceedings-Part II* (Zakopane, Poland, 2014), pp. 627–638
29. W. Miller, J. Throop, B. Upchurch, Pattern recognition models for spectral reflectance evaluation of apple blemishes. Postharvest Biol. Technol. **14**(1), 11–20 (1998)
30. G. Cwikła, A. Sekala, M. Wozniak, The expert system supporting the design of the manufacturing information acquisition system (MIAS) for production management. Adv. Mater. **1036**, 852–857 (2014)

31. A.F. Costa, G.E. Humpire-Mamani, A.J.M. Traina, an efficient algorithm for fractal analysis of textures, in *SIBGRAPI 2012 (XXV Conference on Graphics, Patterns, and Images)* (Ouro Preto, Brazil, 2012), pp. 39–46
32. T. Tiplica, P. Vandewalle, S. Verron, Identification of apple varieties using acoustic measurements, in *International Metrology Conference CAFMET* (2010), pp. 1–8

A Framework for Quality Evaluation of Edible Nuts Using Computer Vision and Soft Computing Techniques

V. G. Narendra, G. Shiva Prasad, and Ancilla J. Pinto

Abstract Generally, humans use edible nuts as food (edible oils, condiments, spices, or drinks) since prehistoric times. The foods are high in protein, minerals, vitamins, and energy so-called as nutrition for human foods. The requirement for precise, quick, and unbiased quality determination of these features increases with increasing demands of decent quality and safety requirements for food products. To satisfy these criteria, digital perception and soft computing provide alternatives to an adaptive, non-destructive, and economical approach. This study has identified a wide range of applications in the food industry, based on image analysis and processing. The consistency study of meat and seafood, bakery food like bread, pizza, cheese, and biscuits was conducted effectively in computer vision and soft computing. This study has been tested on the quality and characteristics of edible nuts.

Keywords Edible nuts · Computer vision · Image processing · Soft computing · Quality evaluation

1 Introduction

People have been using edible nuts as an important food source since prehistoric times and these are used as food, condiments, edible oils, and spices. They can be considered as the most nutritious of human foods since they are high in protein, oil, energy, minerals, and vitamins. Therefore, there is a growing need to determine these properties in foodstuffs precisely, quickly, and objectively with high expectations of high quality and safety standards for edible nuts. Their main manufacturer is the confectionary industry. Nuts and shelled fruits are mainly used to produce snacks, muesli, bakery ingredients, or oils.

V. G. Narendra (✉) · G. S. Prasad · A. J. Pinto
Department of Computer Science and Engineering, Manipal Institute of Technology, Manipal Academy of Higher Education, Manipal 576104, India
e-mail: narendra.vg@manipal.edu

S. M. Nigdeli et al. (eds.), *Proceedings of 6th International Conference on Harmony Search, Soft Computing and Applications*, Advances in Intelligent Systems and Computing 1275, https://doi.org/10.1007/978-981-15-8603-3_30

A large number of effects during growth, harvesting, and processing have an impact on the quality and sensory properties of nuts and shelled fruits. Sensory evaluation and quality grading require extensive experience and commodity-specific knowledge of nuts and shelled fruits. Hence, this work has huge importance in the field of trade, processing industry, and consumers.

The goal of the grading process is to preserve product quality by removing all the damaging elements by sorting homogenous lots of products according to set grade standards [10]. This distinction is thus referred to as its commercial principles and its application. The grading procedures for farmers are subject to the specifications and requirements of the target market [11]. Agricultural products can be categorized according to their biological, chemical, or physical properties. Moisture content, weight, size, texture, color, shape, and external matter are all physical characteristics. The analysis of the product composition, rancidity, or color may be termed as chemical properties. Biological characteristics are the form and amount of damage to the molds, insect damage, and germination percentage of the plant. The most widely used approach is usually the size and weight grading [11]. Weight classifications are applied by weight for products, and size classifications using sieves or bells or bars [12].

Agricultural products are generally graded according to various quality factors. The quality and value of the products are improved by grading edible nuts by farmers, traders, and consumers [9]. Other factors that influence the consistency of the nuts directly include size, color, kernel damage, moisture content, and impurities. Manual grading at every level of product marketing is time-consuming, obsolete, expensive, and fails to provide a suitable quality measure [9].

Classified goods must be insect-free and odor-free. In general, time, demand, high prices, crop loss, and often incorrect quality measurement [11] are problems associated with grading. Section 1 provides a brief introduction to the classification process and classification. Section 2 discourses some associated work and in Sect. 3, the approach of this research is discussed. The experimental outcomes of the proposed work are presented in Sect. 4.

2 Background

Computer image study is one of the areas in which independent grain material assessment is possible. Edible pistachio [3], rice [7], beans [1], and wheat grain [4] have been tested for high-quality requirements. The sorting systems for the qualitative assessment of the samples analyzed are primarily focused on the study of the geometric parameters of the objects studied [4]. For classification based on statistics [6], neural networks [3], genetic algorithms [2], fuzzy sets [1], and multi-dimensional databases relating shapes of the objects studied are used.

Computing vision is a science that creates an algorithmic and theoretical basis. Using system computing [13], useful information is obtained and automatically analyzed from a displayed image. The computer vision is being increased in terms of quality control, sorting, and processing in agriculture due to benefits such as economics, precision, and objective features such as size, form, color, and texture in their ability to provide numerical data [14]. However, for many agricultural products, these features are not specifically mathematical. Due to the need to provide a large number of classification elements [15, 16], the natural variability of these products makes identification and classification extremely difficult and computer-intensive [16]. To simplify decision-making in computer vision, it has a direct advantage in the classification process to incorporate the artificial neural network. A computer vision has been developed in several types of research [17–21] to identify agricultural products using artificial neural networks.

While work in the field of computer vision in recent years has advanced well, computer vision is still in use in the context of automating coffee bean determination. Soedibyo [22] has developed integrated imagery and artificial neural systems to coffee bean sorting. To obtain five quality measures that suit the quality criteria for green coffee, namely length, perimeter, area, defect area, and green color index, the green coffee images have been analyzed. The possibility of classifying coffee beans with its properties such as colors, shape, and size has been studied by Carillo [23]. Food and agricultural product quality assurance is mainly carried out by a visual inspection performed by trained experts. If a large number of samples are analyzed, the evaluation is expensive, time-consuming, and subjective.

The study aimed to investigate the applicability of the procedure, the quality evaluation framework for food products with computer vision, and soft computing techniques. The creation of a classifier was therefore evaluated based on the morphological, color, and textural characteristics of the pictures taken with a digital camera. The methodology suggested is highly robust and can be used to classify foods like cashews, peanuts, and almonds.

3 Materials and Methods

3.1 Data Set

The FoodCast Research Image Database (FRID) attempts to standardize food-related images like edible nuts (cashews, almonds, and peanuts). All 640 × 480 pixel images are stored in the image file format of jpg in data sets. This study includes 360 images from food products and classifies them into edible nuts of 120 images mentioned before. Figure 1 shows the sample data set.

Fig. 1 Sample data set of the edible nuts

3.2 Feature Extraction

The extraction of features is an extremely important process. We used the segmented images of different FRID data set categories. A method of extracting features, such as morphological characteristics (length, width, area, and perimeter), color, and texture, was then developed. To calculate luminance and chrominance, the HSL color space extracts the color attributes of food product categories. The calculated colors are as follows [24].

i. Luminance (L): the "achromatic" component is defined but typically the brightness of an image.
ii. Chrominance (C): the image color details are usually expressed as two components of color difference.
iii. The dominant color: Hue (H).
iv. Distance color metric (ΔE): the metric color difference.
v. Mean (μ): the average total of each color component of the image.
vi. Standard deviation (σ): the average distance from the perceived mean the color component in the image.

3.3 HSL Color Space

Two major aspects depend on the HSL color space. Firstly, it is the color spectrum perceived by humans, and the chrominance components are hue and saturation distinguishing them from the luminosity. Distinct color values for the colors in this space come close to a white color with required saturation. Color purity measurements are taken by hue (H), white color is measured by saturation (S) in a specific color, and color brightness is determined by lightness (L) in this color space [24]. From each sample, we have extracted 14 features mentioned in Table 1.

Table 1 From each sample, the extracted color features by using HSL color space

Sl. No.	Measurement	Code	Sl. No.	Measurement	Code
1	Hue component mean	μ_{HSL_H}	8	Saturation component range	r_{HSL_S}
2	Saturation component mean	μ_{HSL_S}	9	Luminance component range	r_{HSL_L}
3	Luminance component mean	μ_{HSL_L}	10	Chroma of HSL	C_{HSL}
4	The hue component standard deviation	σ_{HSL_H}	11	Hue component color distance metric	ΔE_{HSL_H}
5	The saturation component standard deviation	σ_{HSL_S}	12	Saturation component color distance metric	ΔE_{HSL_S}
6	The luminance component standard deviation	σ_{HSL_L}	13	Luminance component color distance metric	ΔE_{HSL_L}
7	Hue component range	r_{HSL_H}	14	HSL color distance metric	$\Delta \mathrm{E}_{HSL}$

3.4 Texture

On the other hand, the texture is an inherent feature of an image, which correlates with pixel arrangement roughness, granulation, and regularity. The content aspect of an image can be used to segment, classify, and interpret images. Haralick [25] suggested some useful texture features, such as energy, entropy, contrast, and homogeneity. Calculation of texture properties to be carried out using data is expressed as follows in the co-occurrence matrix [25, 26]:

1. Entropy: The randomness level of the intensity distribution could be calculated using this characteristic.

$$\text{Entopy} : -\sum_{i_1}\sum_{i_2} p(i_1, i_2) \log p(i_1, i_2) \tag{1}$$

2. Energy: This feature measures the intensity pair concentration in the co-occurrence matrix.

$$\text{Energy} : \sum_{i_1}\sum_{i_2} p^2(i_1, i_2) \tag{2}$$

3. Contrast: This is used to determine the strength difference between image intensity.

$$\text{Contrast} = \sum_{i_1}\sum_{i_2} (i_1, i_2)^2 p(i_1, i_2) \tag{3}$$

4. Homogeneity: It is a contrast that measures a homogeneous characteristic of the variation of intensity within the image.

$$\text{Homogenity} = \sum_{i_1}\sum_{i_2}\frac{p(i_1, i_2)}{1 + |i_1 - i_2|} \tag{4}$$

Notation p indicates the probability of the above equations. It shows the value of the element in the co-occurrence matrix from zero to one, while i_1 and i_2 show some intensity of the neighboring intensities. This neighboring pair acts as a line number and column number, respectively, in the co-occurrence matrix.

This analysis measures the texture properties for each sample. Color texture features are extracted by computing the HSL color component texture features. For each color component, four texture features are determined to reflect a certain image sample with a vector feature. This represents $4 + 14 + (4 \times 3) = 30$ features of each image. To avoid the drawback of major differences between the magnitude of the feature, each feature is normalized before any machine learning method is implemented. Standardization varies from 0 to 1.

3.5 Feature Selection and Feature Model

The optimal feature model is defined as a subset of relevant data-representing features. For this study, the optimized feature model set is composed of the principal component analysis (PCA). This is an orthogonal transformation procedure that transforms a collection of samples of possible correlated features into a set of values of linearly unrelated functions known as principal components. This process is a statistical one. This technique also describes a new space feature in which data is expressed effectively. The most possible way of doing this is to use the first three principal components representing the highest variability of the data [27]. Thus, the first three principal components are the second feature in the Almond, Cashew nut, and Peanut classification category.

3.6 Soft Computing Techniques

The best classifier is dependent on the data and its capability to perceive the relations between the features when classifying patterns. Turning their particular parameters on the output of the different classifiers can be checked as a general approach. In this study, various soft computing techniques (Backpropagation Neural Network—BPNN and Probabilistic Neural Network—PNN) and an optimum classification model have also been described.

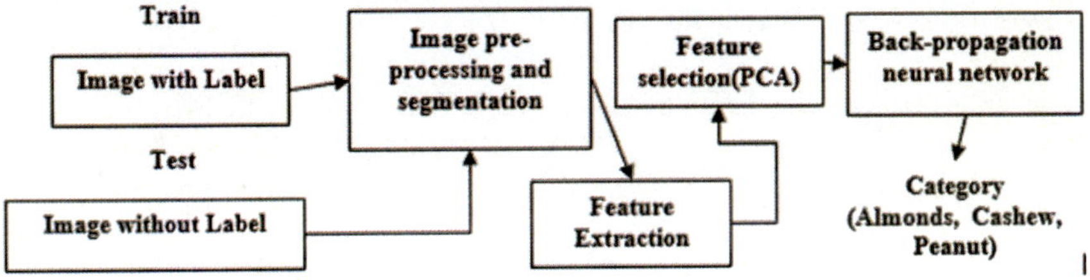

Fig. 2 Schematic Representation of fruit and vegetable classification using BPNN

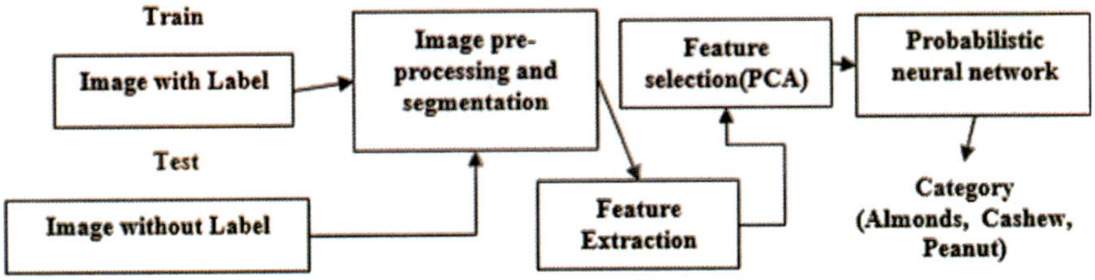

Fig. 3 Schematic Representation of fruit and vegetable classification using PNN

For the classification of varieties, almond, cashew nut, and peanut multilayer feed forward neural networks use backpropagation training (see Fig. 2) and PNN (see in Fig. 3). For preparation, testing, and improvement of the basic classification parameters, a methodology based on the combination of a tenfold cross-validation process is implemented.

The network weights are adapted after appropriate training to determine the overall performance of the model and are used for cross-validation. Just one hidden layer in the network is used to reduce the BPNN's training time. In an exhaustive search for 1 to 50 nodes, the number of neurons in the hidden layer is determined. The 28 nodes in the hidden layer neural network had the lowest standard deviation error and a decent degree of stability. While designing BPNN models, both the hidden and the output layers are used as transition functions for a linear structure of the input layer and a nonlinear hyperbolic tangent function. For the momentum learning law, the learning rate is 0.9.

The PNN is based on the Bayes decision rule and uses the Gaussian Parzen window to estimate the probability density functions (pdf). It allows the pdf calculation to have a constant distribution size equal to the average width of the Gaussian window. Spread parameters or smoothing factors are equivalent to the standard Gaussian Parzen window deviation. The correct selection of the spread parameter relies on the decent output of the PNN. For this analysis, we analyzed the characteristics of 3 categories, such as Almonds, Cashew nuts, and Peanut, for the development of PNNs. For training and testing the characteristics of the sample, which belongs to any type, we used the empiric spread parameter as constant values (0.93).

Table 2 Classification results for sample category

Category	BPNN		PNN	
	Training	Test	Training	Test
	Accuracy in %	Accuracy in %	Accuracy in %	Accuracy in %
Almonds	89.34	93.72	90.45	90.58
Cashew nuts	94.79	93.90	94.30	91.90
Peanuts	92.57	92.00	88.27	89.23

The sets of data are divided into two randomly selected data sets, as such an additional guard against overfitting: 50% of the data used for training and 50% for testing. For the configuration and development of the BPNN and PNN models, the neural network toolbox for the 2019B software is used.

4 Results and Discussions

The color and texture characteristics are developed for the classification experiments. In total there are 360 images (Almonds—120; Cashews—120; Peanut—120). From each category, 60 as training and 60 as testing were considered and taken randomly. For training and testing purposes, tenfold cross-validation was considered. The proportion of training data and testing data is between 90 and 10% per fold. The study deals with the identification of food products in the almond, cashew nut, and peanut categories. Table 2 displays the outcomes from the category of the tests.

The predictive accuracy obtained with BPNN is obtained for Almonds (89.34%), Cashew nuts (94,79%), and Peanuts (92,57%) for the training set. The accuracy of the test results was Almonds (93.72%), Cashew nuts (93.90%), and Peanuts (92.00%). Likewise, for Almonds (90.45%), Cashew nuts (94.3%), and Peanuts (88.27%), a prediction of the training set obtained using PNN was accurate. The accuracy of the test set was Almonds (90.58%), Cashew nuts (91.90%), and Peanuts (89.23%), respectively.

5 Conclusions

Overall color and texture features have been extracted from every sample image as shown in the recognition of categorized images. The research was confined only to Almonds, Cashew nuts, and Peanuts; more studies are needed for more specific edible nuts. The very high precision and prediction accuracy of the tests allowed us to establish food standard measurement and classification systems for edible nuts.

Acknowledgments The authors of Manipal Institute of Technology, Manipal Academy of Higher Education, Manipal-576104, India, are very grateful to the Department of Computer Science and Engineering for providing excellent laboratory equipment for this project.

References

1. Y. Chtioui, S. Panigrahi, L.F. Backer, Rough sets theory as a pattern classification tool for quality assessment of edible beans. Trans. ASAE **42**(4), 1145–1152 (1999)
2. Y. Chtioui, D. Bertrand, D. Barba, Feature selection by genetic algorithm. Application to seed discrimination by Artificial Vision. J. Sci. Food Agric. **76**, 77–86 (1998)
3. A. Ghazanfari, J. Irudayara, A. Kusalik, Grading pistachio nuts using a neural network approach. Trans. ASAE **39**(6), 2319–2324 (1996)
4. A. Kubiak, L. Fornai, Interaction between geometrical features and technological quality of polish wheat grain. Polish J. Food Nutr. **7**/48(3), 160–167 (1994)
5. Z. Mikrut, Recognition of objects normalized in Log-polar space using Kohonen networks. In: *2nd International Proceedings on Image and Signal Processing and Analysis* (Zagreb, Pula, Croatia, 2001), pp. 308–312
6. R. Ohsawa, T. Tsutsumi, H. Uehara et al., Quantitative evaluation of common buckwheat (Fagopyrum esculentum Moench) kernel shape by elliptic Fourier descriptor. Euphtica **101**, 175–183 (1998)
7. N. Sakai, S. Yonekawa, A. Matsuzaki, Two-dimensional image analysis of the shape of rice and its application to separating Varieties. J. Food Eng. **27**, 397–407 (1996)
8. E.L. Schwartz, Spatial mapping in the primate sensory projection: analytic structure and the relevance to perception. Biol. Cybern. **25**, 181–194 (1977)
9. A.B. Fashina, A. Saleh, F.B. Akande, Development and evaluation of a groundnut in-shell grader. Agric. Sci. **6**, 217–223 (2015)
10. T.B. Whitaker, J.W. Dorner, M. Lamb, A.B. Slate, The effect of sorting farmers' stock peanuts by size and color on partitioning aflatoxin into various shelled peanut grade sizes. Peanut Sci. **32**, 103–118 (2005)
11. I.M. Crawford, *Agricultural and Food Marketing Management* (Food & Agriculture Organization of the United Nations, Rome, 1997)
12. J.W. Chapin, J.S. Thomas, *Peanut Money-Maker Production Guide—2004* (Clemson Extension and South Carolina Peanut Board, Blackville, 2004)
13. T.L. Su, F.C. Kung, Y.L. Kuo, Application of back-propagation neural network fuzzy clustering in textile texture automatic recognition system. In: *International Proceedings on International Conference on Wavelet Analysis and Pattern Recognition* (Hongkong, 2008)
14. Y.R. Chen, K. Chao, M.S. Kim, Machine vision technology for agricultural applications. Comput. Electr. Agric. **236**(2–3), 173–191 (2002)
15. N.S. Visen, J. Paliwal, D.S. Jayas, N.D.G. White, Specialist neural networks for cereal grain classification. Biosyst. Eng. **82**(2), 151–159 (2002)
16. M. Markou, S.S. Singh, Color texture analysis of natural scene using neural networks. In: *The International Proceedings on IEEE/INNS Joint Conference on Neural Networks* (Hawaii, 2002)
17. X. Chen, Y. Xun, W. Li, J. Zhang, Combining discriminant analysis and neural networks for corn variety identification. Comput. Electron. Agric. **71**(1), 48–53 (2010)
18. P.M. Granitto, H.D. Navone, P.F. Verdes, H.A. Ceccatto, Weed seeds identification by machine vision. Comput. Electron. Agric. **33**(2), 91–100 (2002)
19. N. Kondo, U. Ahmad, M. Mitsuji, H. Murase, Machine vision-based quality evaluation of Iyokan orange fruit using neural network. Comput. Electron. Agric. **29**(1–2), 135–147 (2002)

20. R. Urena, F. Rodriquez, M.A. Berenquel, Machine vision system for seeds quality evaluation using fuzzy logic. Comput. Electron. Agric. **32**(1), 1–20 (2001)
21. D.W. Soedibyo, K.B. Seminar, U. Ahmad, I.D.M. Subrata, The development of automatic coffee sorting system based on image processing and artificial neural network. In: *International Proceedings on Asian Federation for Information Technology in Agriculture* (Bogor, 2010), pp. 272–275
22. E. Carillo, A.A. Penaloza, Artificial vision to assure coffee-excelso beans quality. In: *The Proceedings on EATIS* (Prague, 2009), pp. 1–2
23. Standar Nasional Indonesia (SNI) 01-2907-1999. Biji Kopi (Coffee Bean). Badan Standarisasi Nasional (1999)
24. S. Da-Wen, Computer vision technology for food quality evaluation. In: *Food Science and Technology*, 1st edn (Elsevier Inc, 2008)
25. R.M. Haralick, K. Shanmugan, I. Dinstein, Textural features for image classification. IEEE Trans. Syst. Man Cybern. SCM-3 **6**, 610–621 (1973)
26. A. Fadhil, An automatic identification system of human skin irritation. TELKOMNIKA Indonesian J. Electr. Eng. **8**(3), 255–264 (2010)
27. F. Kurtulmus, H. Unal, Discriminating rapeseed varieties using computer vision and machine learning. Expert Syst. Appl. **42**, 1880–1891 (2015)

Discovery of Spatial Patterns of Types of Cooking Fuels Used in the Districts of India Using Spatial Data Mining

B. M. Ahamed Shafeeq and Zahid Ahmed Ansari

Abstract The key objective of this research paper is to apply data processing techniques to research the patterns of cooking fuels utilized in India. In a couple of years, the applications of spatial data analytical techniques have spread to most of the fields. The spatial data repository is growing day by day exponentially. The main intention of this paper is to identify the clusters of cooking fuel usage and to seek out the correlation between the different types of cooking fuels used. Every household requires energy to satisfy most of the essential requirements. Household air pollution from cooking with polluting fuels is recognized as a major cause of health hazards. There is poor knowledge of the health effects of prolonged exposure to smoke from unclean cooking fuels among residents. Residents are either unaware or neglect the health impacts of persistent exposure to smoke from polluting cooking fuels, and this leads to serious concerns of indoor air pollution. The local body has to be very proactive in guiding/educating the residents on the health effects of the use of unclean indoor cooking fuel. The administration can use the result of the study to scale back or to seek out the alternate resources of less-polluting cooking fuels. Spatial data processing is a powerful method to extract very much essential and useful information from the large spatial databases.

Keywords Spatial data mining · Autocorrelation · Exploratory spatial data analysis · Spatial clustering

---

B. M. A. Shafeeq (✉)
Department of Computer Science & Engg., Manipal Institute of Technology, Manipal Academy of Higher Education, Manipal, India
e-mail: Ahamed.shafeeq@manipal.edu

Z. A. Ansari
Department of Computer Science & Engg. P.A. College of Engineering, Mangalore, India
e-mail: zahid.pace@gmail.com

S. M. Nigdeli et al. (eds.), *Proceedings of 6th International Conference on Harmony Search, Soft Computing and Applications*, Advances in Intelligent Systems and Computing 1275, https://doi.org/10.1007/978-981-15-8603-3_31

1 Introduction

According to Sashi Shekar et al., “Spatial data mining is the process of discovering interesting and previously unknown, but potentially useful patterns from large spatial datasets” [1]. Spatial data is the data related to the geographical region. The spatial database is the collection of location and its related nonspatial features. The explosive growth of spatial data and therefore the extensive use of spatial databases stress the necessity for automatic extraction of hidden spatial information [2]. Modern spatial analysis focuses on computer-based techniques due to the massive amount of spatial data available in the spatial database. The amount of data obtained from satellite images, computer tomography or other scientific equipment is huge and is growing rapidly.

In spatial information, mining examiners utilize topographical or spatial data to create business insight and other valuable outcomes. This requires specific techniques and resources to urge geographical data into relevant and useful formats. Challenges involved in spatial data processing include identifying patterns or objects that are relevant to the questions that drive scientific research.

According to the United Nations, “half of the world’s population, almost all living in developing countries, use solid fuels for heating and cooking” [3]. There is a very high usage of coal and biomass in developing countries compared to the developed countries. In China, solid fuel is utilized in three out of four households, and in a majority of the rural households as their basic source of energy for food preparation. About 60% of households in India use solid fuels to satisfy their domestic needs. The polluted gas produced by this may result in various health hazards [4]. In the last couple of decades, the Indian government has introduced many schemes/subsidies to encourage the transition from polluting fuels to cleaner fuels like liquid petroleum gas (LPG). A proper study has not been conducted, on what level these plans have influenced overall emissions from the utilization of alternate fuels. The transition to LPG cooking from other sources will have a tremendous positive impact on the environment. The spatial clusters are often formed with the conceptualization of places via measuring co-occurrence of tags at different distance scales [5]. The results show that a conceptualization of place is unveiled by tag co-occurrences at a suitable distance scale. Spatial autocorrelation is an important spatial statistical method which provides spatial structures and patterns of regional variables [6]. The study results demonstrate that statistical analysis of spatial autocorrelation can well reveal the energy usage patterns, and it’s important for the government to make the decision of scaling down more polluting cooking oil or try alternate ones.

The purpose of Spatial co-location pattern mining is to check the coexistence of spatial features in a geographical area. The co-location pattern mining is one of the important tasks in the field of spatial data mining [7]. The research is conducted over various types of data sets to show the effectiveness and feasibility of the method.

The key goal of the current study is to comprehend (a) the spatial clusters of different types of cooking fuel usage among the districts of India and (b) Spatial correlation of different types of cooking fuel usage such as LPG, Firewood, Charcoal, Kerosene and Biogas in the districts of India.

2 Methodology

2.1 *Exploratory Spatial Data Analysis (ESDA)*

The exploratory spatial data analysis technique (ESDA) is a set of procedures designed for analyzing and visualizing spatial patterns and relations. It is used for various tasks such as finding local relations or spatial outliers and detecting patterns of spatial association, groups or hot spots [8]. In the analysis of Spatial patterns, the spatial autocorrelation technique plays a vital role. Spatial autocorrelation indices measure the spatial resemblance between values of the same variable in different places in space [9]. ESDA is applied to explore the spatial dependency and spatial diversity of different types of cooking fuels used in 594 districts of India. The attributes considered from each of the 594 districts are different types of cooking fuel usage such as LPG, Firewood, Charcoal, Kerosene and Biogas. The cooking fuel usage in different districts of India for the year 2011 is used for the study and is obtained from the census of India website: "https://censusindia.gov.in/2011census/HLO/HL_PCA/Houselisting-housing-HLPCA.html". The global statistics can be determined by computing the spatial autocorrelation index [10]. The weight matrix is created using the neighborhood relation. The Moran's I value is computed to check global autocorrelation using GeoDa software [11, 12]. To identify the pattern, the spatial clusters of different types of cooking fuel usage such as LPG, Kerosene, Biogas, Cow dung cake, Charcoal and crop residue are formed.

2.1.1 Spatial Weights Matrix

The basic requirement of analysis of spatial information is the availability of a weight matrix. The proper selection of the weight matrix is a little tedious job in the analysis of spatial data [12]. Below is the general structure of the weight matrix W of spatial data.

$$\begin{bmatrix} w_{11} \ w_{12} \cdots w_{1n} \\ w_{21} \ w_{22} \cdots w_{2n} \\ \cdots\cdots\cdots \\ w_{n1} w \cdot \cdots w_{nn} \end{bmatrix}$$

The 'n' indicates the total number of regions (districts). If districts i and j are adjacent regions, then $w_{ij} = 1$ else $w_{ij} = 0$. The value of self-distance is zero in the matrix, i.e. $w_{ii} = 0$ for $i = 1, 2, \ldots, n$.

2.1.2 Global Spatial Autocorrelation

It is known from the study that there exists a higher resemblance between referred entities, when the two features of different entities show some similarity and are not arbitrarily distributed [13]. By using Moran's I statistic, the global spatial autocorrelation is computed [12]. The Moran's I statistic is given below:

$$I = \frac{n\sum_i\sum_j W_{ij}(x_i - \overline{x})(x_j - \overline{x})}{\left(\sum_i\sum_j w_{ij}\right)\sum_i(x_i - \overline{x})^2} \quad (1)$$

The total number of regions are indicated by 'n'; w_{ij} is ijth element in the matrix; x_i is the value of region i; $\overline{x}$ is the average of all observation values. The value of Moran's I is always between −1 and 1. The value very close to 1 indicates that similar entities are grouped (Higher value districts bounded by Higher value districts—HH or Lower value districts surrounded by Lower value districts—LL), and a value close to −1 signifies that different entities are grouped (HL—Higher value region surrounded by Lower value region or Lower value districts surrounded by Higher value districts). The value close to zero of Moran's I signifies that there is an arbitrary cluster or no spatial autocorrelation.

2.1.3 Local Spatial Autocorrelation

Moran's Scatter Plot: Moran's scatter plot is obtained using GeoDa software for all the features, i.e. LPG, Kerosene, Biogas, Cow dung cake, Charcoal and crop residue. In the study, the districts of India are considered as the basic study element and the nonspatial attributes such as LPG, Kerosene, Biogas, Cow dung cake, Charcoal and crop residue are used to understand the spatial distribution among the districts of India.

2.2 Clustering

Clustering is one of the most popular tasks in the field of Knowledge discovery research for spatial and nonspatial databases. In clustering, all the similar entities will be put into the same group and dissimilar entities will be put into different groups. Due to the availability of the large spatial database, the challenges in spatial data are increasing day by day. It is very much essential to address the issues and meet the demands of commercial communities. The huge explosion in spatial data emphasizes the importance of developing the algorithms for mining large data sets. It is expensive and more importantly impractical for users to check spatial data in depth and search for useful patterns or associations among a huge amount of data. The clustering is applied to the target data followed by the analysis of local and global spatial autocorrelation. The idea is to consider every feature and split the districts into three groups. The groups can be considered as regions with different ranges of values, i.e. low-, medium- and high-usage level groups for all the features considered. The clustered districts of the different features are compared against each other for their impact and distribution. The k-means algorithm technique available in GeoDa software is used for clustering. The k-means clustering is simple to understand and easy to implement. The k-means clustering is widely used by researchers and industrial applications. The required number of clusters must be mentioned in advance. The distance measure used in this clustering is Euclidean distance. The clusters can be made very compact by minimizing intra-cluster distance and maximizing inter-cluster distance [14]. The process may be repeated for a fixed number of epochs to get better results.

The k-means clustering is used for the discovery of spatial patterns.

1. The preliminary k cluster centers are selected randomly from the data set D
2. Reiterate step 3 and step 4.
3. Each data point is assigned to the cluster based on a distance measure.
4. Compute the average value of a data point in each cluster to form a new cluster center.
   $\mathrm{v_i} = \frac{\sum_{i=1}^{c_i} x_i}{c_i}$
   where c_i is the number of data objects in the ith cluster.
5. Repeat step 2, if there are changes in the cluster center.

3 Results

Table 1 below gives the score of Moran's Index for the LPG usage of the 594 regions (districts). The result obtained are based on the 1000 permutations.

Table 1 indicates that there is a positive correlation on the usage of LPG among the districts of India.

The scatter plot (Fig. 1) shows the correlation of usage of LPG of all the districts (594 districts). It indicates that 79.5% of the districts show a positive spatial association and there is a low negative association among the remaining 31.5% of the districts. More than half of the districts are positively associated. The plot reveals that there is a close association of similar districts (First quadrant 31.1% and Third quadrant 48.4%) and it is shown by 79.5% of Indian districts.

Table 1 Moran's statistics for LPG usage

Attribute	Moran's I value	Standard deviation value	Expected value of Moran's I $E(I)$
LPG usage	0.516	0.0259	−0.0017

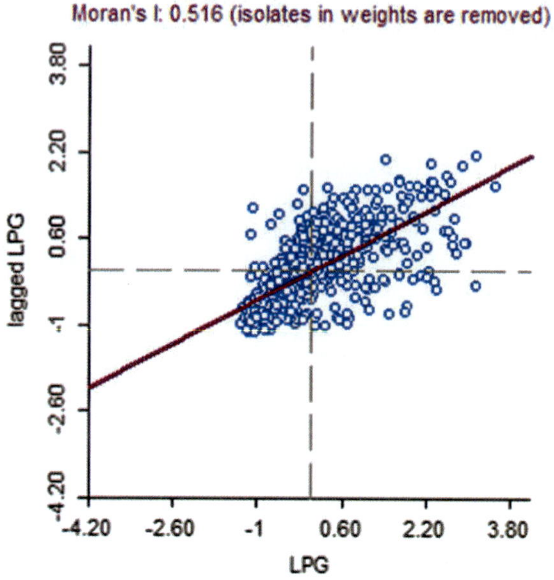

Fig. 1 Moran's scatter plot—LPG

The scatter plot (Fig. 2) shows the correlation of usage of Firewood among the districts (594 districts) for Firewood usage. It is observed that most of the Indian districts show a positive spatial association. The plot shows that 78.95% of

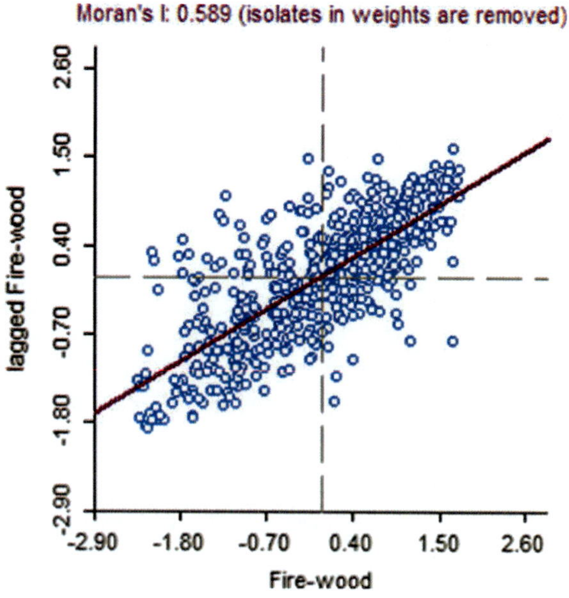

Fig. 2 Moran's scatter plot—Firewood

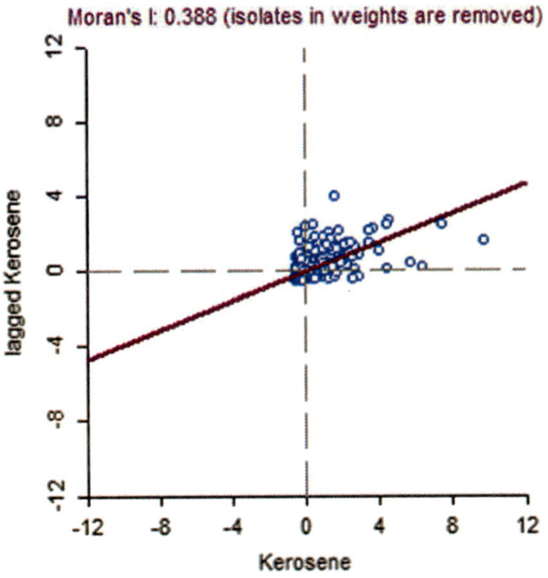

Fig. 3 Moran's scatter plot—Kerosene

Indian districts have an association of similar values (Third quadrant—Low to Low relationship—37.05% and in First quadrant—High to High relationship—41.9%).

The scatter plot (Fig. 3) shows the correlation of usage of Kerosene among the districts (594 districts). There is a high correlation for low usage among the neighboring districts of India (59.59%). The high usage correlation between the neighboring districts are very low (19%).

The scatter plot (Fig. 4) shows the correlation of usage of Crop residue among the districts (594 districts). It can be seen that most of the Indian districts show a positive spatial association. The majority of the districts (89.17%) shows the association of similar values (Low-Low quadrant—63.45% and in High-High quadrant—25.72%).

The scatter plot (Fig. 5) shows the correlation of usage of Charcoal among the districts (594 districts). It is observed that most the Indian districts show a positive spatial association. The association of similar values is shown in 85.61% of districts (Low-Low quadrant—73.43% and in quadrant High-High quadrant—12.18%).

The scatter plot (Fig. 6) shows the correlation of usage of Biogas among the districts (594 districts). It is observed that most the Indian districts show a positive spatial association. The association of similar values is shown in 82.57% of the districts (Low-Low quadrant—58.3% and in quadrant High-High quadrant—24.19%).

Figure 7 shows the three clusters of LPG usage of 594 regions. The cluster shows only 12% of districts use LPG as a major cooking fuel, 32% of districts moderate usage and 56% of the districts use LPG as a secondary source. This is an alarming sign for the governing body to take necessary actions. The indoor pollution created by the cooking fuel will lead to many health hazards.

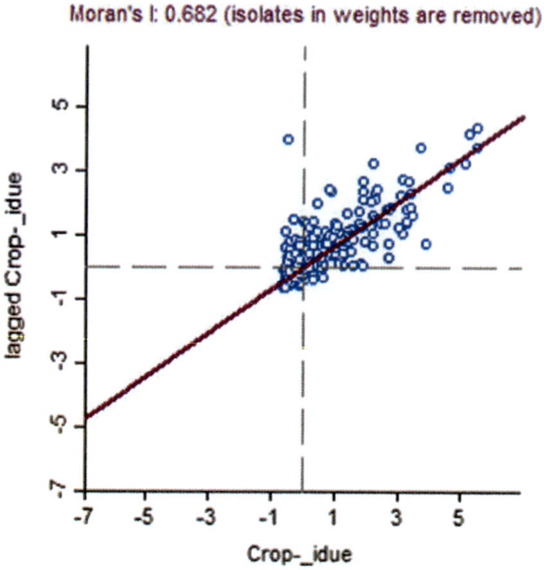

Fig. 4 Moran's scatter plot—Crop residue

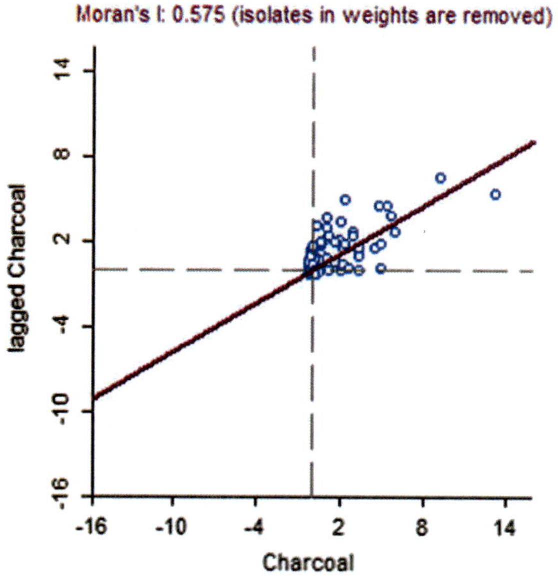

Fig. 5 Moran's scatter plot—Charcoal

Figure 8 shows the clusters of Firewood usage in the districts of India. The cluster map shows that38% of the districts of India use Firewood as primary cooking fuel. This must be brought down to reduce air pollution and save forests. The smoke produced from burning wood will cause a number of severe respiratory and cardiovascular health problems.

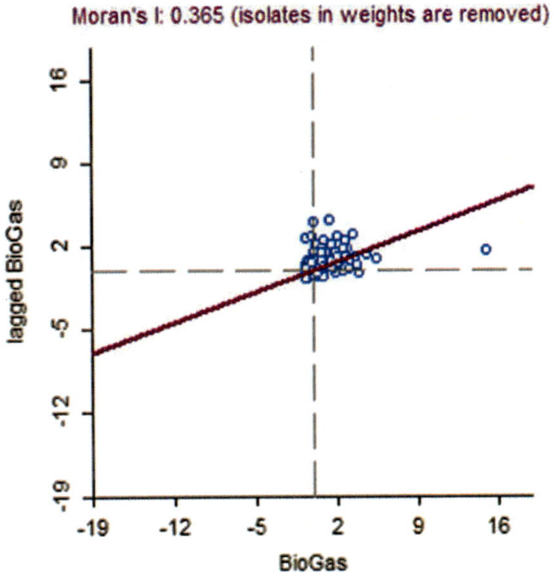

Fig. 6 Moran's scatter plot—Biogas

Figure 9 shows that only 7% of the districts use more kerosene. The usage of Kerosene is very low in the districts of India.

Figure 10 shows that only 6% of the districts use Cow Dung cake as its primary source. The usage of cow dung is found in only a few districts of Uttar Pradesh and Himachal Pradesh. The burning of dung cakes produces a lot of arsenic in fumes and it leads to many life-threatening diseases.

Figure 11 shows the clusters of usage of Crop residue in the districts of India. Only 7% of the districts use Crop residue as their primary source. There is a huge amount of agricultural wastes comprising crop residues produced by leading agro-based companies. The usage of Crop residue as a cooking fuel will lead to many health problems and indoor pollution. According to S. Bhuvaneshwari et al., "The main adverse effects of crop residue burning is the emission of greenhouse gases that contributes to the global warming" [14].

Figure 12 shows that the clusters with more than 70% of households use polluting cooking fuels in more than 60% of the districts. Switching to LPG from other polluting cooking fuels will enormously reduce air pollution and save the forest.

4 Discussion

This experimental analysis discovers the spatial pattern of various types of cooking fuel usage in the districts of India. The analysis focuses on spatial autocorrelation of cooking fuel usage and on identifying the patterns of cooking fuel usage among the

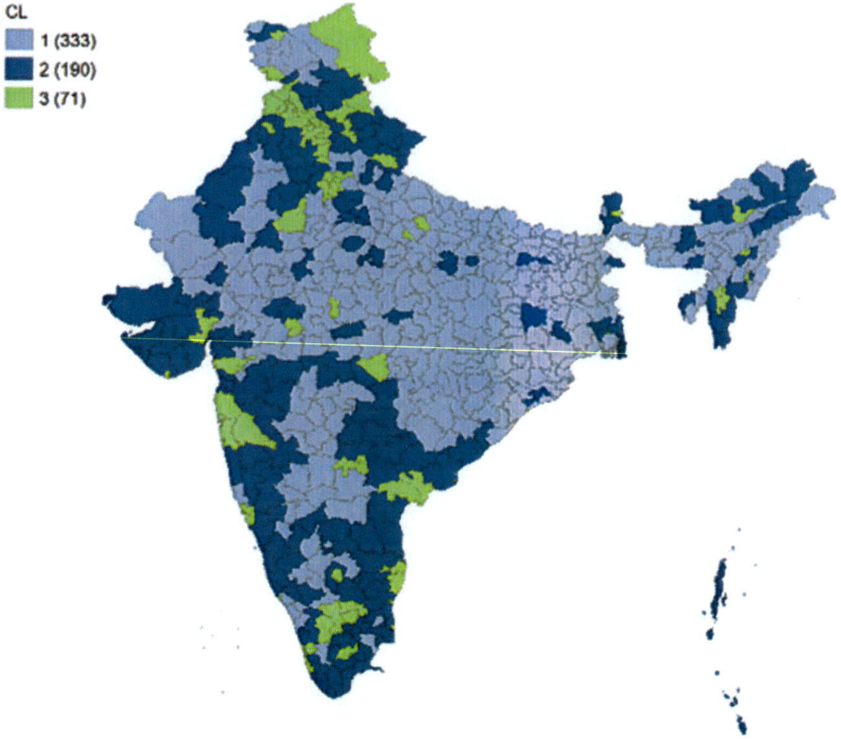

Fig. 7 LPG usage cluster

districts. The different kinds of cooking fuel usage clusters are created. The study is conducted on health survey Uttar Pradesh and it is very much useful in understanding the prevalence of acute respiratory infections [15]. The article is focused on different types of cooking fuels such as LPG, Kerosene, Biogas, Cow dung cake, Charcoal and crop residue. The empirical analysis signifies that there are differences in inter-district cooking fuel sources and the occurrence of severe respiratory infections is significant.

The GeoDa software is used to obtain the scatter plot of usage of different types of cooking fuels in the districts of India. Scatter plot shows a positive correlation for almost all types of cooking fuel usage in the districts. But a few districts show a negative correlation for the usage pattern. The cluster map of LPG usage shows that LPG is the main source of cooking fuel in more than 60% of households among the 20% of districts in India. There are no similar patterns among the usage of LPG and Biogas. In order to reduce indoor air pollution, these two cooking fuels can be made available to every household. More or less similar patterns are found in the usage of cow dung cake and crop residue. Firewood is the main soure of cooking fuel in more than 80% of households among the 40% of districts. The heavy usage of Firewood

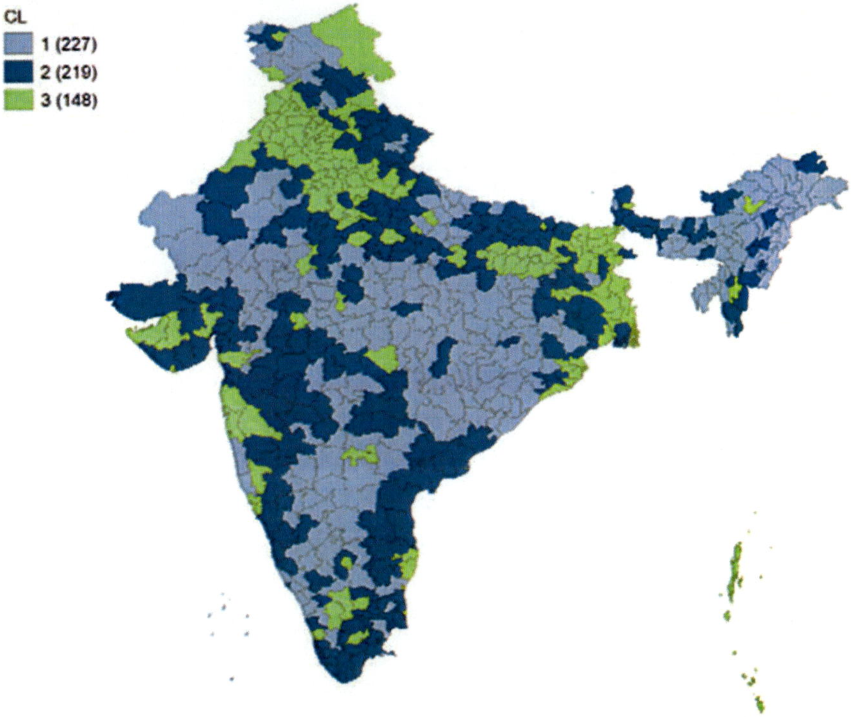

Fig. 8 Firewood usage cluster

leads to many health and environmental issues [16]. The high usage is found in the districts of central and northeastern states of India.

The empirical analysis is limited to check the spatial correlation and to identify the clusters of the districts for the cooking fuel usage such as LPG, Firewood, Cow dung Cake, Kerosene, Biogas, Charcoal and Crop residue.

5 Conclusion

The experimental results of cooking fuel usage among the districts show that there is a positive autocorrelation. The usage of kerosene among the districts shows a moderate positive correlation. The study also reveals that there is positive local autocorrelation of polluting cooking fuel usage between the districts. The spatial pattern shows that there is less spatial correlation among the districts with the different types of cooking fuels used. The spatial distribution of cooking fuel usage in the districts identifies (a) clusters of usage of polluting and non-polluting cooking fuels, (b) Global spatial autocorrelation of the usage of different cooking fuels and (c) Local spatial autocorrelation of usage of different cooking fuels. Moran's scatter plot and Index are

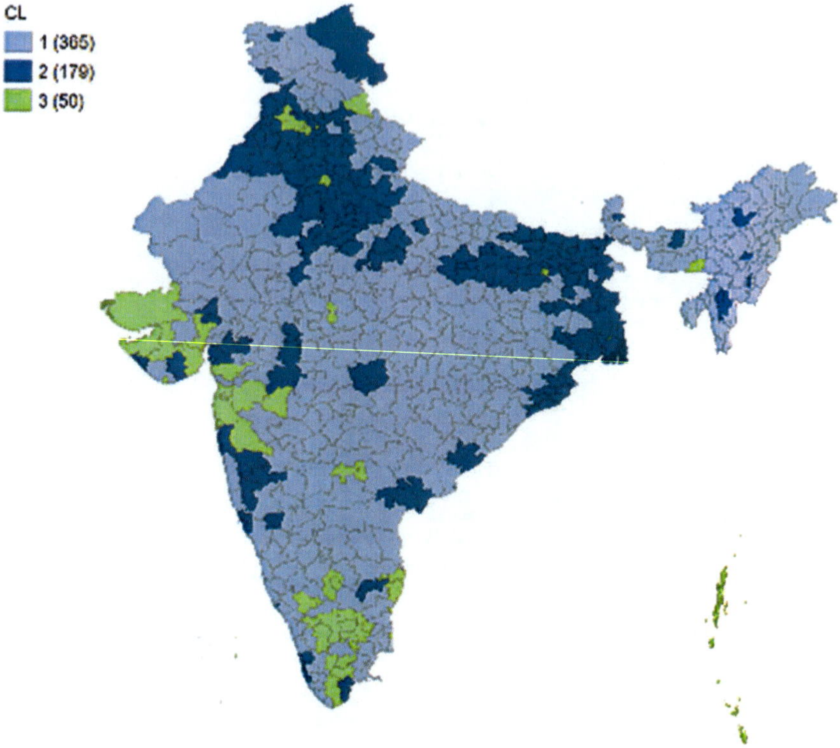

Fig. 9 Kerosene usage cluster

emerged as a helpful application in learning and discovering spatial patterns of usage of cooking fuels in the districts of India. ESDA reveals that there is a positive global spatial autocorrelation on cooking fuel usage among the districts. Moran scatter plot is the one of the best tools to identify positive and negative spatial autocorrelation. The global level analysis reveals that there is a homogeneous relationship and shows heterogeneous relationship in the local analysis. The heterogeneous regions (districts) found in the scatter plot (Low–High and High–Low) in the local analysis is very low.

The outcome will not reveal the causal relation of the cooking fuel usages in different parts of India. The main reason may be the abundant availability of the particular cooking fuels in that region. The increment in the usage of LPG will save the forest and reduce the emission of carbon dioxide. The patterns are identified for each attribute to check the spatial distribution of cooking fuel usage.

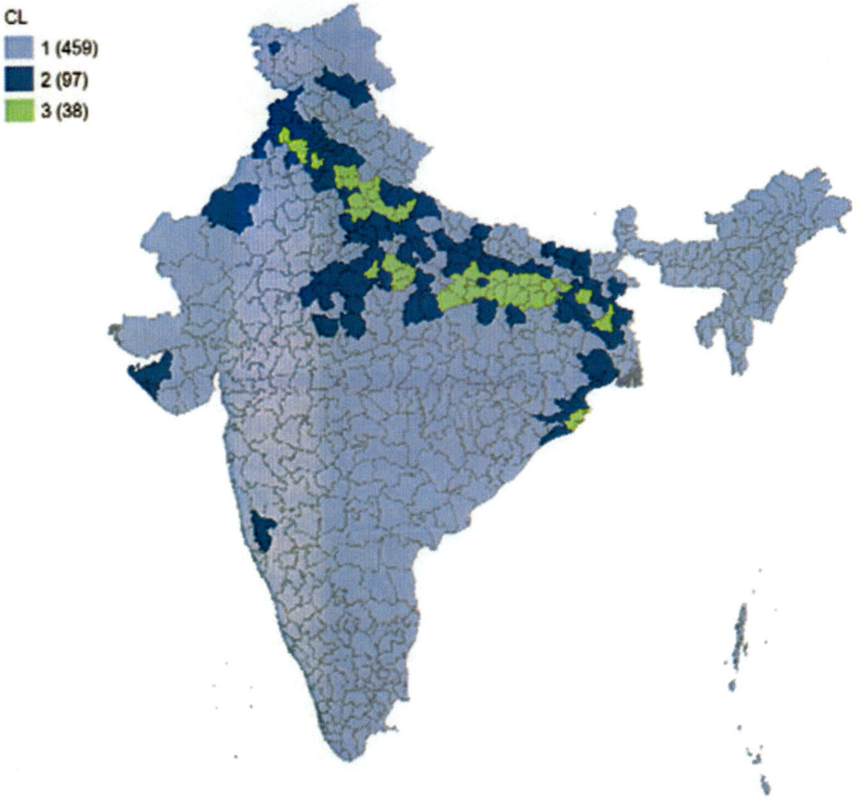

Fig. 10 Cow Dung cake usage cluster

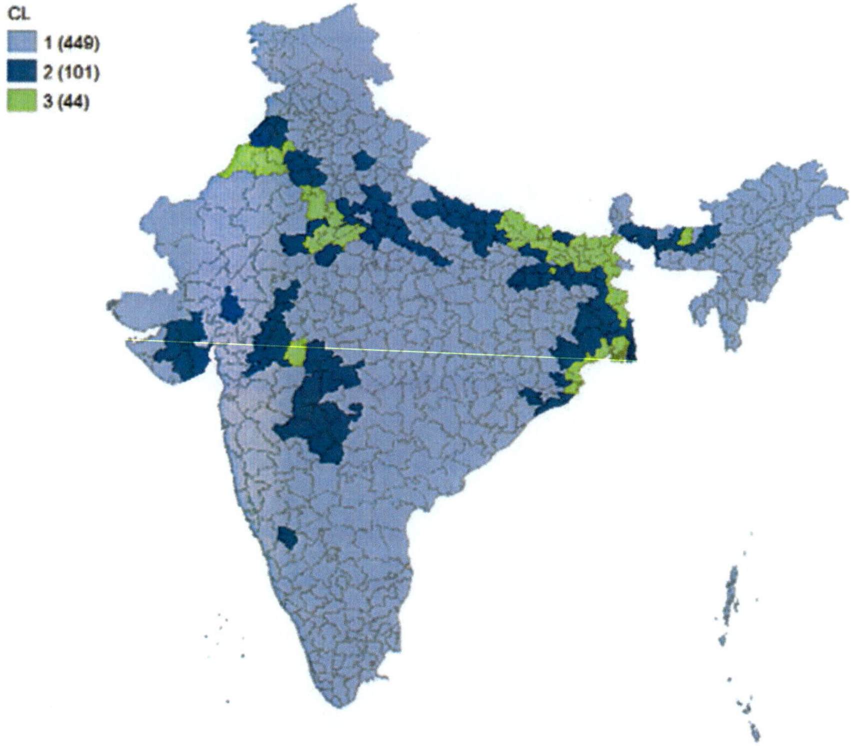

Fig. 11 Crop residue usage cluster

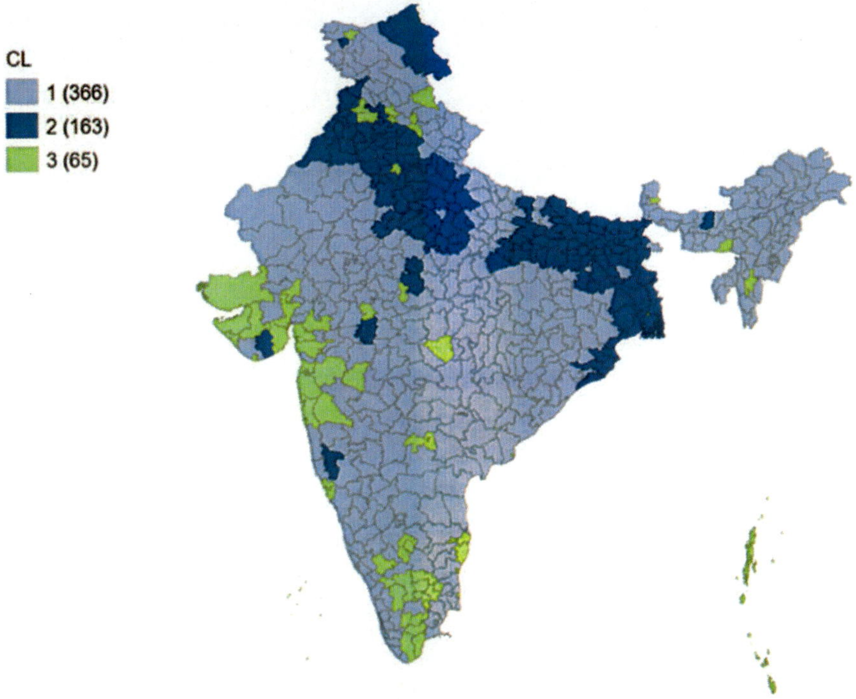

Fig. 12 Clusters of districts with highly polluted cooking fuels used

References

1. S. Shekar, P. Zhang, Y. Huang, *Spatial Data Mining: Data Mining and Knowledge Discovery Handbook* (Springer, Boston, 2005)
2. P. Haggett, A.D. Cliff, A. Frey, *Location Analysis in Human Geography 2: Locational Methods* (Wiley, New York, 1977)
3. H.D. Hosgood, Q. Lan, Indoor air pollution attributed to solid fuel use for heating and cooking and cancer risk. J. Earth Syst. Environ. Sci. (2011), pp. 633–635
4. D. Singh, S. Pachauri, H. Zerrifi, Environmental payoffs of LPG cooking in India. Environ. Res. Lett. **12**(11) (2017)
5. D.-P. Deng, T. Ruey Chuang, R. Lemmens, Conceptualization of place via spatial clustering and co–occurrence analysis. In: *International Workshop on Location Based Social Networks* (Seattle, USA, 2009)
6. M. Yang, J. Ma, P. Jia, Y. Pu, G. Chen, The use of spatial autocorrelation to analyze changes in spatial distribution patterns of population density in Jiangsu Province China. In: *International Conference on Geoinformatics, Shanghai China* (2011)
7. J. Duan, L. Wang, X. Hu, The effect of spatial autocorrelation on spatial co-location pattern mining. In: *International Conference on Computer, Information and Telecommunication Systems (CITS)* (Dalian China, 2017)
8. A.D. Cliff, J.K. Ord, *Spatial Autocorrelation* (Pion, London, 1973)
9. A.D. Cliff, J.K. Ord, *Spatial Processes: Models and Applications* (Pion, London, 1981)
10. W. Yuanfei, He. Honglin, *Spatial Data Analysis Method* (Science Press, Beijing, 2007)

11. L. Anselin, P.A. Longley, M.F. Goodchild, D.J. Maguire, Interactive techniques and exploratory spatial data analysis. In: *Geographical Information Systems, Principles, Technical Issues, Management Issues and Applications* (Wiley, 1999), pp. 253–266
12. L. Anselin, Local indicators of spatial association—LISA. Geogr. Anal. **27**, 93–115 (1995)
13. L. Anselin, *Spatial Econometrics: Methods and Models* (Kluwer Academic Publishers, Dordecht, 1998)
14. S. Bhuvaneshwari, S. Hettiarchchi, H. Meegoda, N. Jay, Crop residue burning in india: policy challenges and potential solutions. Int. J. Environ. Res. Public Health 16(5), 1–19 (2019)
15. S.K. Patel, S. Patel, A. Kumar, Effects of cooking fuel sources on the respiratory health of children: evidence from the annual health survey, Uttar Pradesh, India. J. Public Health (169), 59–68(2019)
16. J. Jung, M. Huxham, Firewood usage and indoor air pollution from traditional cooking fires in Gazi Bay, Kenya. Int. J. Student Res. (11) (2018)

An Interactive Approach of Rule Mining and Anomaly Detection for Internal Risks

Kun Liu, Yunkun Wu, Wenting Wei, Zhonghui Wang, Jiaqi Zhu, and Hongan Wang

Abstract How to prevent internal risks to the information system, especially for undefined risks, is a great challenge. A reasonable approach is to mine the behavior rules of internal staff on historical data through various data mining algorithms and then use the behavior rules to detect abnormal behaviors. However, in practice, risk control officers are often not familiar with data mining technologies, so it is hard to make them effectively choose and adapt these algorithms to find internal risks. In this paper, we propose an interactive approach for behavior rule mining and anomaly detection. Firstly, we express behavior rules and abnormal behaviors as complex events uniformly to accommodate different mining algorithms. Then, the internal staff's history behavior logs generated during production are used for mining behavior rules. Next, mined behavior rules are applied to new logs for anomaly detection. Finally, the detected abnormal behavior will be reported to the risk control officer for evaluation, and the feedback will be used for improving mining and detection settings to form a gradual and interactive process. The experiments on the real production

This work was supported by the National Key R&D Program of China (2018YFC0116703).

K. Liu (✉) · Y. Wu
University of Chinese Academy of Sciences, Beijing 100049, China
e-mail: liukun315@mails.ucas.ac.cn

K. Liu · Y. Wu · J. Zhu · H. Wang
Institute of Software Chinese Academy of Sciences, Beijing 100190, China
e-mail: zhujq@ios.ac.cn

W. Wei
China Development Bank, Beijing 100031, China

Z. Wang
State Grid Liaoning Electric Power Supply Co. Ltd., Shenyang 110006, China

S. M. Nigdeli et al. (eds.), *Proceedings of 6th International Conference on Harmony Search, Soft Computing and Applications*, Advances in Intelligent Systems and Computing 1275, https://doi.org/10.1007/978-981-15-8603-3_32

data show that the approach is effective and efficient to detect abnormal behavior and can be used to prevent internal risks of the information system of big corporations such as banks.

Keywords Internal risks · Behavior rule mining · Anomaly detection · Complex events

1 Introduction

The information system often faces various risks and threats, and organizations need to use various methods to strengthen the security of information systems [1]. The security threats of the information system are considered to be generated by four sources: external threats (human-made, non-human-made) and internal threats (human-made, non-human-made) [2]. Organizations use tools such as anti-virus software, firewalls, and intrusion detection methods to prevent external human-made threats, and establish disaster recovery centers [3] to prevent non-human-made threats such as natural disasters. For internal threats, the misbehavior of internal staff with legitimate identities can often cause worse damage [4] and is considered to be the most significant vulnerability [5] of information systems. To this end, organizations often formulate information security policies [6] to prevent internal risks and use information security awareness [7] to ensure the implementation of information security policies.

Nevertheless, these policies are based on identified risks, and they cannot prevent undefined risks [8]. Undefined risk is the type of risk that can hardly be identified in advance and impossible to detect directly. A reasonable method is to mine the normal behaviors of internal staff through data mining on historical data and detect internal staff's improper behaviors compared to the normal ones [9]. Due to the complexity of mining algorithms and different mining configuration requirements, the mining of behavior rules of internal staff and the detection of abnormal behaviors are usually performed by technicians who are proficient in data mining. Risk control officers of the big organizations in the industry who are concerned about the risks of internal staff have rich business experiences and understanding of security policies. They may not know much about the details of the mining algorithms. Moreover, with the development of the business, the behaviors of internal staff will also change, and people who are familiar with the business will be required to adjust the settings of mining and detection algorithms. Therefore, how to bridge the gap between business experience and mining technology, so that risk control officers can smoothly use multiple data mining algorithms to detect improper behaviors efficiently, has become a great challenge.

In this paper, we propose an interactive approach to rule mining and anomaly detection for internal risks. First, we unify the behavior of internal staff and establish a complex event representation model of behavior rules and abnormal behaviors. Then, we use classical behavior rule mining algorithms to discover the normal behavior rules of internal staff on historical behavior records. Next, corresponding detection algorithms are used to detect abnormal behaviors that deviate from the behavior rules. Finally, risk control officers evaluate behavior rules and abnormal behaviors, and the evaluation results are used as feedback to update the configuration settings of mining and detection algorithms.

Section 2 describes the existing methods for detecting internal threats. Section 3 introduces our approach. Section 4 shows our experimental results on actual operating data.

2 Related Work

For internal risks, researchers have proposed many countermeasures. For example, facing known risks, Li and Liu [10] used rule engines to detect violations of rules and regulations in behavior records. Hu et al. [11] compared Role-Based Access Control (RBAC) with security policies to find out differences between roles and processes for anomaly detection. In terms of undefined risks, machine learning and data mining techniques are wildly used [9]. For supervised machine learning, Lin et al. [12] focused on the analysis of multi-domain behavior patterns and used DBN (deep belief networks) to extract multi-domain behavioral features and identified risk behaviors. Meng et al. [13] used Long Short-Term Memory Recurrent Neural Network (LSTM-RNN) to classify insider behaviors and detected risk behaviors. With unsupervised data mining, Eberle et al. [14] used graph-based anomaly detection to represent normal and abnormal data as a graph to detect abnormal behaviors. Parveen et al. [15] combined graph-based anomaly detection with one class SVM for better performance. Chen et al. [16] used community discovery to analyze the access logs of the collaborative system and find individuals who have deviated from the community behavior. Mining will consider all deviations from the behavior rules as abnormal, which will make threat detection encounter a lot of false alarms. Therefore, it is necessary to apply multiple types of information combined with business experience to ensure the effectiveness of threat detection.

In order to mine the behavior rules of internal staff and detect abnormal behaviors, many existing data mining algorithms can be utilized and adapted. Frequent itemset mining [17] can identify frequently occurring behavior rules, and detect possibly risky behaviors such as occasional and significant changes through comparison in behavioral frequency. Frequent sequence mining [18] can represent normal behaviors that occur in a particular order and detect behaviors that are inconsistent in order. Association rule mining [19] can represent a set of behaviors with frequent associations and can detect situations where the correlations are missing. Rare sequence mining [20] can find behaviors that frequently occur for some individuals but are very rare overall. We used the above mining algorithms in this paper.

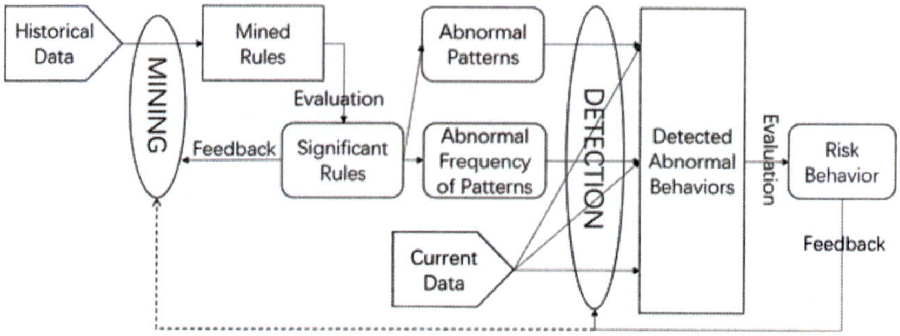

Fig. 1 The framework of interactive approach to rule mining and anomaly detection

3 Rule Mining and Anomaly Detection

The framework of our interactive approach to rule mining and anomaly detection for internal risks is shown in Fig 1.

First, the behavior rules of internal staff are mined through historical data. Then, the mined behavior rules are evaluated by risk control officers. The evaluation results are used as feedback to adjust behavior rule mining algorithms on the one hand and to select which behavior rules should be transformed for anomaly detection on the other. The transformed abnormal rules include two types: abnormal patterns and abnormal frequency of patterns. Next, the current data is detected to find abnormal behaviors in three ways: detect abnormal patterns, compare with the abnormal frequency of patterns, and mine abnormal patterns directly without comparing to any behavior rule. Finally, the detected abnormal behaviors are evaluated by risk control officers, and the evaluation results are used again as feedback to adjust the detection algorithms.

3.1 Event Representation of Behavior Rules and Abnormal Behaviors

Various internal control processes of modern organizations faithfully record the activities of internal staff, which are very helpful to analyze various behavioral characteristics of internal staff and detect potential threats through various methods. However, the content format recorded by various systems is different. For example, the access control system records the time and location information of internal staff, and the operation log records the internal staff's login and operation commands as well as operating parameters. Since all the information needs to be mined and detected together, we require a unified representation for these records. Borrowing the ideas from complex event presentation [21], we take the behavior records of internal personnel collected by each system as primitive events and express them as a quadruple

(Staff, Time, EventType, EventAttribute)

The event attributes of the access control system include whether the authorization was successful, the location of the access control device, the area to be controlled, etc.

If there are no emergencies, then there will be no significant deviation in the daily work of the stable operation organization. Although the internal staff's behavior rules for normal production operations are not necessarily written in the regulation of the organization, they usually reflect a normal and reasonable operational process. Behaviors that deviate from the rules of normal operating behaviors often mean that there are incidents or threatening actions, no matter what, it is worth paying attention to.

We classify the behavior rules and the corresponding abnormal rules for anomaly detection into two types:

- **Complex Event Patterns**. Similar to the regulations, the behavior of internal staff needs to be consistent with them, and any inconsistent behavior will be considered as abnormal behavior.
  The same as the detection of violation of regulations [10], we represent this as NormalPattern = (Premise, PremiseConstraint, Conclusion, ConclusionConstraint), and use Anomaly = (Premise $\wedge$ PremiseConstraint) $\wedge$ $\neg$(Conclusion $\wedge$ ConclusionConstraint) as a condition for anomaly detection.
- **Frequency of Patterns**. Such as the number of times an operation or a pattern occurs in a unit of time. The deviation of the behavioral frequency of internal staff will be considered as abnormal behavior.
  We represent this type as (NormalPattern, $Freq_{TimeUnit}$, Support), and use (Pattern, $Freq_{TimeUnit}$, FreqTolerance, Support, SupportTolerance) for abnormal behavior detection. $Freq_{TimeUnit}$ stands for the counts of Pattern occurring per time period, and Support stands for the percentage of transactions (sessions) where the pattern occurs.

In addition to the above two kinds of abnormal behaviors corresponding to the behavior rules, some mined patterns are anomaly themselves, and we are directly mined by the rare sequence mining algorithm.

3.2 *Behavior Rule Mining*

When mining behavior rules, risk control officers need to (1) determine the type of behavior rules to be mined and select the corresponding mining algorithm, (2) decide whether the execution method is one-time mining or periodic mining, (3) set appropriate mining parameters, (4) select the data range to be mined like data time range, internal staff range, event and event attribute range (for example, whether the mining of the operation logs focuses only on the operation commands themselves or

Table 1 Mining algorithms and mined behavior rules

Mining algorithm	Rule type	Rule name
Frequent sequence	Behavior pattern frequency patterns	Behavior sequence behavior sequence frequency
Frequent itemset	Frequency patterns	Behavior frequency
Association rule	Behavior pattern	Associated behavior

also includes command parameters). Different mining algorithms can mine different types of behavior rules, as shown in Table 1.

The following are some examples of mined behavior rules.

- Behavior sequence: Behaviors that occur in a particular sequences, and using it as a behavior rule means that any instance in a different order will be regarded as abnormal behavior. For example, system hardware maintenance needs three steps: shutdown, confirm shutdown, startup. This sequence must be followed, and the wrong order may cause a system failure. Behavior sequence will be represented like
  - Premise: shutdown $\wedge$ confirm $\wedge$ startup
  - PremiseConstraint: ((shutdown, confirm, startup) $\in$ TimeWindow) $\wedge$ (shutdown.time $<$ confirm.time)
  - Conclusion: null
  - ConclusionConstraint: shutdown.time $<$ confirm.time $<$ startup.time
- Behavior Sequence Frequency: Frequency of behaviors that occur in a certain sequence. Taking it as a behavior rule means that the frequency of any behavior in the same sequence should be within a reasonable interval. Taking the sequence above, for example, too rare or especially too frequent occurrences often means an anomaly. Behavior sequence frequency will be represented like
  - Pattern:
    - Premise: shutdown $\wedge$ confirm $\wedge$ startup
    - PremiseConstraint: ((shutdown, confirm, startup) $\in$ TimeWindow) $\wedge$ (shutdown.time $<$ confirm.time $<$ startup.time)
    - Conclusion: null
    - ConclusionConstraint: null
    - $\text{Freq}_{\text{TimeUnit}}$: OncePerDay
    - Support: 5%
- Behavior Frequency: Same as behavior sequence frequency, but the sequences is irrelevant.
  - Pattern:
    - Premise: shutdown $\wedge$ confirm $\wedge$ startup
    - PremiseConstraint: ((shutdown, confirm, startup) $\in$ TimeWindow)
    - Conclusion: null
    - ConclusionConstraint: null

- $\text{Freq}_{\text{TimeUnit}}$: OncePerDay
- Support: 5%

- Correlated Behavior: Conclusion behavior that should occur when the premise behavior occurs. The lack of correlation may imply abnormal behavior. For example, it may be abnormal to fail in starting up after shutting down in hardware maintenance. Correlated behavior will be represented like
  - Premise: shutdown $\wedge$ confirm
  - PremiseConstraint: (shutdown, confirm) $\in$ (TimeWindow) $\wedge$ (shutdown.time $<$ confirm.time)
  - Conclusion: startup
  - Constraints: (startup $\in$ TimeWindow) $\wedge$ (shutdown.Time $<$ confirm.time $<$ startup.time)

3.3 *Abnormal Behavior Detection*

When detecting abnormal behavior, different detection methods need to be applied according to different requirements. We have three different ways of detecting abnormal behavior:

1. To detect abnormal behavior through abnormal patterns. Similar to the detection method of regulations, the method of complex event detection will be used to detect any abnormal behavior that does not conform to the behavior rules but satisfy the corresponding abnormal rules.
2. To detect the abnormal behavior frequencies through the abnormal frequency of patterns. Risk control officers need to (1) determine the behavior frequency of which to detect, and select the corresponding algorithm, (2) set the execution method to be one-time or periodic, (3) set appropriate detection parameters such as frequency tolerance and support tolerance, (4) choose the range of objects to be detected, such as internal staff range and time range, but keep event and event attribute unchanged as behavior rules.
3. To perform abnormal behavior mining directly without relying on normal behavior rules. At present, we only use a rare sequence mining algorithm to detect the behavior characteristics of internal staff or a group of internal staff that are significantly different from the whole. The mining process can also be regarded as a detection process and is performed after fitting various settings according to requirements.

3.4 Result Evaluation and Feedback

Risk control officers at first evaluate the behavior rules obtained through data mining and mark whether they are reasonable and can be used for the detection of abnormal behavior. Then the IT technicians can adjust the mining configuration based on the feedback. Similarly, the risk control officer will also evaluate the detected abnormal behavior, determine whether further processing is required based on the severity of the abnormal behavior, and the evaluation results will be fed back to adjust the detection settings.

Specifically, the control officers rate each behavior pattern (from meaningless to very useful) and abnormal behavior (from entirely harmless to very dangerous) within a range from 0 to 10. The scoring results are used to calculate the effectiveness of the mining and detection settings. At present, we calculate the effect of different settings by calculating the average score over the mined rules or detected behaviors. Risk control officers are also allowed to directly modify the behavior rules based on experience and business needs, making them more in line with the actual situation of the organization. The different effectiveness obtained with different settings guides the risk control officer to choose a more suitable mining and detection setting.

The mining process and the detection process, as well as their evaluation, will be executed gradually and interactively shown in Algorithm 1. During the operation

Algorithm 1 Rule Mining, Anomaly Detection, and Feedback

$i \leftarrow 1$
repeat ▷ repeat during business
  $R_i \leftarrow$ `Mine`(m_i, MS_i, E_i, MO_i) ▷ mine behavior rules R_i using algorithm m_i with settings MS_i from events E_i and staff in MO_i.
  $RUS_i \leftarrow$ `EvaluateRule`(Ri) ▷ evaluate R_i and get a mining score RUS_i.
  $j \leftarrow 1$
  while need to detect **do** ▷ abnormal behavior detection
    $DR_{ij} \leftarrow$ `Select`(Ri), $DR_{ij} \subseteq R_i$ ▷ select rules DR_{ij} to detect.
    for all $r \in DR_{ij}$ **do**
      $A_{ijr} \leftarrow$ `Detect`$(d_{ijr}, DS_{ijr}, r, DO_{ijr})$ ▷ detect abnormal behaviors A_{ijr} through behavior rule r and detection algorithm d_{ijr} with settings DS_{ijr} from staff in DO_{ijr}.
      $AUS_{ijr} \leftarrow$ `EvaluateAnomaly`(A_{ijr}) ▷ evaluate A_{ijr} and get a detection score AUS_{ijr}.
    end for
    $DS_{i(j+1)r} \leftarrow$ `AdjustDetect`$(\{AUS_{ijr}\}, \forall(r))$ ▷ adjust next detection
    $j \leftarrow j+1$
  end while
  $MS_{(i+1)} \leftarrow$ `AdjustMine`$(\{RUS_i, AUS_{ijr}\}, \forall(r, j))$ ▷ detection scores AUS_{ijr} are combined with mining score RUS_i to adjust next mining.
  $i \leftarrow i+1$
until business ends

of organization, the behavior rules (R) of internal staff are mined (`Mine`) multiple times numbered by i. The risk control officer evaluates (`EvaluateRule`) each

Table 2 Summary of mined behavior rules using 2016 data

Mining algorithm	Number	Comments
Frequent itemset	35	Mostly about accessing certain tables or files
Frequent sequence	10	Stop a process, do something and then restart
Association rule	10	Some are the same as frequent sequence

set of behavior rules (R). And then, they select (`Select`) the behavior rules for detection. For selected behavior rules (DR), anomaly detection (`Detect`) will be performed multiple times numbered by j. The risk control officer also evaluates (`EvaluateAnomaly`) each set of the anomaly. The evaluation results (AUS) of abnormal behavior, on the one hand, are directly used to adjust the detection settings (`AdjustDetect`). On the other hand, they are combined with evaluation results (RUS) of behavior rules to improve the mining settings (`AdjustMine`).

4 Case Study

We applied our approach using real production data at a data center of a bank. We used some staff's 2016 annual production operation and maintenance operation logs (about 480,000 records after cleaning). As a critical step, we invited the risk control officer of the bank to evaluate the results of mining and detection. For behavior rules mining, we used the three algorithms as introduced in Sect. 3.2, frequent itemset, frequent sequence, and association rule to mine the behavior rules. The mined number and explanations of behavior rules are roughly shown in Table 2.

Here are some example of behavior rules:

- Frequent Sequence
  - Pattern = (shutdown, crsctl, startup), Support = 5.6%
  - Pattern = (stopapa, ps, startapa), Support = 7.6%
  - Pattern = (horcmshutdown.sh, horcmstart.sh), Support = 59.3%

  - Association Rule
  - Premise = (stopapa, ps), Conclusion = (startapa)
  - Premise = (su − grid, @sqlfile), Conclusion = (sqlplus)
  - sqlfile must be executed in sqlplus enviroment.

As we can see, these behavior rules focus on some meaningful command combinations, such as the stopped service will eventually be started, and some commands need to be executed in a specific environment. These behavior rules are not rigid security policies, but they are essential for safe production.

Table 3 Summary of detected abnormal behavior using 2016 data

Abnormal type	Algorithm	Number	Comments
Frequency of patterns	Frequent itemset	449	Frequency change in a certain month
Behavior patterns	Association rule	342	Missing correlated operation
Mined anomaly	Rare sequence	331	Rare in global, frequent in individuals

We performed the corresponding detection on all the mined rules. The detected number and explanations of abnormal behavior are roughly shown in Table 3, according to the types of abnormal behaviors, here are some example of the anomaly:

- Frequent Itemset:
  - File or table access frequency changed.
  - We detected a sudden increase in the frequency of 426 access for files and 23 for data tables. For example, a Global Cash Management (GCM) table was accessed by one staff 14 times in August compared with nearly no access in other months.
  - It is caused by a sudden change in operating frequency, especially when many abnormalities are concentrated in December, which is reasonable for a bank.
  - Association Rule:
    - missing start service command
    - For example, with behavior rule Premise = (stopapa, ps), Conclusion = (startapa), we detected a staff with 3 instances missing correlated operation startapa.
    - If the service is not started after the service is stopped, then there will be problems. After inspection, we found that there is no failure caused by no start of the service, maybe because of the time windows setting, and we consider this as feedback for adjusting further detection.
    - missing environment change command
    - For example, with behavior rule Premise = (su − grid, @sqlfile), Conclusion = (sqlplus), we detected another staff with 6 instances missing correlated operation sqlplus.
    - Executing commands in the wrong environment can be hazardous behaviors, such as forgetting to switch users or forgetting to enter the SQL environment. The investigation by the risk control officers revealed that such acts were mostly careless. Although they did not cause any failures, carelessness when operating in a production environment was a terrible sign, and risk control officers should take action to prevent this from happening.

We got much help from the risk control officers during the experiment. For example, when mining association rules for a specific group of internal staff, we initially set the threshold of minimum support to 0.01 and minimum confidence to 0.5. Such mining settings let us get a lot of meaningless behavior rules. They carefully checked the rules that mined, found some meaningful behavior rules, and introduced some business details in their daily production. After several rounds of optimization, we finally set the mining of association rules to thresholds with minimum support of 0.5 and minimum confidence of 0.85, and get an excellent result of behavior rules.

Due to the bank's strict management of internal staff and the effective implementation of security policies, we have not been able to detect threats from internal staff that have caused substantial risks. However, it is still significant to discover potential risks in internal staff; on the one hand, our approach complements the security policy with some feasible rules and regulations through the mining of behavior rules. On the other hand, we also detect some careless behaviors, although these behaviors did not cause any failure, for the production environment of banks, such careless behaviors should be corrected.

5 Conclusion

To effectively use multiple data mining algorithms to detect the risk behavior of internal staff, this paper proposes an interactive risk behavior mining and detection approach. This method represents the behavior of insiders as complex events, uses multiple mining and detection approach in combination, and utilizes risk control officers' feedback to adjust mining and detection settings. The experiments show that the behavior rules and abnormal behaviors of insiders can be effectively mined and detected, and the feedback mechanism can eventually improve the mining and detection results.

This paper pays much attention to how multiple algorithms can work together. However, the processing of feedback information, for now, remains very simple and worth further studying in the following aspects: the content of feedback is limited and straightforward, and the association between mining and detection is not considered in the feedback. In future work, we will focus on feedback processing, use more data collected during operation to establish a more suitable interactive feedback framework, and consider using machine learning and other methods to adjust mining and detection settings based on feedback results automatically. Our goal is that insider threat detection can be performed more effectively and automatically, so that the threats can be prevented before they happen.

References

1. E. Dubois, P. Heymans, N. Mayer, R. Matuleviius, A systematic approach to define the domain of information system security risk management, in *Intentional Perspectives on Information Systems Engineering* (Springer, Berlin, Heidelberg, 2010), pp. 289–306
2. K.D. Loch, H.H. Carr, M.E. Warkentin, Threats to information systems: today's reality, yesterday's understanding. Mis. Q. 173–186 (1992)
3. P. Fallara, Disaster recovery planning. IEEE potentials **23**(5), 42–44 (2004)
4. M. Warkentin, R. Willison, Behavioral and policy issues in information systems security: the insider threat. Eur. J. Inf. Syst. **18**(2), 101–105 (2009)
5. P. van Kessel, Cybersecurity regained: preparing to face cyber attacks. Ernst & Young Global Limited (2017). http://www.ey.com/Publication/vwLUAssets/ey-cybersecurity-regained-preparing-to-face-cyber-attacks/$FILE/ey-cybersecurity-regained-preparing-to-face-cyber-attacks.pdf
6. S. Barman, *Writing Information Security Policies* (New Riders Publishing, 2001)
7. S. Bauer, E.W. Bernroider, K. Chudzikowski, Prevention is better than cure! Designing information security awareness programs to overcome users' non-compliance with information security policies in banks. Comput. Secur. **68**, 145–159 (2017)
8. B. Von Solms, R. Von Solms, The 10 deadly sins of information security management. Comput. Secur. **23**(5), 371–376 (2004)
9. I.A. Gheyas, A.E. Abdallah, Detection and prediction of insider threats to cyber security: a systematic literature review and meta-analysis. Big Data Analytics **1**(1), 6 (2016)
10. Z. Li, K. Liu, An event based detection of internal threat to information system, in *The 5th International Conference on Harmony Search, Soft Computing and Applications (ICHSA 2019)* (Springer, 2019), pp. 44–53
11. N. Hu, P.G. Bradford, J. Liu, Applying role based access control and genetic algorithms to insider threat detection, in *ACM Southeast Regional Conference: Proceedings of the 44 th annual Southeast regional conference*, vol. 2006 (2006), pp. 790–791
12. L. Lin, S. Zhong, C. Jia, K. Chen, Insider threat detection based on deep belief network feature representation, in *2017 International Conference on Green Informatics (ICGI)* (IEEE, 2017), pp. 54–59
13. F. Meng, F. Lou, Y. Fu, Z. Tian, Deep learning based attribute classification insider threat detection for data security, in *2018 IEEE Third International Conference on Data Science in Cyberspace (DSC)* (IEEE, 2018), pp. 576–581
14. W. Eberle, J. Graves, L. Holder, Insider threat detection using a graph-based approach. J. Appl. Secur. Res. **6**(1), 32–81 (2010)
15. P. Parveen, N. Mcdaniel, Z. Weger, J. Evans, B. Thuraisingham, K. Hamlen, L. Khan, Evolving insider threat detection stream mining perspective. Int. J. Artif. Intell. Tools **22**(05), 1360013 (2013)
16. Y. Chen, S. Nyemba, B. Malin, Detecting anomalous insiders in collaborative information systems. IEEE Trans. Dependable Secure Comput. **9**(3), 332–344 (2012)
17. G. Grahne, J. Zhu, Fast algorithms for frequent itemset mining using fp-trees. IEEE Trans. Knowl. Data Eng. **17**(10), 1347–1362 (2005)
18. M.J. Zaki, SPADE: an efficient algorithm for mining frequent sequences. Mach. Learn. **42**(1–2), 31–60 (2001)
19. C. Zhang, S. Zhang, *Association Rule Mining: Models and Algorithms* (Springer, 2002)
20. J. Zhu, K. Wang, Y. Wu, Z. Hu, H. Wang, Mining user-aware rare sequential topic patterns in document streams. IEEE Trans. Knowl. Data Eng. **28**(7), 1790–1804 (2016)
21. D. Luckham, *The Power of Events*, vol. 204 (Addison-Wesley, Reading, 2002)

Wear Particles Classification Using Shape Features

Mohammad Shakeel Laghari, Ahmed Hassan, and Mubashir Noman

Abstract Wear debris provides important information about the machine condition that can be used to prevent the loss of expensive machinery. This information is crucial as it portrays the condition of the machines and can be used to predict early failure of the machinery that can prevent a major loss to the industry. Wear debris or particles produced in different parts of machine vary in shape, size, color, and texture. These characteristic features can be utilized to identify the type of wear debris. Human experts are extremely efficient in recognizing such objects; however, wear judgments are occasionally based on their specific perceptions. The goal is to look beyond the personal opinions and bring consistency in judging and recognizing wear particles. Keeping in view the above findings, this study focuses on the identification of wear particles by using shape-based features only. Different shape features, which include the Histogram of Oriented Gradients (HOG), Rotation Scale Translation (RST) invariant features, solidity, aspect ratio, circularity, and Euler number, are extracted and used to train three classification models of Support Vector Machine (SVM), k-Nearest Neighbors (kNN), and discriminant analysis (DA) classifier. The performance of the classifiers are compared with each another and classification of debris based on shape features is analyzed.

Keywords Wear debris · Particles classification · Shape-based features · Histogram of oriented gradients · RST invariant features

1 Introduction

Computer vision is one of the emerging fields that is influencing the automation industry since last few decades. It is assisting humans in a broad range of application areas including security, transportation, medical diagnostics, space science,

M. S. Laghari (✉) · A. Hassan · M. Noman
Department of Electrical Engineering, United Arab Emirates University, P. O. Box: 15551, Al Ain, United Arab Emirates
e-mail: mslaghari@aueu.ac.ae

S. M. Nigdeli et al. (eds.), *Proceedings of 6th International Conference on Harmony Search, Soft Computing and Applications*, Advances in Intelligent Systems and Computing 1275, https://doi.org/10.1007/978-981-15-8603-3_33

visual inspection systems, etc. Analysis of wear particles is one of the prime application areas of visual inspection systems that can provide essential information about machines condition. This information is crucial as it depicts the condition of the machines, which can be used to predict the early failure of the machinery to prevent the major loss to the industry.

Monitoring of machine wear is performed by using techniques such as ultrasound and X-rays. Wear particles are contained in the lubrication oil that can be segregated for analysis by using various methods. One of these techniques use glass slides to deposit the particles making it ready for viewing. Another approach makes use of different sized filters to separate wear particles. Another technique to separate wear debris from oil is *Ferrography*. In this technique, particles are separated and laid out on a transparent substrate according to their size [1]. In addition to the above-mentioned techniques, Magnetic Chip Detector (MCD) is another method used to extract particles. This approach uses magnetic plugs that are small removable units fitted with a powerful permanent magnet and are placed in appropriate positions in the machine. Metallic wear particles stick to the plugs that are spread out on a slide for analysis [2]. The separated particles are then identified by using high specification microscopes.

The characteristic parameters of wear particles play an important role in their identification. These parameters are categorized into six distinct attributes of shape, surface texture, color, edge details, size, and thickness ratio. A typical particle is classified and identified by using most of the six attributes; however, shape is one of the most important attributes that can be used for better classification results. Commonly used shape features are aspect ratio, circularity, rectangularity, convexity, Euler number, area, perimeter, etc.

Manual analysis of wear debris is very time consuming and tedious; therefore, automated analysis is very useful for the experts to make key decisions in reasonable times. Wear particles are classified in many types; however, in this investigation, only six types of cutting, fatigue, fiber, severe sliding, non-metallic, and water are included and analyzed. Each particle type possesses some distinguishable characteristics, e.g., cutting edge particles are elongated and curved, fiber particles have high aspect ratio whereas water particles are typically spherical and both of these particles may have holes in them, fatigue particles are thick and irregular shaped, etc. Shape parameters like aspect ratio, circularity, solidity, Euler number, Rotation Scale Translation invariant features, and Histogram of Oriented Gradients are used to classify the wear particles.

2 Related Work

Several methods exist in literature to analyze and recognize wear debris. Peng et al. have used deep neural network to identify fatigue, oxide, and spherical wear particles considering the case of overlapping particles. They have used transfer-learning approach by using Inception-v3 model to extract features and classify the particles.

The last few layers of Inception-v3 are replaced with one fully connected layer followed by the classification layer [3].

Li et al. have trained a single-layered feedforward neural network by using Extreme Learning Machine technique to classify the wear particles based upon the shape, texture, and color features [4].

Peng et al. have proposed a hybrid search-tree discriminant technique to discriminate oxide, fatigue, sliding, cutting and spherical wear particles. They have integrated multiclass Support Vector Data Description (SVDD), k-means and Support Vector Machine (SVM) to make a three-level search tree model classify the wear debris [5].

Wu et al. have proposed an online fluid monitoring scheme for particles analysis. They have used background subtraction and thresholding to segment the region of interest. The segmented particles are then analyzed from multi-view images by using the features like height aspect ratio, sphericity, spatial diameter, etc. [6].

Laghari et al. have proposed the Fourier descriptors based technique for the analysis of particles profile [7, 8].

Laghari et al. have used the shape and edge details of the wear particles to devise an interactive image analysis system [9].

Stachowiak et al. have worked on automated classification of abrasive, adhesive, and fatigue wear debris by using surface texture and shape features. They have used Discrete Wavelet Transform (DWT), statistical co-occurrence features to represent the surface texture. Finally, a linear SVM classifier is trained to classify these particles [10].

Peng et al. have monitored the machinery using wear debris and vibration analysis for fault diagnosis [11].

Yuan et al. proposed a radial concave deviation based method to determine feature parameters that are relevant to the size and shape of the wear particles [12]. They used linear discriminant analysis, quadratic discriminant analysis, naive bayes, and classification and regression trees for particles classification. They concluded that classification and regression trees classifier has better performance as compared to other classifiers by using features based on radial concave deviation method.

The rest of the paper is organized as follows. The next section discusses the methodology of the proposed scheme. In Sect. 4, experimentation results of the methodology are explained and Sect. 5 summarizes the conclusions.

3 Methodology

The proposed methodology is shown in Fig. 1, and is described as follows:

Object segmentation is the first step to extract its features; therefore, it is required to segment the particles in the images. A dataset of particles is prepared that contains grayscale images with good contrast between foreground and background regions. Consequently, simple global thresholding can be used to segment these particles from the background. Although, thresholding separates the particles agreeably well; however, it also produces some noise in the form of tiny dots. This noise occurring in

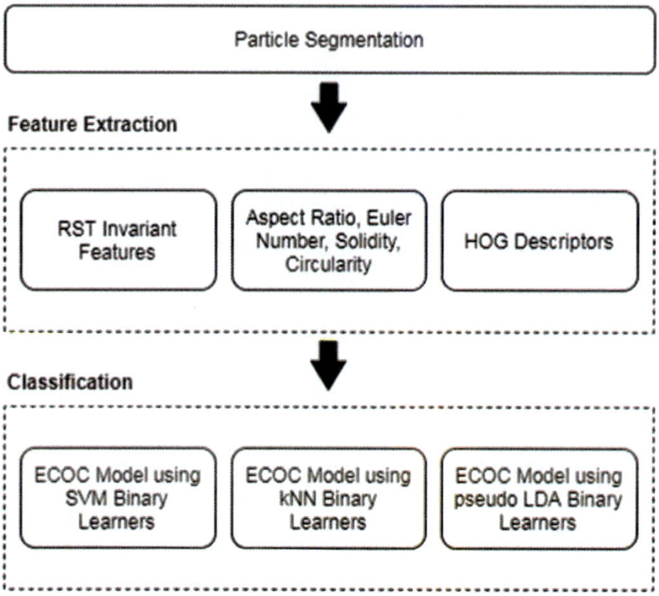

Fig. 1 Block diagram of methodology

some images is due to blurred edges along the boundary region and low resolution of some images. The noise is typically removed by finding the connected components in the images and retain only the largest connected component.

The following step involves computing the particle's shape features. The earlier mentioned four features are selected because they are one of the notable descriptors to characterize and describe the shape of an object. Similarly, HOG features are also significant that can be used to describe the structural shape of an object for detection purposes. HOG technique counts occurrences of gradient orientation in restricted portions of an image. To compute the RST invariant features, it is required to calculate the central moments and hence it is computed as:

$$\mu_{ij} = \Sigma_x \Sigma_y \left(x - x'\right)^i \left(y - y'\right)^j I(x, y) \quad (1)$$

where x and y represent the coordinates of the pixel, and $I(x, y)$ represents the intensity of the pixel. The central moments of the image are only translation invariant.

They are also required to be made rotation and scale invariant. The scale invariance can be achieved as below.

$$\eta_{\mathrm{ij}} = \mu_{\mathrm{ij}} / \mu_{00}^{(1+(\mathrm{i}+\mathrm{j})/2)} \quad (2)$$

After achieving the scale invariance, central moments can be made rotation invariant by using the equations as below [13]:

$$I_1 = \eta_{20} + \eta_{02} \tag{3}$$

$$I_2 = (\eta_{20} - \eta_{02})^2 + 4\eta_{11}^2 \tag{4}$$

$$I_2 = (\eta_{20} - \eta_{02})^2 + 4\eta_{11}^2 \tag{5}$$

$$I_3 = (\eta_{30} - 3\eta_{12})^2 + (3\eta_{21} - \eta_{03})^2 \tag{5}$$

$$I_4 = (\eta_{30} + \eta_{12})^2 + (\eta_{21} + \eta_{03})^2 \tag{6}$$

$$\begin{aligned} I_5 = {} & (\eta_{30} - 3\eta_{12})(\eta_{30} + \eta_{12})\left[(\eta_{30} + \eta_{12})^2 - 3(\eta_{21} + \eta_{03})^2\right] \\ & + (3\eta_{21} - \eta_{03})(\eta_{21} + \eta_{03})\left[3(\eta_{30} + \eta_{12})^2 - (\eta_{21} + \eta_{03})^2\right] \end{aligned} \tag{7}$$

$$I_6 = (\eta_{20} - \eta_{02})\left[(\eta_{30} + \eta_{12})^2 - (\eta_{21} + \eta_{03})^2\right] + 4\eta_{11}(\eta_{30} + \eta_{12})(\eta_{21} + \eta_{03}) \tag{8}$$

$$\begin{aligned} I_7 = {} & (3\eta_{21} - \eta_{03})(\eta_{30} + \eta_{12})\left[(\eta_{30} + \eta_{12})^2 - 3(\eta_{21} + \eta_{03})^2\right] \\ & - \eta_{30} - 3\eta_{12})(\eta_{21} + \eta_{03})\left[3(\eta_{30} + \eta_{12})^2 - (\eta_{21} + \eta_{03})^2\right] \end{aligned} \tag{9}$$

The next task involves calculating the shape feature like aspect ratio, Euler number, etc. To calculate aspect ratio of the particle, minor and major axis length is computed, and their ratio is used as the aspect ratio. Another shape feature is circularity that is given as

$$\text{Circularity} = (4 \times \text{pi} \times \text{area})/\left(\text{perimeter}^2\right) \tag{10}$$

Therefore, area and outer perimeter of the particle is estimated to compute circularity. Euler number is found by estimating the number of holes in the particle and subtracting the holes count from a value one. Solidity is given as the ratio of area of the particle and convex area of the particle; therefore, the convex area is required to be calculated first, which can be computed by finding the convex hull of the particle and then calculating its area.

HOG features are based upon local intensity gradient orientation that makes them good shape descriptors. Before computing HOG features, segmented particles are aligned on the major axis to avoid the effect of rotation. The alignment of the particles is necessary because rotated particles can have different gradient orientations that may result in dissimilar descriptor vector, which can affect the accuracy of the classifier. After aligning particles, image gradients are computed in the horizontal and vertical directions. Then, magnitude and phase of the gradients are calculated as

$$\text{Magnitude} = \left(S_x^2 + S_y^2\right)^{0.5} \tag{11}$$

$$\text{theta} = \text{arc tan}\left(S_y / S_x\right) \tag{12}$$

As default parameters are used for computing HOG, therefore, images are divided into blocks of size 16 × 16. Each block contains four 8 × 8 cells. Orientation binning is then performed by using nine evenly spaced unsigned bins. Each block is normalized and arranged into a vector. Finally, descriptor vector is made by concatenating all the block descriptors into a single vector.

After computing features of the particles, a multiclass Error-Correcting Output Codes (ECOC) model is trained by using linear SVM, k-Nearest Neighbors and pseudo linear Discriminant Analysis (DA) binary learners. SVM is used due to the optimal margin gap between the separating hyperplanes that give more robustness against outliers. Similarly, kNN can provide reasonable accuracy as it groups similar objects quite well. The measure of separation is also good for discriminant analysis based classifier. The coding scheme used for ECOC model is one versus one and the multiclass ECOC model predicts the output class as

$$k' = \arg\min_{k}[(\Sigma_{l=1\,to\,L}|m_{kl}|\,g(m_{kl},\ s_l))/(\Sigma_{l=1\,to\,L}|m_{kl}|)] \tag{13}$$

For linear SVM learners, ECOC model uses Hinge loss function whereas quadratic loss function is used for kNN and pseudo linear DA learners.

$$\text{Hinge loss} = \max(0,\ 1 - y_i s_i) \tag{14}$$

$$\text{Quadratic loss} = (1 - y_i(2s_i - 1))^2/2 \tag{15}$$

4 Experimentation Results

The software tools used for the experimentation are Matlab 2015a, and Leica's Quantimet Computer Vision and Image Processing tools.

The detail of the dataset and experimentation results are described below.

4.1 Dataset

The dataset used in this paper contains six types of wear particles including severe sliding, cutting edge, non-metallic, fatigue, water, and fiber particles. This dataset only contains shape information of the wear particles. Some images of the dataset are shown in Fig. 2.

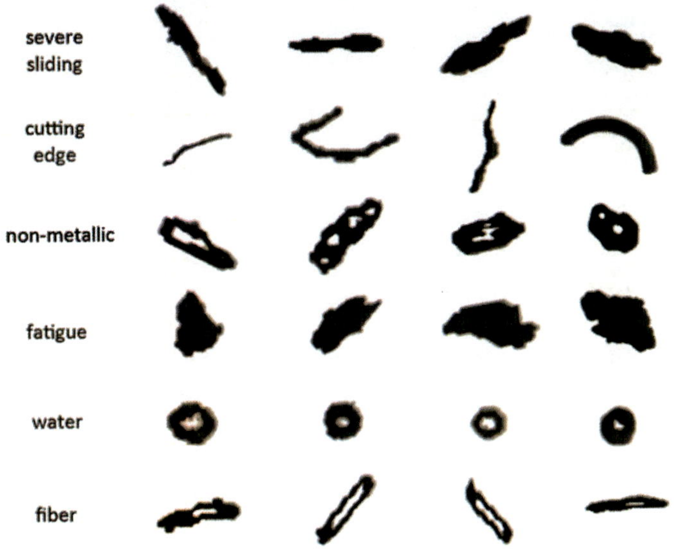

Fig. 2 Dataset images

4.2 *Testing Procedure*

Different combinations of features and classifiers are used to test the results of the proposed method. The performance of different feature sets and classifiers is evaluated by using cross-validation accuracy. Firstly, five different feature sets are made, i.e., first feature set includes only RST invariant features, second feature set includes only HOG features, third feature set consists of shape parameters (aspect ratio, solidity, Euler number, circularity), fourth feature set contains combination of shape parameters and RST features, and fifth feature set is composed of shape parameters and HOG features.

After making five feature sets, three models of multiclass ECOC classifier are trained by using linear SVM, kNN, and pseudo linear DA binary learners, respectively. The performance of the classifiers is determined by using the classification error and accuracy. The trained models are cross-validated by using tenfold cross-validation scheme to estimate the accuracy of the classifier. Then, k-fold loss is estimated from the cross-validated model. The k-fold loss corresponds to the average classification error in all the folds, i.e., the proportion of observations that are misclassified by the classifier. Finally, the accuracy of each model is obtained by subtracting the k-fold loss from a value of '1' and the results are shown in Table 1.

The table shows that ECOC model trained by using the third feature set (shape parameters only) has reasonable accuracy whereas other feature sets have low accuracy. It is also observed that when RST and HOG features are combined with the shape parameters, the accuracy of the classifier is decreased. The accuracy becomes considerably low when shape parameters are combined with HOG features. One of

Table 1 Results

Feature set	Accuracy %		
	Linear SVM	KNN	Pseudo linear DA
RST invariant features	22.2	16.7	33.3
HOG	22.2	27.8	16.7
Shape parameters	94.4	94.4	61.1
Shape parameters + RST features	83.3	88.9	61.1
Shape parameters + HOG	27.8	44.4	16.7

the reasons is that the HOG features comprises a major portion of the fifth feature set and the classification error of HOG features is high which results in almost similar performance to the second feature set. The high classification error of HOG is because wear particles have many variations in the shape within the same class that is affecting the local features like HOG. Although, HOG is a good descriptor to represent the structural shape of an object but in the case of wear particles, there is a lack of shape similarity due to the irregularity in the boundary of the particles. This irregularity between the boundaries of the particles within same class results in different orientation histograms. As a result, classifier is unable to generalize the shape similarity of the particles. Conversely, shape parameters are depicting the overall trend of the shape, i.e., elongation, convexity, number of holes, etc. These global properties of the shapes have insignificant variability; therefore, shape parameters are better in classifying the wear particles.

On the other hand, linear SVM and kNN have almost similar results whereas models based on linear DA learners exhibit very low accuracy. The reason for this low accuracy is that the linear DA assumes that all groups are identically distributed and by using all the data to estimate the covariance matrices that can be prone to outliers. In contrast, SVM and kNN do not make any assumptions about the data and give relatively better accuracy.

5 Conclusion

In this paper, automated classification of six types of wear particles is presented. It is observed that shape parameters, i.e., aspect ratio, solidity, circularity, etc., are good features to categorize wear debris when images lack texture and color features. It is also observed that shape features that are based on local intensity gradients do not perform well with wear debris classification. The reason may be that wear particles have many variations in the shape within the same class whereas parameters like aspect ratio, solidity, etc., can generalize well with the overall shape of the particles.

References

1. E.R. Bowen, D. Scott, W. Seifert, V.C. Westcott, Ferrography. Int. J. Tribol. **6**, 109–115 (1976)
2. A.C. Cumming, Condition monitoring today and tomorrow—an airline perspective. In: *1st International Conference COMADEN*, vol. 89 (Birmingham, UK, 1989)
3. P. Peng, J. Wang, Wear particle classification considering particle overlapping. Wear **422–423**, 119–127 (2019)
4. Q. Li, T. Zhao, L. Zhang, W. Sun, X. Zhao, Ferrography wear particles image recognition based on extreme learning machine. Electr. Comput. Eng. 2017, Article ID 3451358 (2017)
5. Y. Peng, T. Wu, G. Cao, S. Huang, H. Wu, N. Kwok, Z. Peng, A hybrid search-tree discriminant technique for multivariate wear debris classification. Wear **393–393**, 152–158 (2017)
6. T. Wu, Y. Peng, S. Wang, F. Chen, N. Kwok, Z. Peng, Morphological feature extraction based on multiview images for wear debris analysis in on-line fluid monitoring. Tribol. Trans. **60**(3), 408–418 (2016)
7. M.S. Laghari, T.R. Soomro, G.A. Khuwaja, The use of Fourier Descriptors to recognize particle profile. In: *4th IEEE International Conference on Modeling, Simulation, and Applied Optimization* (Kuala Lumpur, 2011), pp. 1–6
8. M.S. Laghari, Wear particle profile analysis by using Fourier Analyses. In: *5th IEEE International Workshop on Signal Processing and its Applications* (Sharjah, 2008)
9. M.S. Laghari, F. Ahmed, J. Aziz, Wear particle shape & edge detail analysis. In: *2nd IEEE International Conference on Computer and Automation Engineering* (Singapore, 2010), pp. 122–125
10. G.P. Stachowiak, G.W. Stachowiak, P. Podsiadlo, Automated classification of wear particles based on their surface texture and shape features. Tribol. Int. **41**, 34–43 (2008)
11. M.K. Hu, Visual pattern recognition by moment invariants. Proc. IRE Trans. Inf. Theory **8**(2), 179–187 (1962)
12. W. Yuan, K.S. Chin, M. Hua, G. Dong, C. Wang, Shape classification of wear particles by image boundary analysis using machine learning algorithms. Mech. Syst. Signal Process. **72–73**, 346–358 (2016)
13. R.C. Gonzalez, R.E. Woods, *Digital Image Processing*, 4th edn. (Pearson, London, 2018)

Smart Academic Guidance for Institutional Students

Mohammad Shakeel Laghari, Ahmed Dirir, and Mubashir Noman

Abstract Smart time conflict solvers and event time managers are some of the essential tools for educational institutions that can help students to attend events, which might time conflict with a lecture or other activities to gain advantage of the opportunities enriching their professional benefits. Normally, an institutional timetable may encounter certain time conflicts that remain undiscovered until the time of students' registration. Therefore, a smart and an innovative approach in terms of a software package is implemented and described in this paper. The developed software uses smart and efficient searching methods for achieving an optimum time match between any number or group of students and under any specified constraints. Consequently, the devised event organizer can find an optimum time to allocate for the event by finding an optimum time at which the majority of the intended people are free for the event. In addition, students who struggle with time conflicts between two or more courses can also find the most suitable times in which their time conflicts are resolved. The software searches for classes by locating the time and courses intersections between all the students who are associated with such conflicts. These time conflict processes are difficult to handle manually and may be prone to errors.

Keywords Time conflict solver · Software package · Educational institution · Time table · Event management

1 Introduction

Timetabling is an essential process for an educational organization, particularly in universities and institutions, due to scheduling of large number of time-based events and lectures to be planned between great numbers of participants. Usually, such organizations resort to external services to handle these issues. Three related online

M. S. Laghari (✉) · A. Dirir · M. Noman
Department of Electrical Engineering, United Arab Emirates University, P. O. Box: 15551, Al Ain, United Arab Emirates
e-mail: mslaghari@uaeu.ac.ae

S. M. Nigdeli et al. (eds.), *Proceedings of 6th International Conference on Harmony Search, Soft Computing and Applications*, Advances in Intelligent Systems and Computing 1275, https://doi.org/10.1007/978-981-15-8603-3_34

services of DegreeWorks, Jadawil, and Google form are currently in use at the United Arab Emirates University (UAEU).

DegreeWorks is a web-based service that allows students to create study plans with the help of their academic advisors. This process confirms the course planners to indicate the exact number of students who will register specific courses in all their planned semesters. However, it does not provide an automated scheduling procedure and a class-time conflict solver. Hence, a student has to resolve time conflicts manually in this service, which is both time consuming and difficult to perform [1].

Jadawil is another service that is currently in use by UAEU students. It is a website that was developed by an Electrical Engineering alumnus. It provides a reasonable number of suitable time arrangements for any given set of courses. This is useful when there are many course selection choices for a student and helps in deciding which set of courses to consider. However, it does not communicate the students' course data to one another and neither does it provide solutions to the time conflicts for the events. Hence, the students are not much benefited from this kind of service [2].

A third tool in use for event time scheduling by the University is Google Form. It is a service that allows collection of course data from the participants. Data collection is performed by sending a form link in which they are asked to register for a certain event or workshop by specifying their available times. After the data is collected, the organizers manually plan appropriate time slots to suit the majority of students [3].

The use of Google form seems logical, however, it has many flaws. Primarily, asking the students to click on a link to fill in the necessary information greatly decreases the potential number of interested students. Moreover, in the case of large numbers of students' registrations, the organizers find it difficult to choose ideal times based on the vast amount of data provided because it becomes too laborious to sort such issues. A more frequent scenario for the students is that multiple events may overlap with course lectures that cannot be avoided. As a result, the students may loose a chance to attend many important events that could benefit them in their professional carrier.

2 Survey of Existing Solutions and Their Limitations

2.1 DegreeWorks

DegreeWorks is an online service provided to the student to create study plans. The main page of DegreeWorks is shown in Fig. 1. It allows academic advisors to follow up with the courses, study plans, and other academic records associated with their students. Moreover, students can easily keep record of the remaining courses required to graduate and hence, it provides better and more reliable study plans. Students can review any of their previously prepared plans because DegreeWorks keeps a record of all plans in a database. It also provides the assigned time for any

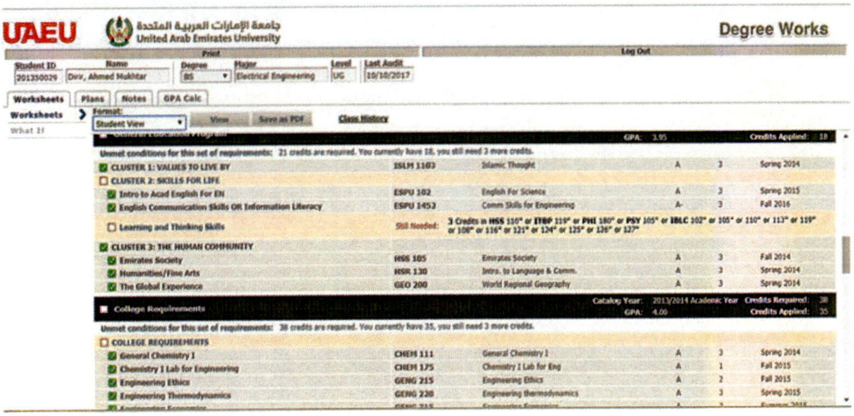

Fig. 1 UAEU DegreeWorks homepage

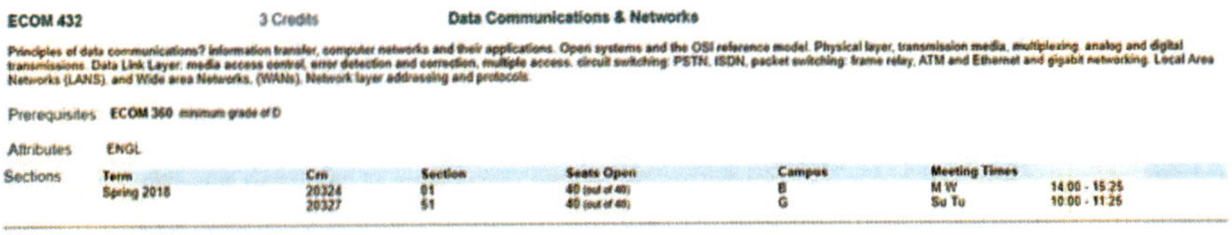

Fig. 2 Course details as shown from DegreeWorks

course with all its sections details. Therefore, it is easier to find out all the required courses and their offering times for the next semester course offering. The main disadvantage of DegreeWorks is that a student has to rearrange courses manually especially, whenever there is a time conflict between two or more courses. Although DegreeWorks can manage the student's courses in general, it does not provide any tool to manage the student's classes conflicts. Figure 2 shows the course details as shown by DegreeWorks.

2.2 *Jadawil*

Jadawil is a website that intends to show all suitable time arrangements for any given set of courses. The interfaces shown in Figs. 3 and 4, lists courses supplied by the student, and the corresponding website gives all the possible arrangements. The disadvantage of this service is that it takes into account only the set of given courses; hence, any attempt to solve a time conflict between two or more students cannot be achieved as the website does not know what courses are selected by other students. Jadawil does not give alternatives if certain time conflicts between two or

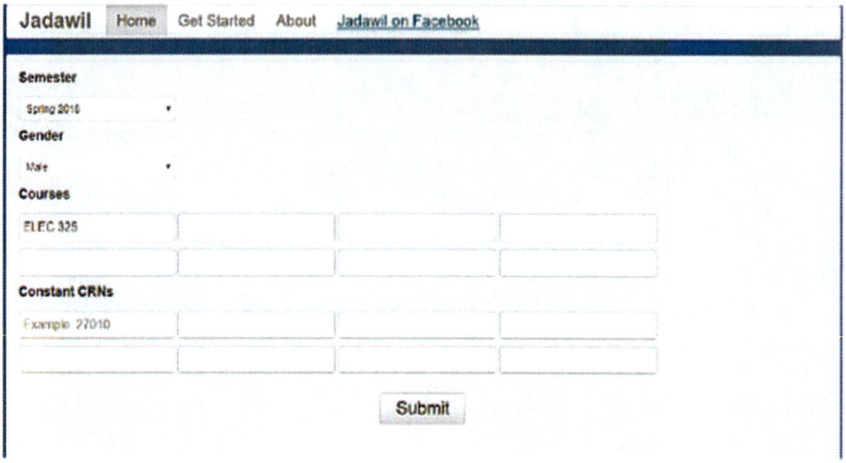

Fig. 3 Jadawil main interface

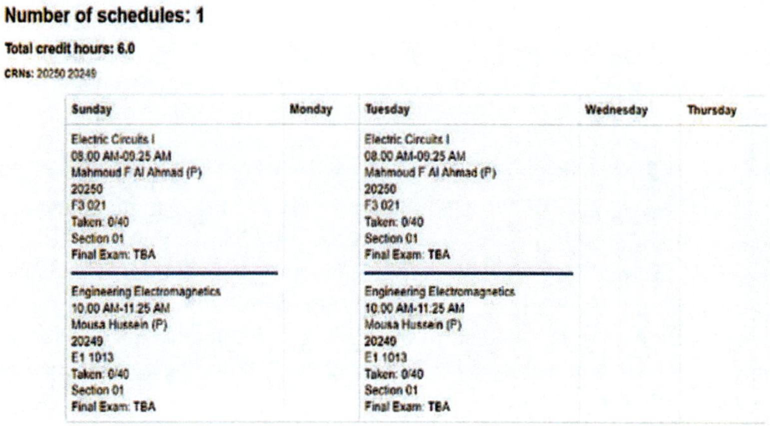

Fig. 4 Jadawil created schedule interface

more courses given in a specific set is encountered. Therefore, the student has to change the set of courses every time until no time conflicts are achieved.

2.3 *Google Form*

The Students' Activities Department (SAD) frequently uses this Google service. Since SAD has no records of student's courses or timetables for a specific semester, they cannot help them in events' organization. A form link is sent to the students to fill in their available times, and, then the SAD manually sorts through the received data to find out the best time suitable to most students. The two disadvantages here are that not enough number of students fill the form, and this technique cannot be applied to large numbers of students since finding the paramount time is extremely difficult. The *Tutorial Center* at the university is also using this technique to sort out the courses needed for tutorial purposes.

3 Related Work

Le Kang and George M. White proposed an approach based on logic programming to resolve the constraints in timetabling [4].

Victor A. Bardadym discussed the problems related to academic timetabling and proposed a classification for timetabling requirements, solution methods, mathematical models, data representation, and interface design [5].

Elmohamed et al. compared the annealing techniques to tackle the problem of academic class scheduling and concluded that best results were obtained by using simulated annealing with adaptive cooling and reheating as a function of cost [6].

S. M. Selim proposed the modified Almond's algorithm to construct the university faculty timetable [7].

S. Kadry and B. Ghazal described a solution for exam scheduling by using Maximal Independent Set, Multi Criteria, and Heuristic Graph [8].

Lopez et al. proposed a system that automatically generates course assignment for teachers and class scheduling by using mathematical modeling and data mining techniques [9].

V. J. Kadam and S. S. Yadav briefly described the problems related to academic timetable scheduling in [10].

J. E. Guo and H. X. Zhang presented a special laboratory timetable algorithm and conflict solving method in [11].

Dufournaud et al. addressed the conflict issue in timetabling and proposed a solution that uses a utility function to resolve the scheduling conflicts [12]. F. Guo and H. Song introduced a reinforcement learning algorithm-based timetable-scheduling model in [13].

4 Methodology and Framework

4.1 Bitwise Operation

In computer science, a bitwise operation operates on one or more bit patterns at the level of their individual bits. It is directly supported by the processor and operates fast to manipulate values for comparisons and calculations.

A bitwise AND takes two equal-length binary numbers and performs the logical AND operation on each pair of the corresponding bits, by multiplying them. Thus, if both bits in the compared position are 1, the bit in the resulting binary representation is 1 ($1 \times 1 = 1$); otherwise, the result is 0 ($1 \times 0 = 0$, $0 \times 1 = 0$, and $0 \times 0 = 0$). For example:

$$0111(\text{decimal } 7)$$
$$\text{AND}0011(\text{decimal } 3)$$
$$= 0011(\text{decimal } 3)$$

This is regularly called bit masking. The bitwise AND may be used to clear selected bits (or flags) of a register in which each bit represents an individual Boolean state. This technique is an efficient way to store a number of Boolean values using as smaller memory as possible.

4.2 Bitwise AND Operation Implementation for Student Data Storage

Database management is important to any software that has multiple fields and many levels of data. It is also dependent on the way the database is organized, and on the implication of the data type to the operation of the software. The main test in this project was to figure out how the students free times and schedules are saved and optimized in such a way that it required a minimum number of processes to visualize or compare to other free times.

The process starts by dividing the 24 h day to 30 min time slots resulting in 48-time slots array to represent different time periods within a day. Starting from the most significant digit (left most digit) or time slot, this array of times is filled with '0's or '1's depending on the intended task. Storing a '1' in the time slot represents student's free times, whereas a '0' corresponds to a busy or scheduled time slot. The most significant slot corresponds to 00:00; the next one corresponds to 00:30 and so on. For instance, if only the most significant time slot is filled with a '1', and the rest are filled with '0's, then the student is said to be free from 00:00 to 00:30 (12:00 am to 12:30 am).

The next step is to convert this array of '0' and '1' to its integer number equivalent in such a way that the integer number accepts a typical 32-bit binary number. Therefore, the morning period of a 24-h whole day is divided to start at 00:00 am and to end up at 12:00 noon, which corresponds to the time slots from the most significant slot up to the 24th time slot (moving from most significant toward the least significant). The evening period starts at 12:00 noon and ends up at 00:00 am. Notice that starting at 8:00 am in the morning is more convenient; however, it is found that starting at 00:00 makes the process of converting the user input representing the free time to binary number is easier. The conversion process (Fig. 5) takes place as follows:

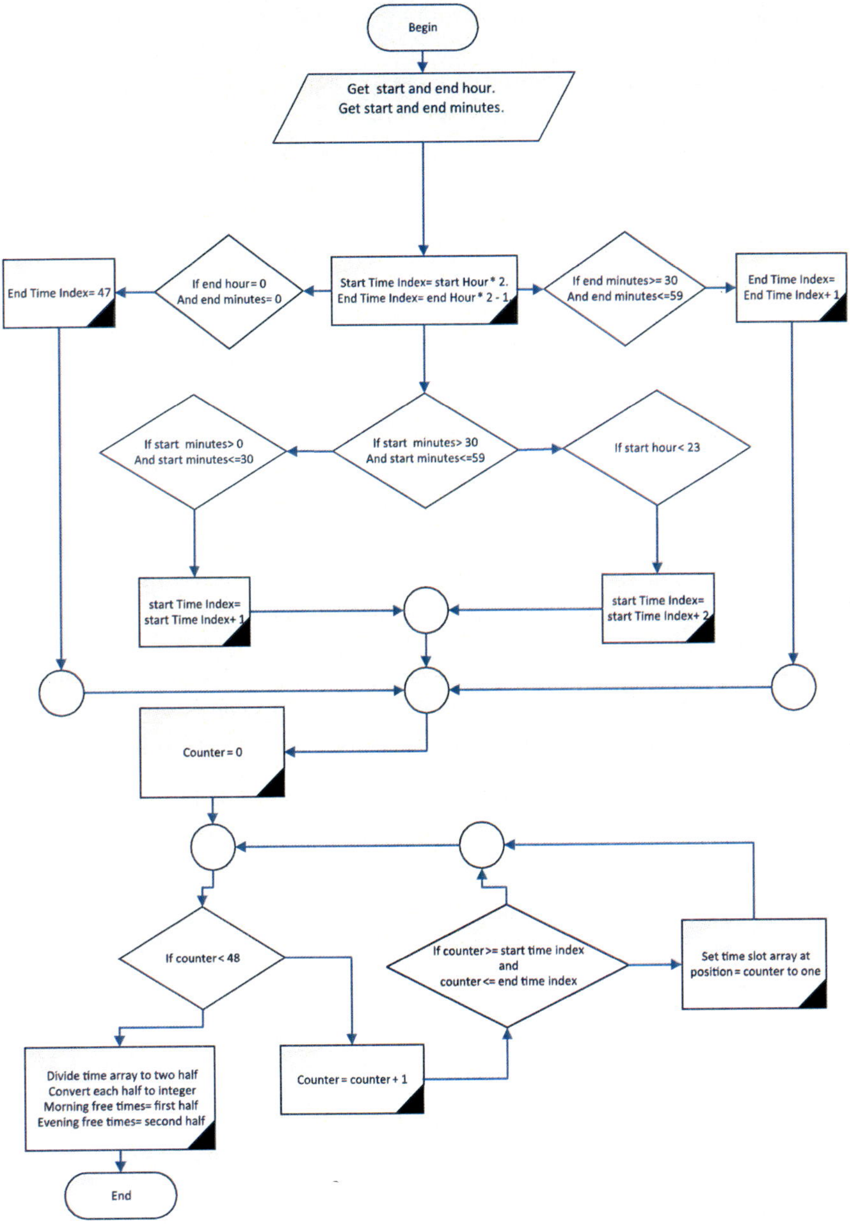

Fig. 5 Flow chart of the conversion algorithm

1. Get the start time and end time from the user. The format is (hour: minutes).
2. The first time slot corresponds to 00:00, from an array point of view, the index of this time slot is zero. So, if starting hour is multiplied by two, the exact start index is known in the times array. So, for any starting hour with minutes = 00. The start index = start hour x 2. Alternatively, if the minutes are between 1 and 30, then it is assumed that the student is unavailable for this time slot. A '1' is added to the calculated start index, and the start index = start hour x 2 + 1. If the starting minutes is somewhat between 31 and 59. Then, a 2 is added to start time so the next hour is taken as a starting point.
3. For the end time index, it is the same except that a '1' is subtracted from the calculated index. For instance, if the student decides that the end time is 10:48, then the actual end time to the system is 10:30, because an assumption is made from the beginning that the student is either available or unavailable for the whole time slot (30 min.). Similarly, if the student end minutes are between 1 and 29, then the end time is the starting of the same hour, but with minutes = 0, nothing is added.

Ex: If student start time is 08:39 and end time is 10:17.

Start index = 8 x 2 = 16, add 2 so start index = 18 which corresponds to 09:00.

End index = 10 x 2 = 20. Nothing is added, so end index = 20 which corresponds to 10:00. If end time is 10:42, end index = 10 x 2 = 20, then add 1 to make end index = 21 which represents 10:30. All the time slots in the times array which falls between the start index and end index are associated as '1', and others as '0's. So for starting time of 08:39 and end time of 10:17.

000000000000000000110000000000000000000000000000

Morning free times = 000000000000000000110000 = 48.

4.3 *Student Information Interface*

The second phase of the project involves designing a simple graphical user interface (GUI) so that student data can be entered in a more convenient way. The software inputs the student name, ID, email address, mobile number, college, and department. The same GUI is used to get instructor data. An example of the data set to be used to prove the concept of timetabling consists of nine courses, which are Calculus I, ESPUI, Chemistry, Physics, Biology, Finance, Economics, Marketing, and Supply Chain. There are five available instructors, and their names, courses and classes times are shown in Fig. 6. There are 15 students and their names and courses are shown in Table 1, consequently, their classes can be sorted out by matching Table 1 and Fig. 6. The GUI is shown in Fig. 7. Keep in mind, that the student free times are stored in the system; other times are assumed to be occupied either by courses or by other activities. The created database is shown in Fig. 8.

Week Schedule (From Sunday to Thursday)

30/10/2017	31/10/2017	01/11/2017	02/11/2017	03/11/2017
Chemistry from8 to 10 (Yasser). Calculus from8 to 10 (Omar). Marketing from8 to 10 (Yazan). Economics from10 to 12 (Yazan) ESPU I from12 to 2 (Mike).	Physics from8 to 10 (Adel). Supply chain from8 to 10 (Yazan). Finance from10 to 12 (Yazan). Biology from10 to 12 (Yasser).	Chemistry from8 to 10 (Yasser). Calculus from8 to 10 (Omar). Marketing from8 to 10 (Yazan). Economics from10 to 12 (Yazan) ESPU I from12 to 2 (Mike).	Physics from8 to 10 (Adel). Supply chain from8 to 10 (Yazan). Finance from10 to 12 (Yazan). Biology from10 to 12 (Yasser).	No classes

Fig. 6 Weekly schedule for the courses

Table 1 Student names and associated courses

Student names	Courses
Ahmed	Calculus I, ESPUI
Ali	ESPUI
Anas	ESPUI, Economics
Bassel	Calculus I, ESPUI
Bashaar	Chemistry, Physics
Baraa	Chemistry
Dirir	Calculus I, ESPUI, Physics
Fahad	Calculus I, ESPUI
Falah	Calculus I
Faisal	Physics, Biology
Hamad	Chemistry, Biology
Humaid	Physics, Biology
Jassem	Finance, Economics, Marketing, Supply Chain
Mohammed	ESPUI, Calculus I
Mubarek	Finance, Economics, Marketing, Supply Chain

5 Applications to Real-World Problems

5.1 *Event Organization*

All over the world, institutions typically organize several events across the academic year, which are mainly students-based events with students as the major audience. As mentioned earlier, the UAEU uses Google form to ask the students for their available times, and then a manual matching process takes place. This manual process can be automated in such a way that an optimal event timing can be found; hence, attendance to the events is maximized. The algorithm to do with this automatic matching is as follows:

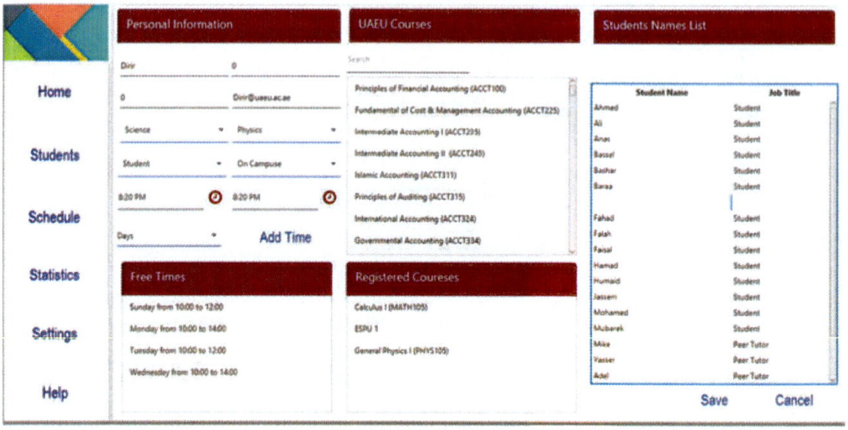

Fig. 7 Registered student (Dirir) data is shown

	StudentName	StudentID	MobileNumber	Email	College	Department	Locations	JobTitle	Courses
1	Ahmed	0	0	Ahmed@uaeu.ac.ae	Science	Mathematical Sciences	On Campuse	Student	{Calculus I (MATH105)}{ESPU 1}
2	Ali	0	0	Ali@uaeu.ac.ae	Business and Economics	Accounting	On Campuse	Student	{ESPU 1}
3	Anas	0	0	Anas@uaeu.ac.ae	Business and Economics	Economics and Finance	On Campuse	Student	{ESPU 1}{Principles of Economics (ECON110)}
4	Bassel	0	0	Bassel@uaeu.ac.ae	Science	Mathematical Sciences	On Campuse	Student	{ESPU 1}{Calculus I (MATH105)}
5	Bashar	0	0	Bashar@uaeu.ac.ae	Science	Physics	On Campuse	Student	{General Chemistry I (CHEM111)}{General Physics I (PHYS105)}
6	Baraa	0	0	Baraa@uaeu.ac.ae	Science	Chemistry	On Campuse	Student	{General Chemistry I (CHEM111)}
7	Dirir	0	0	Dirir@uaeu.ac.ae	Science	Physics	On Campuse	Student	{Calculus I (MATH105)}{ESPU 1}{General Physics I (PHYS105)}
8	Fahad	0	0	Fahad@uaeu.ac.ae	Science	Mathematical Sciences	On Campuse	Student	{Calculus I (MATH105)}{ESPU 1}
9	Falah	0	0	Falah@uaeu.ac.ae	Science	Mathematical Sciences	On Campuse	Student	{Calculus I (MATH105)}
10	Faisal	0	0	Faisal@uaeu.ac.ae	Science	Biology	On Campuse	Student	{General Physics I (PHYS105)}{Basic Biology I (BIOC100)}
11	Hamad	0	0	Hamad@uaeu.ac.ae	Science	Biology	On Campuse	Student	{Basic Biology I (BIOC100)}{General Chemistry I (CHEM111)}{General Physics I (PHYS105)}
12	Humaid	0	0	Humaid@uaeu.ac.ae	Science	Biology	On Campuse	Student	{Basic Biology I (BIOC100)}{General Physics I (PHYS105)}
13	Jassem	0	0	Jassem@uaeu.ac.ae	Business and Economics	Economics and Finance	On Campuse	Student	{Principles of Economics (ECON110)}{Principles of Financial Accounting (ACCT100)}{Principles of Mark
14	Mohamed	0	0	Mohamed@uaeu.ac.ae	Science	Mathematical Sciences	On Campuse	Student	{Calculus I (MATH105)}{ESPU 1}
15	Mubarek	0	0	Mubarek@uaeu.ac.ae	Business and Economics	Economics and Finance	On Campuse	Student	{Principles of Economics (ECON110)}{Principles of Financial Accounting (ACCT100)}{Supply Chain Mana
16	Mike	0	0	Mike@uaeu.ac.ae	Business and Economics	Accounting	On Campuse	Peer Tutor	{ESPU 1}
17	Yasser	0	0	Yasser@uaeu.ac.ae	Science	Chemistry	On Campuse	Peer Tutor	{Basic Biology I (BIOC100)}{General Chemistry I (CHEM111)}
18	Adel	0	0	Adel@uaeu.ac.ae	Science	Physics	On Campuse	Peer Tutor	{General Physics I (PHYS105)}
19	Yazan	0	0	Yazan@uaeu.ac.ae	Business and Economics	Economics and Finance	On Campuse	Peer Tutor	{Principles of Marketing (MKTG200)}{Supply Chain Management & Operations (SCML200)}{Principles of I
20	Omar	0	0	Omar@uaeu.ac.ae	Science	Mathematical Sciences	On Campuse	Peer Tutor	{Calculus I (MATH105)}

	SundayMorning	SundayEvening	MondayMorning	MondayEvening	TuesdayMorning	TuesdayEvening	WednesdayMorning	WednesdayEvening	ThursdayMorning	ThursdayEvening
1	15	0	255	15728640	15	0	255	15728640	0	0
2	255	0	255	15728640	255	0	255	15728640	0	0
3	240	0	255	15728640	240	0	255	15728640	0	0
4	15	0	255	15728640	15	0	255	15728640	0	0
5	15	15728640	15	15728640	15	15728640	15	15728640	0	0
6	15	15728640	255	15728640	15	15728640	255	15728640	0	0
7	15	0	15	15728640	15	0	15	15728640	0	0
8	15	0	255	15728640	15	0	255	15728640	0	0
9	15	15728640	255	15728640	15	15728640	255	15728640	0	0
10	255	15728640	0	15728640	255	15728640	0	15728640	0	0
11	15	15728640	0	15728640	15	15728640	0	15728640	0	0
12	255	15728640	0	15728640	255	15728640	0	15728640	0	0
13	0	15728640	240	0	240	0	240	0	0	0
14	15	0	255	15728640	15	0	255	15728640	0	0
15	0	15728640	0	15728640	0	15728640	0	15728640	0	0
16	15	0	255	0	15	0	255	0	0	0
17	15	15728640	240	0	15	15728640	240	0	0	0
18	255	15728640	15	15728640	255	15728640	15	15728640	0	0
19	0	15728640	0	15728640	0	15728640	0	15728640	0	0
20	15	15728640	255	15728640	15	15728640	255	15728640	0	0

Fig. 8 Student's data as stores in the database

1. Identify the group of students to find the best time match for them.
2. Specify the time constraints for the search (starting time and ending time and day).
3. Apply the filter.

The software retrieves all the specified student's free times, and then do a bitwise operation with the integer number resulting from the conversion of the specified period for search.

"SELECT * FROM StudentInformation WHERE SundayMorning & " + SpecifiedTimePeriod + " > 0 AND College = '" + college + "' AND JobTitle = 'Student'".

This statement returns the students who are free in the specified time. Only one statement is needed to find out the results quickly and accurately. This algorithm can be applied to any time-matching processes. The only remaining thing is to add more constraints to find out the optimum time for any given set of students. The algorithm is tested on three cases.

1. Based on Table 1, the best time for science students is to search from all science students who are taking one of these courses, Calculus, ESPU, Biology, Physics, and Chemistry. If the search for the best times is for Sunday from 8:00 am to 12:00 noon, then the searched results from 10:00 am to 12:00 noon. As shown in Fig. 9, Segment 2, all science students are listed starting with Ahmed.

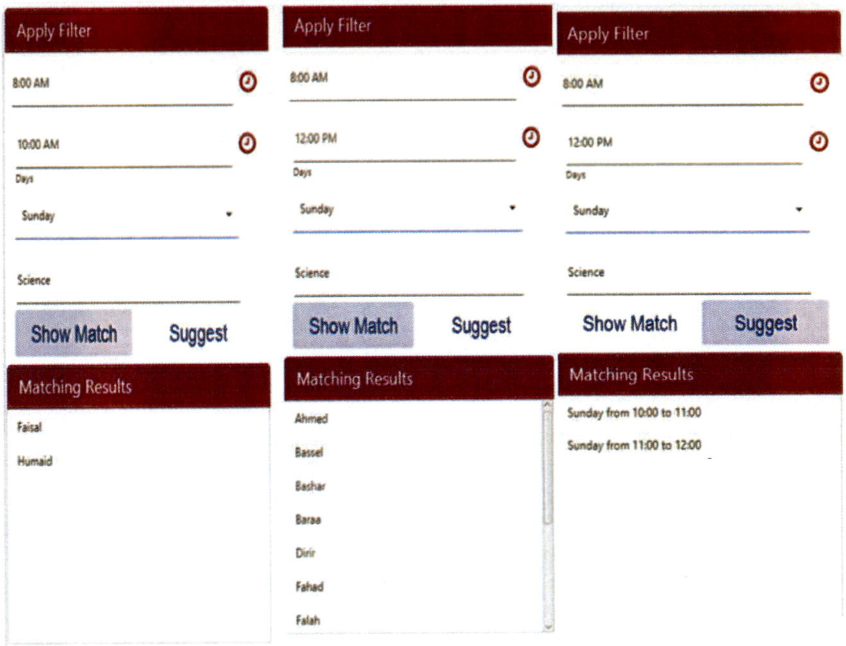

Fig. 9 Test cases for optimal time matching

2. If the time constraint is changed to be from 8:00 to 10:00 am, then only two students show up in the results (Faisal and Humaid). As shown from Table. 1, these two are the only available students within the specified time period. Figure 9, Segment 1, is showing the results of this search.
3. For the given time period which is from 8:00 to 10:00 am and selecting only science students, the best times is shown in Fig. 9, Segment 3, which is Sunday from 10:00 am to 12:00 noon.

5.2 *Time Conflict Alternatives*

A time conflict arises when a Biology student is interested in taking a Finance class. Assume that the Finance class cannot be changed; therefore, the Biology class time needs to be changed. In order to do this and since Biology students are taking Chemistry and Physics, and similarly Physics students are taking ESP and Calculus as well, the class times of Chemistry, Physics, Calculus, and ESPU are all taken into consideration. The proposed system seeks for all the courses taken by the students registered in the Biology class, and the system tries to find out the free time where all the students of these courses are free.

Chemistry, Calculus, Biology and Physics are called student-intersected courses. The search result is presented in Fig. 10. As expected, the system finds out that 10:00 am to 12:00 noon on Sunday is the best time alternative for the Biology class.

6 Conclusion

This paper details the use of an innovative and adaptive event organizer that can help institutional in particularly university students to attend events that conflict with their lectures or other activity timings in order to enrich their professional benefits. It summarizes the use of efficient searching methods for achieving an optimum time match between any group of students and under any specified constraints in order to avoid time conflict processes that are both challenging to handle manually and prone to errors. The results of this research could assist the universities in such issues.

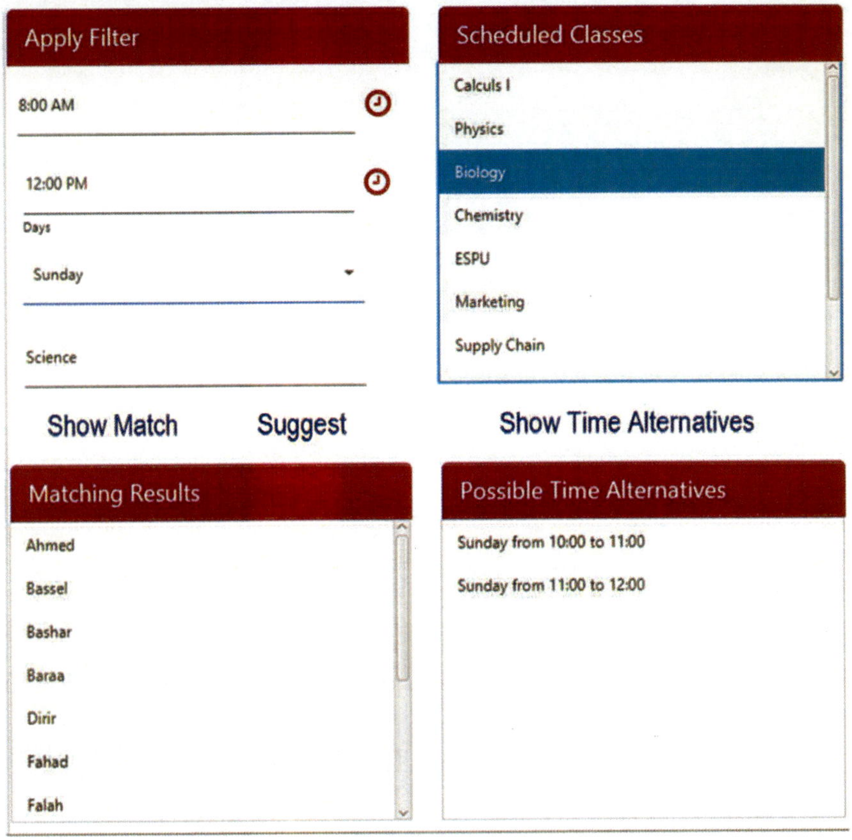

Fig. 10 Time alternative for Biology class

References

1. DEGREE WORKS, https://it.stonybrook.edu/services/degree-works
2. JADAWIL, https://jadawil.assil.me/
3. GOOGLE FORMS, https://docs.google.com/forms/d/e/1FAIpQLSegJ1bgM0UbTjrp9Yt-BqSCvqn7ecE-zKfbg1V_Gnq8q2jf9w/viewform
4. L. Kang, G.M. White, A logic approach to the resolution of constraints in timetabling. Eur. J. Oper. Res. **61**(3), 306–317 (1992)
5. V.A. Bardadym, Computer-aided school and university timetabling: the new wave, in *Practice and Theory of Automated Timetabling PATAT 1995, LNCS*, eds. by E. Burke, P. Ross, vol. 1153 (Springer, Berlin, Heidelberg, 1996)
6. M.A.S. Elmohamed, P. Coddington, G. Fox, A comparison of annealing techniques for academic course scheduling, in *Practice and Theory of Automated Timetabling II PATAT 1997, LNCS*, eds. E. Burke, M. Carter, vol. 1408 (Springer, Berlin, Heidelberg, 1998)
7. S.M. Selim, An algorithm for constructing a university faculty timetable. Comput. Educ. **6**(4), 323–332 (1982)

8. S. Kadry, B. Ghazal, New proposed design to solve examination timetable problem, in *3rd International Proceedings on Modern Trends in Science, Engineering and Technology* (Dubai, 2015)
9. D. Calle-López, J. Cornejo-Reyes, F. Pesántez-Avilés, V. Robles-Bykbaev, M. Rodas-Tobar, C. Vásquez-Vásquez, A university administration system to automatically assign courses to teachers and support the design of timetables through mathematical modeling and restrictions analysis, in *Proceedings on IEEE World Engineering Education Conference* (IEEE Xplore, Buenos Aires, 2018), pp. 1–5
10. V. Kadam, S. Yadav, Academic timetable scheduling: revisited. Int. J. Res. Sci. Eng. April/2016, 417–423 (2016)
11. J.E. Guo, H.X. Zhang, The research of special laboratory timetable algorithm and solving conflicting method. Appl. Mech. Mater. **220–223**, 3064–3067 (2012)
12. C.M. Dufournaud, G. McBoyle, J.T. Quinn, Harrington, schedule conflict resolution in a small faculty, in *IEEE International Conference on Systems, Man and Cybernetics 2003, SMC'03 Conference Proceedings* (Washington, DC, 2003), pp. 4401–4406
13. F. Guo, H. Song, Research and application of data-mining technique in timetable scheduling, in *2nd International Conference on Computer Engineering and Technology* (Chengdu, 2010), pp. V1-409–V1-412

Grey Wolf Optimizer with Crossover and Opposition-Based Learning

Shitu Singh and Jagdish Chand Bansal

Abstract Grey wolf optimizer (GWO) is a relatively new optimizer in the field of swarm intelligence. It is based on the leadership hierarchy and hunting behavior of grey wolves in nature. Due to less number of parameters and ease of implementation, it has gained significant interest among the researchers of different fields. However, in some cases, the insufficient diversity of wolves prone it toward the local optima and degrades the performance of GWO. To resolve this issue, a new Grey wolf optimizer with crossover and opposition-based Learning (GWO-XOBL) is proposed. In the GWO-XOBL, two strategies have been introduced to get a proper balance between exploration and exploitation. First, the crossover is employed to increase the population diversity of GWO. Then, opposition-based learning is applied to improve the convergence rate and to avoid the stagnation. To investigate the effectiveness of the proposed algorithm, it is tested on 13 well-known standard benchmark problems. Statistical tests of the numerical results demonstrate that the performance of the proposed GWO-XOBL is significantly better than GWO and other compared nature-inspired algorithms.

Keywords Grey wolf optimizer · Uniform crossover · Opposition-based learning · Global optimization

1 Introduction

Nature-inspired algorithms are the most popular strategies that are used to solve complex real-life optimization problems. Swarm intelligence (SI) is one of the interesting nature-inspired algorithms. It mimics the collective and cooperative behavior of social creatures. Swarm-based optimization techniques find a solution by a collab-

S. Singh · J. Chand Bansal (✉)
Department of Mathematics South Asian University, New Delhi, India
e-mail: jcbansal@sau.ac.in

S. Singh
e-mail: singh.shitu93@gmail.com

S. M. Nigdeli et al. (eds.), *Proceedings of 6th International Conference on Harmony Search, Soft Computing and Applications*, Advances in Intelligent Systems and Computing 1275, https://doi.org/10.1007/978-981-15-8603-3_35

orative hit and trial method. Some popular techniques based on swarm intelligence are particle swarm optimization (PSO) [1], ant colony optimization (ACO) [2], spider monkey optimization (SMO) [3], and grey wolf optimizer (GWO) [4], etc. The GWO algorithm is introduced by Mirjalili et al. [4] in 2014 which simulates the leadership behavior and hunting procedure of grey wolves pack. The GWO search process, like other SI-based algorithms, starts with randomly generated population of grey wolves. Each wolf represents a candidate solution and is updated through the searching process. The pack's members in GWO are divided into four types of the wolves: alpha, beta, delta, and omega, where the alpha represents the first-best solution, beta is a second-best solution, delta is the third-best solution, and remaining solutions are omega. In GWO, the omega wolves update their positions with the help of alpha, beta, and delta. GWO has been applied to solve various real-life applications for example GWO and DE have been used for the single and multi-objective optimal flow problem [5], surface waves analysis [6], multi-layer perceptron (MLP) [7], feature selection in intrusion detection system [8], the binary GWO wrapper approach was applied in [9] for cancer classification on gene expression data, path planning tasks [10], and GWO has been hybridized with crossover and mutation to solve four problems of economic dispatch [11], in which GWO shows its potential as on very efficient algorithm. In the literature, researchers have some attempts to prevent from stagnate to local optima. For example—In [12], an improved version of GWO is proposed to enhance the exploration ability. In [13], opposition-based learning is integrated into GWO to reduce the problem of stagnation at local optima. There are many other modifications that have been done in the literature [14–17] to improve the performance of GWO algorithm.

In this paper an improved variant of GWO, Grey wolf optimizer with crossover and opposition-based Learning (GWO-XOBL) is proposed with the help of two strategies, known as uniform crossover [18] and opposition-based Learning [19]. The crossover is employed to increase the population diversity of GWO and opposition-based learning is applied to improve the convergence rate and to avoid the stagnation so that an appropriate balance between exploration and exploitation can be established. In order to verify the performance of the GWO-XOBL, 13 well-known benchmark functions are considered as test problems.

The organization of rest of the paper is as follow: Section 2 describes brief overview of the basic GWO. In Sect. 3, the proposed GWO-XOBL is discussed in detail. The experimentation and validation of the proposed algorithm on benchmark test problems are performed in Sect. 4 and the conclusions of the work is presented in Sect. 5.

2 An Overview of Grey Wolf Optimizer Algorithm

This section briefly presents the mathematical model of the GWO algorithm.

2.1 *Social Behavior*

In order to mathematically model the social behavior of wolves, the best solutions are considered as alpha (α), second best as beta (β), and third-best solution as delta (δ) while rest of the solutions are assumed as omega (ω) solutions. The (ω) wolves (solutions) are iteratively improved by α wolf (solution), β wolf (solution), and δ wolf (solution).

2.2 *Encircling Prey*

When the wolves do hunting, they tend to encircle their prey. The mathematical form of encircling behavior of grey wolves is given as follows:

$$Y^{(t+1)} = Y^t_{\text{prey}} - P \cdot Q \tag{1}$$

$$P = 2 \cdot \mu \cdot r_1 - \mu \tag{2}$$

$$Q = |b \cdot Y^t_{\text{prey}} - Y^t| \tag{3}$$

$$b = 2 \cdot r_2 \tag{4}$$

where Y^t_{prey} is the position of the prey (α, β, and δ wolves) at tth iteration, $Y^{(t+1)}$ and Y^t are the position of the wolf at $(t+1)$th and tth iteration, respectively. P and b are the coefficient vectors, r_1 and r_2 are random numbers in [0,1]. μ is a linearly decreasing vector from 2 to 0 over iterations and expressed as

$$\mu = 2 - 2\left(\frac{t}{T}\right) \tag{5}$$

T denotes the maximum number of iterations.

2.3 Hunting

In the hunting process of wolves, α, β, and δ are the three best solutions so far saved then the other search agents (ω wolves) update their positions according to the current best position. These situations are expressed in the following mathematical equation:

$$Y^{(t+1)} = \frac{Y_1^t + Y_2^t + Y_3^t}{3} \tag{6}$$

where Y_1, Y_2, and Y_3 are calculated as

$$Y_1^t = Y_\alpha^t - P_1 \cdot Q_\alpha \tag{7}$$

$$Y_2^t = Y_\beta^t - P_2 \cdot Q_\beta \tag{8}$$

$$Y_3^t = Y_\delta^t - P_3 \cdot Q_\delta \tag{9}$$

where

$$Q_\alpha = |b_1 \cdot Y_\alpha^t - Y^t|$$

$$Q_\beta = |b_2 \cdot Y_\beta^t - Y^t|$$

$$Q_\delta = |b_3 \cdot Y_\delta^t - Y^t|$$

and Y_α^t, Y_β^t, and Y_δ^t are the positions of α, β, and δ wolves at tth iteration. b_1, b_2 and b_3 denotes the random vectors in the range $[0, 2]$ and Y^t represent the position of the grey wolf at tth iteration.

3 Proposed Method

In the nature-inspired algorithms, exploration and exploitation are two conflicting features and an appropriate balance is needed between them for better performance of the algorithm. In the GWO, it has been observed that all the individuals attract toward the α, β, and δ leading wolves. This nature sometimes creates the problem of insufficient exploration and, therefore, the algorithm may converge prematurely. In order to overcome this problem, in this paper a new GWO called GWO-XOBL is proposed. In this algorithm, the uniform crossover is used to increase the population diversity, and the OBL is used to improve the convergence rate and to avoid the stagnation at local optima. The new solution $Z^{(t+1)}$ is obtained as explained in the following crossover operator.

$$Z^{(t+1)} = \begin{cases} Y^{(t+1)}, & (\text{rand} \leq \text{pcr} \quad \text{or} \quad j = j_{\text{rand}}) \\ X^t, & \text{otherwise.} \end{cases} \tag{10}$$

where pcr is the crossover rate (pcr) and is fixed to 0.5 in the proposed algorithm. rand is the uniformly distributed random number in [0, 1]. j_{rand} is a randomly chosen index in the range [0, dim], dim is the dimension of the problem. After the uniform crossover, a greedy search approach given by Eq. (11) decides that whether the obtained vector Z will survive in the next iteration or not.

$$X^{(t+1)} = \begin{cases} Z^{(t+1)}, & f(Z^{(t+1)}) \leq f(X^{(t)}) \\ X^t, & \text{otherwise} \end{cases} \tag{11}$$

where f is the objective function. Equation (11) is described for the minimization problem which can be maximized if the problem is maximization. Moreover, to improve the convergence rate and to avoid the stagnation we have applied opposition-based learning on there best solutions (α, β, and δ wolves). The opposite solutions corresponding to best solutions are calculated by Eq. (12).

$$\overline{X}_L = (lb + ub) - X_L \tag{12}$$

where $\overline{X}_L$ is the opposite solution corresponding to leading wolves (α, β, and δ wolves) and X_L denotes the solution corresponding to α, β, and δ wolves. lb and ub are the lower and upper bounds of the decision variable, respectively. After that, all the updated wolves are sorted according to their fitness values and then the last 3 wolves having worst fitnesses are replaced by the opposite solutions of (α, β, and δ wolves). The proposed algorithm GWO-XOBL is presented in Algorithm 1.

4 Results and Discussion

In this section, the proposed GWO-XOBL is benchmarked on a set of 13 classical benchmark problems. These benchmark problems are presented in Table 1. In Table 1, dim denotes the dimension of the problems, Range refers the range of the decision variables, and $f_{\min}$ refers optimal value of the problem. The problems are divided into two different categories namely, unimodal and multimodal. $f1 - f7$ are unimodal and $f8 - f13$ are multimodal benchmark functions. Parameters which are used in the proposed algorithm are as follows:

Number of grey wolf (N) = 50
Maximum number of iteration (T) = 1000
Maximum number of function evaluation = $N \times T$
Number of independent runs= 30

Table 1 Description of benchmark functions

Function	dim	Range	f_{min}
$f1(x) = \sum_{i=1}^{n} x_i^2$	30	[−100,100]	0
$f2(x) = \sum_{i=1}^{n} \lvert x_i^2 \rvert + \prod_{i=1}^{n} \lvert x_i \rvert$	30	[−10,10]	0
$f3(x) = \sum_{i=1}^{n} \left(\sum_{j-1}^{i} x_j \right)^2$	30	[−100,100]	0
$f4(x) = \max_i \{\lvert x_i \rvert, 1 \le i \le n\}$	30	[−100,100]	0
$f5(x) = \sum_{i=1}^{n-1} \left[100 \left(x_{i+1} - x_i^2 \right)^2 + (x_i - 1)^2 \right]$	30	[−30,30]	0
$f6(x) = \sum_{i=1}^{n} ([x_i + 0.5])^2$	30	[−100,100]	0
$f7(x) = \sum_{i=1}^{n} i x_i^4 + random[0, 1)$	30	[−1.28,1.28]	0
$f8(x) = \sum_{i=1}^{n} -x_i \sin \left(\sqrt{\lvert x_i \rvert} \right)$	30	[−500,500]	−418.9829 · dim
$f9(x) = \sum_{i=1}^{n} \left[x_i^2 - 10 \cos (2\pi x_i) + 10 \right]$	30	[−5.12,5.12]	0
$f10(x) = -20 \exp(-0.2\sqrt{\frac{1}{n} \sum_{i=1}^{n} x_i^2}) - \exp \left(\frac{1}{n} \sum_{i=1}^{n} cos\ (2\pi x_i) \right) + 20 + e$	30	[−32,32]	0
$f11(x) = \frac{1}{4000} \sum_{i=1}^{n} x_i^2 - \prod_{i=1}^{n} \cos \left(\frac{x_i}{\sqrt{i}} \right) + 1$	30	[−600,600]	0
$f12(x) = \frac{\pi}{n} \left\{ 10 \sin (\pi y_1) + \sum_{i=1}^{n-1} (y_i - 1)^2 \left[1 + 10 \sin^2(\pi y_{i+1}) \right] + (y_n - 1)^2 \right\} + \sum_{i=1}^{n} u(x_i, 10, 100, 4)$ $y_i = 1 + \frac{x_i + 1}{4} u(x_i, a, k, m) = \begin{cases} k(x_i - a)^m & x_i > a \\ 0 - a & < x_i < a \\ k(-x_i - a)^m & x_i < -a \end{cases}$	30	[−50,50]	0
$f13(x) = 0.1 \left\{ \sin^2(3\pi x_1) + \sum_{i=1}^{n} (x_i - 1)^2 \left[1 + \sin^2(3\pi x_i + 1) \right] + (x_n - 1)^2 \left[1 + \sin^2(2\pi x_n) \right] \right\} + \sum_{i=1}^{n} u(x_i, 5, 100, 4)$	30	[−50,50]	0

Algorithm 1 Grey Wolf Optimizer with Crossover and Opposition-based Learning (GWO-XOBL) algorithm

Initialize randomly grey wolf positions say (x_i); (i=1,2...n)
Evaluate the fitnesses at x_i say $f(x_i)$; (i=1,2...n)
Select α, β, and δ wolf
Initialize μ, P, b, and crossover rate (pcr)
$t = 0$
while ($t \leq$ maxiter & feval $\leq$ maximum no of feval) **do**
  for each grey wolf **do**
    Update the positions of grey wolves using Eq. (6)
    Apply Crossover between current solution and updated solution as described in Eq. (10)
    calculate fitness of the updated solutions
    perform greedy selection as defined in Eq. (11)
  end for
  Select new α, new β, and new δ from the wolf pack
  Apply OBL on new α, new β, and new δ and obtain α^*, β^*, and δ^* wolves
  Sort the population and replace three worst fitness with α^*, β^*, and δ^*
  Update α, β, and δ wolves of the wolf pack
  $t = t + 1$
end while
return the α

The obtained results in terms of the average, best value, maximum value, and standard deviation(STD) on these benchmark problems are reported in Table 2. Table 2 also consists of the results obtained by differential evolution (DE) [18], biogeography-based optimization (BBO) [20], GWO [4], and GWO-XOBL. In Table 2, the better results obtained by considered algorithms are highlighted by bold font. In $f1$, $f2$, $f3$, $f4$, $f5$, $f9$, $f10$, and $f11$, results obtained by GWO-XOBL are better than DE, BBO, and GWO in terms of the average, best value, maximum value, and standard deviation(STD) and in $f1$, the GWO-XOBL achieves the global optimal value (0). In $f7$, GWO performs better than DE, BBO, and GWO-XOBL and DE is better than BBO, GWO, and GWO-XOBL in $f6$, $f8$, $f12$, and $f13$. The significance of the difference in the performance can be verifed through the statistical test Wilcoxon signed-rank test at 5% level of significance, which has been performed in Table 3. In Table 3, the sign "+" is used to present that the proposed GWO-XOBL is significantly better, " =" indicates that the proposed GWO-XOBL is similar and "−" is used to present that the proposed GWO-XOBL is worse than DE, BBO, or GWO algorithms. From the results, it can be concluded that GWO-XOBL is a better optimizer.

Table 2 Comparison of results of DE, BBO, GWO, and GWO-XOBL

Benchmark function	Algorithm	Average	Best value	Maximum	STD
$f1$	**DE**	8.10E-10	2.92E-12	1.79E-11	4.54E-12
	BBO	5.95E+00	2.25E+00	1.36E+01	2.98E+00
	GWO	8.83E-77	1.19E-79	2.00E-75	3.63E-76
	GWO-XOBL	**0.00E+00**	**0.00E+00**	**0.00E+00**	**0.00E+00**
$f2$	**DE**	1.11E-07	7.47E-08	1.48E-07	1.79E-08
	BBO	8.24E-01	5.66E-01	1.23E+00	1.78E-01
	GWO	5.45E-45	1.16E-45	1.98E-44	4.39E-45
	GWO-XOBL	**1.09E-177**	**3.15E-181**	**1.43E-176**	**0.00E+00**
$f3$	**DE**	1.62E+04	1.04E+04	2.11E+04	2.93E+03
	BBO	8.80E+03	4.97E+03	1.42E+04	2.26E+03
	GWO	7.54E-18	1.73E-24	1.61E-16	2.93E-17
	GWO-XOBL	**1.58E-211**	**2.47E-233**	**4.73E-210**	**0.00E+00**
$f4$	**DE**	3.44E+00	2.61E+00	4.21E+00	3.91E-01
	BBO	6.22E+00	3.88E+00	9.23E+00	1.12E+00
	GWO	1.83E-16	6.12E-18	1.77E-15	3.49E-16
	GWO-XOBL	**2.42E-133**	**3.18E-137**	**1.53E-132**	**4.60E-133**
$f5$	**DE**	5.03E+01	3.01E+01	1.07E+02	2.05E+01
	BBO	3.86E+02	1.96E+02	9.38E+02	1.98E+02
	GWO	2.63E+01	2.51E+01	2.79E+01	7.17E-01
	GWO-XOBL	**2.55E+01**	**2.48E+01**	**2.70E+01**	**6.03E-01**
$f6$	**DE**	**7.24E-12**	**3.68E-12**	**1.41E-11**	**2.61E-12**
	BBO	5.93E+00	3.15E+00	1.14E+01	2.45E+00
	GWO	2.97E-01	4.55E-06	7.53E-01	2.45E-01
	GWO-XOBL	5.19E-02	1.60E-03	2.54E-01	1.01E-01
$f7$	**DE**	2.62E-02	1.63E-02	3.96E-02	5.24E-03
	BBO	2.57E-02	6.87E-03	4.39E-02	9.67E-03
	GWO	**6.24E-04**	**1.98E-04**	**1.17E-03**	**2.41E-04**
	GWO-XOBL	1.60E-03	2.59E-04	3.20E-03	7.66E-04
$f8$	**DE**	**-1.26E+04**	**-1.26E+04**	**-1.25E+04**	**3.00E+01**
	BBO	-1.26E+04	-1.26E+04	-1.25E+04	5.66E+00
	GWO	-6.47E+03	-8.53E+03	-3.76E+03	1.02E+03
	GWO-XOBL	-9.05E+03	-1.03E+04	-8.07E+03	5.74E+02
$f9$	**DE**	1.80E-02	1.11E-03	6.03E-02	1.81E-02
	BBO	2.35E+00	1.17E+00	4.10E+00	7.70E-01
	GWO	4.30E+00	0.00E+00	1.36E+01	5.06E+00
	GWO-XOBL	**0.00E+00**	**0.00E+00**	**0.00E+00**	**0.00E+00**
$f10$	**DE**	7.82E-07	5.16E-07	1.01E-06	1.20E-10
	BBO	1.16E+00	7.06E-01	1.92E+00	2.62E-01
	GWO	7.99E-15	7.99E-15	7.99E-15	0.00E+00
	GWO-XOBL	**4.44E-15**	**4.44E-15**	**4.44E-15**	**0.00E+00**
$f11$	**DE**	1.98E-10	2.72E-11	1.67E-09	3.06E-10
	BBO	1.05E+00	1.02E+00	1.10E+00	1.97E-02
	GWO	3.16E-03	0.00E+00	2.31E-02	6.23E-03
	GWO-XOBL	**0.00E+00**	**0.00E+00**	**0.00E+00**	**0.00E+00**
$f12$	**DE**	**2.02E-13**	**8.92E-14**	**3.73E-13**	**6.49E-14**
	BBO	4.34E-02	5.10E-03	1.85E-01	4.80E-02
	GWO	2.17E-02	6.26E-03	6.20E-02	1.35E-02
	GWO-XOBL	2.40E-03	3.74E-05	1.91E-02	4.60E-03
$f13$	**DE**	**1.61E-12**	**5.08E-13**	**4.33E-12**	**8.87E-13**
	BBO	3.01E-01	1.34E-01	6.45E-01	1.14E-01
	GWO	2.97E-01	8.51E-06	8.01E-01	1.98E-01
	GWO-XOBL	7.18E-02	6.26E-04	2.06E-01	6.57E-02

Table 3 p-values obtained by Wilcoxon signed-rank test

Benchmark function	Statistical results	DE	BBO	GWO
$f1$	**p-values**	1.73E-06	1.73E-06	1.73E-06
	conclusion	+	+	+
$f2$	**p-values**	1.73E-06	1.73E-06	1.73E-06
	conclusion	+	+	+
$f3$	**p-values**	1.73E-06	1.73E-06	1.73E-06
	conclusion	+	+	+
$f4$	**p-values**	1.73E-06	1.73E-06	1.73E-06
	conclusion	+	+	+
$f5$	**p-values**	1.11E+01	1.73E-06	4.73E-06
	conclusion	=	+	+
$f6$	**p-values**	1.73E-06	1.73E-06	8.47E-06
	conclusion	−	+	+
$f7$	**p-values**	1.73E-06	1.73E-06	3.41E-05
	conclusion	+	+	-
$f8$	**p-values**	1.73E-06	1.73E-06	1.73E-06
	conclusion	−	−	+
$f9$	**p-values**	1.73E-06	1.73E-06	2.00E-03
	conclusion	+	+	+
$f10$	**p-values**	1.73E-06	1.73E-06	1.21E-07
	conclusion	+	+	+
$f11$	**p-values**	1.73E-06	1.73E-06	1.11E+01
	conclusion	+	+	=
$f12$	**p-values**	1.73E-06	1.92E-06	2.35E-06
	conclusion	−	+	+
$f13$	**p-values**	1.73E-06	5.22E-06	1.97E-05
	conclusion	−	+	+

5 Conclusion

In this paper, an improved version of the classical GWO (GWO-XOBL) has been proposed which is based on uniform crossover and opposition-based learning. The crossover is employed to increase the population diversity of GWO and opposition-based learning is applied to improve the convergence rate and to avoid the stagnation. The proposed GWO-XOBL has been evaluated on 13 classical benchmark problems. The obtained results are compared with DE, BBO, and GWO algorithms. The comparison illustrate that the proposed GWO-XOBL algorithm is better optimizer than the classical GWO and other compared algorithms.

References

1. J. Kennedy, R. Eberhart, Particle swarm optimization, in *Proceedings of ICNN'95-International Conference on Neural Networks*, vol. 4 (IEEE, 1995), pp. 1942–1948
2. M. Dorigo, L.M. Gambardella, M. Birattari, A. Martinoli, R. Poli, T. Stützle, *Ant Colony Optimization and Swarm Intelligence: 5th International Workshop, ANTS 2006, Brussels, Belgium, September 4–7, 2006, Proceedings*, vol. 4150 (Springer, 2006)
3. J.C. Bansal, H. Sharma, S.S. Jadon, M. Clerc, Spider monkey optimization algorithm for numerical optimization. Memetic Comput. **6**(1), 31–47 (2014)
4. S. Mirjalili, S.M. Mirjalili, A. Lewis, Grey wolf optimizer. Adv. Eng. Softw. **69**, 46–61 (2014)
5. A.A. El-Fergany, H.M. Hasanien, Single and multi-objective optimal power flow using grey wolf optimizer and differential evolution algorithms. Electr. Power Compon. Syst. **43**(13), 1548–1559 (2015)
6. X. Song, L. Tang, S. Zhao, X. Zhang, L. Li, J. Huang, W. Cai, Grey wolf optimizer for parameter estimation in surface waves. Soil Dyn. Earthq. Eng. **75**, 147–157 (2015)
7. S. Mirjalili, How effective is the grey wolf optimizer in training multi-layer perceptrons. Appl. Intell. **43**(1), 150–161 (2015)
8. E. Devi, R. Suganthe, Feature selection in intrusion detection grey wolf optimizer. Asian J. Res. Soc. Sci. Humanit. **7**(3), 671–682 (2017)
9. M. Vosooghifard, H. Ebrahimpour, Applying grey wolf optimizer-based decision tree classifer for cancer classification on gene expression data, in *2015 5th International Conference on Computer and Knowledge Engineering (ICCKE)* (IEEE, 2015), pp. 147–151
10. S. Zhang, Y. Zhou, Z. Li, W. Pan, Grey wolf optimizer for unmanned combat aerial vehicle path planning. Adv. Eng. Softw. **99**, 121–136 (2016)
11. T. Jayabarathi, T. Raghunathan, B. Adarsh, P.N. Suganthan, Economic dispatch using hybrid grey wolf optimizer. Energy **111**, 630–641 (2016)
12. J.C. Bansal, S. Singh, A better exploration strategy in grey wolf optimizer. J. Ambient Intell. Humanized. Comput. 1–20 (2020)
13. S. Gupta, K. Deep, An efficient grey wolf optimizer with opposition-based learning and chaotic local search for integer and mixed-integer optimization problems. Arab. J. Sci. Eng. **44**(8), 7277–7296 (2019)
14. A. Saxena, R. Kumar, S. Das, β-chaotic map enabled grey wolf optimizer. Appl. Soft Comput. **75**, 84–105 (2019)
15. S. Dhargupta, M. Ghosh, S. Mirjalili, R. Sarkar, Selective opposition based grey wolf optimization. Expert Syst. Appl. 113389 (2020)
16. S. Gupta, K. Deep, A memory-based grey wolf optimizer for global optimization tasks. Appl. Soft Comput. 106367 (2020)
17. A.A. Heidari, P. Pahlavani, An efficient modified grey wolf optimizer with lévy flight for optimization tasks. Appl. Soft Comput. **60**, 115–134 (2017)
18. R. Storn, K. Price, Differential evolution-a simple and efficient heuristic for global optimization over continuous spaces. J. Global Optim. **11**(4), 341–359 (1997)
19. S. Rahnamayan, H.R. Tizhoosh, M.M. Salama, Opposition-based differential evolution. IEEE Trans. Evol. Comput. **12**(1), 64–79 (2008)
20. D. Simon, Biogeography-based optimization. IEEE Trans. Evol. Comput. **12**(6), 702–713 (2008)

Development of Discrete Artificial Electric Field Algorithm for Quadratic Assignment Problems

Anita, Anupam Yadav, Nitin Kumar, and Joong Hoon Kim

Abstract Quadratic Assignment Problem (QAP) is a problem of facility locations of individual resources. QAP is a proven NP-hard challenging optimization problem and has a large number of real-life applications in diverse fields such as hospital layout problems, machine scheduling, keyboard design, and backboard wiring problem. Artificial electric field optimization (AEFA) is a new metaheuristic optimization algorithm and has achieved great success in continuous optimization problems. This paper presents a discrete artificial electric field algorithm for QAP. Due to the combinatorial nature of QAP, the general operations of AEFA such as particle representations, velocity and position update rules, and subtraction operations are modified. The proposed algorithm is applied to solve the QAP instances taken from the QAP library. The results show the promising performance of the proposed algorithm.

Keywords AEFA algorithm · Quadratic assignment problem · Soft computing optimization

Anita
Department of Mathematics, National Institute of Technology Uttarakhand, Srinagar 246174, Uttarakhand, India
e-mail: annusajwan93@gmail.com

A. Yadav (✉)
Department of Mathematics, Dr. B.R. Ambedkar National Institute of Technology Jalandhar, Jalandhar 144011, Punjab, India
e-mail: anupuam@gmail.com

N. Kumar
Department of Computer Science and Engineering, National Institute of Technology Uttarakhand, Srinagar 246174, Uttarakhand, India
e-mail: nitin@nituk.ac.in

J. H. Kim
School of Civil Environmental and Architectural Engineering, Korea University, Seoul 136-713, South Korea
e-mail: jaykim@korea.ac.kr

S. M. Nigdeli et al. (eds.), *Proceedings of 6th International Conference on Harmony Search, Soft Computing and Applications*, Advances in Intelligent Systems and Computing 1275, https://doi.org/10.1007/978-981-15-8603-3_36

1 Introduction

Quadratic Assignment Problem (QAP) attracts notable attention of researchers due to its wide range of applicability to real-life problems and its complexity to solve them. The real-life problems such as hospital layout problems, bin packing problems, backboard wiring, graph partitioning, keyboard layout problem, and traveling salesman problem can be modeled as a QAP. It was first proposed by Koopmans and Beckmann [1]. This is a problem of finding an optimal allocation of m facility to m destinations. The objective of this problem is to minimize the total cost. The total cost includes the cost of transporting the objects to the facilities and the cost of placing the facilities at the respective locations. Mathematically, this is a problem of finding a permutation π of m facilities to minimize the total cost given as follows:

$$Min\ Z(\pi) = \sum_{i=1}^{m} c_{i\pi(i)} + \sum_{i=1}^{m}\sum_{j=1}^{m} d_{ij} b_{\pi(i)\pi(j)} \tag{1}$$

here m represents the total number of facilities, $D = (d_{ij})$ corresponds to the distance matrix, where d_{ij} indicate the distance between two locations i and j for all $i, j \in \{1, 2, \ldots, m\}$, $B = (b_{ij})$ is the flow material, where b_{ij} is the flow of facility from location i to location j for all $i, j \in \{1, 2, \ldots, m\}$, and $C = (c_{ij})$ is the cost of the facility, where c_{ij} represents the cost of placing the facility from a location i to location j for all $i, j \in \{1, 2, \ldots, m\}$.

The QAPs are the most challenging combinatorial and proven NP-hard optimization problems. The small QAP instances can be solved by the traditional or exact methods such as branch and bound technique, cutting plane, and dynamic programming methods. But out of these only the branch and bound method guarantees an optimal solution [2]. However, these methods fail to find the optimal solution for the large instances of QAP due to their computational limitations. The heuristic and metaheuristics algorithms are the robust tools to solve these problems efficiently with an acceptable computational time. Among these algorithms, genetic algorithm (GA) [3], ant colony optimization (ACO) [4], evolutionary strategy (ES) [5], artificial bee colony (ABC) [6], iterative tabu search (ITS) [7, 8], particle swarm optimization (PSO) [9], simulated annealing (SA) [10], teaching- learning-based optimization (TLBO) [11], and their hybrids [12–15] are some successful optimizers for the QAPs.

The artificial electric field algorithm (AEFA) [16] is a newly designed charged population-based optimization algorithm and is inspired by Coulomb's law of electrostatic force. It was initially proposed for continuous unconstrained function optimization problems [16]. Then it's variant AEFA-C was designed for the constrained function optimization and engineering optimization problems [17]. Further, this is designed for the discrete problems of high-order graph matching [18]. Remarkable results are reported by AEFA and its variants in comparison to the existing optimization algorithms for these problems. In this paper, we present the modified version of discrete AEFA for the QAPs. To fulfill the problem requirements of QAPs, the AEFA algorithm is discretized by redefying the particle's position and velocity representation, velocity, and position update rules of AEFA and subtraction operation. The paper is organized as follows: in Sect. 2, the mechanism of AEFA is presented briefly. The detailed framework of the proposed discrete AEFA for the QAPs is given in Sect. 3. Section 4 demonstrates the computational results and their discussions and Sect. 5 concludes the presented work.

2 Artificial Electric Field Algorithm

AEFA is a charged-population-based metaheuristic algorithm designed for continuous optimization problems. It works on the principle of Coulomb's law of electrostatic force. This electrostatic force is responsible for the movement of the charged particles and the position of the charged particles corresponds to a possible solution to the problem. The details of the algorithm can be extracted from the published article [16]. The following section will address the development of AEFA for QAPs.

3 Development of Discrete Artificial Electric Field Algorithm (DAEFA)

For the first time, the AEFA algorithm is implemented for the quadratic assignment problems (QAPs). For this, we proposed a simple algorithm DAEFA by redefying the existing AEFA algorithm to fulfill the problem requirements of the QAPs. The details of the designing algorithm are given in the following subsections.

3.1 Particle Representation

In AEFA, the particle X_i represents the solution to the continuous problem directly, but for the discrete problem this scheme is simply encoded as follows: each particle is represented by a permutation π, where the dimension and each value of the particle represents a location and the corresponding facility, respectively. For instance, the particle $X_i = \{5\ 1\ 4\ 3\ 2\}$ indicates that the fifth facility is assigned to the first location, the first facility is assigned to the second location, the fourth facility is assigned to the third location, and so on.

3.2 Velocity Updating

The velocity is the important characteristic of AEFA to update the charged particles toward the promising positions. In DAEFA, it is redefined and given by an $m \times m$ matrix. Each element of this velocity indicates the probability of selecting a facility for a particular location. Let the velocity for m instances be given as follows:

$$V_i = \begin{bmatrix} v_{11} & \dots & v_{1m} \\ v_{21} & \dots & v_{2m} \\ \vdots & \ddots & \vdots \\ v_{m1} & \dots & v_{mm} \end{bmatrix}$$

where v_{ij} denotes the probability of assigning the jth facility to the ith location. The velocity of each charged particle is updated by the learning factor (electrostatic force) and the performance of the particles. The learning factor helps to identify the beneficial assignment of a facility to a particular location. These beneficial assignments are identified through the electrostatic force, global and personal best fitness values of the charged particles. For this, the subtraction operation in AEFA is replaced by the $setdiff$ operator. Let X_1 and X_2 be the particle's positions, then the subtraction operation is given by the following equation:

$$setdiff(X_1, X_2) = \{x \in X_1 \text{ and } x \notin X_1 \cap X_2\} \tag{2}$$

To update the particle's velocity, first, calculate the minimum (best) and maximum (worst) cost as follows:

$$best(t) = \min\{Z(X_i)\}, \quad i = \{1, 2, \dots, N\} \tag{3}$$

$$worst(t) = \max\{Z(X_i)\}, \quad i = \{1, 2, \dots, N\} \tag{4}$$

where $Z(X_i)$ represents the cost or objective function value at the assignment X_i and is calculated by Eq. 1. The personal best is evaluated using Eq. 5 of [16]. Then the redefined subtraction operation is used to calculate the difference $(P_j - X_i)$ and the distance R_{ij}, which is explained as follows.

Let the positions X_i and P_j be $X_i = \{2\ 5\ 4\ 3\ 1\}$ and $P_j = \{3\ 5\ 4\ 1\ 2\}$. Then their difference is calculated by the $setdiff$ operator given in Eq. 2 as follows:

$$setdiff(P_j, X_i) = \{3\ \phi\ \phi\ 1\ 2\}$$

where P_j is the position of the personal best fitness of jth particle. Then the permutation vectors for the positions X_i and P_j are transformed into a permutation matrix of size $m \times m$ using Eq. 5.

$$\lambda_{ij}(A) = \begin{cases} 1 & \text{if } A_i = j \\ 0 & \text{otherwise} \end{cases} \tag{5}$$

Based on the above definition (Eq. 5), the permutation matrices of X_i and P_j are given follows:

$$\lambda(X_i) = \lambda(\{2\,5\,4\,3\,1\}) \begin{bmatrix} 0\,1\,0\,0\,0 \\ 0\,0\,0\,0\,1 \\ 0\,0\,0\,1\,0 \\ 0\,0\,1\,0\,0 \\ 1\,0\,0\,0\,0 \end{bmatrix}$$

$$\lambda(P_j) = \lambda(\{3\,5\,4\,1\,2\}) \begin{bmatrix} 0\,0\,1\,0\,0 \\ 0\,0\,0\,0\,1 \\ 0\,0\,0\,1\,0 \\ 1\,0\,0\,0\,0 \\ 0\,1\,0\,0\,0 \end{bmatrix}$$

$$\lambda\left(setdiff\,(P_j, X_i)\right) = \begin{bmatrix} 0\,0\,1\,0\,0 \\ 0\,0\,0\,0\,0 \\ 0\,0\,0\,0\,0 \\ 1\,0\,0\,0\,0 \\ 0\,1\,0\,0\,0 \end{bmatrix}$$

Finally, we calculate the electrostatic force and the acceleration for each particle to update their velocities using the equations of AEFA.

3.3 Position Updating

In the context of QAP, the position updating rule of AEFA is updated to select a facility from m available facilities for each location. For this, first, the updated velocity V_{i+1} is converted into a probability matrix as follows:

$$n(V_{i+1}) = \begin{bmatrix} \frac{v_{11}}{\sum_{j=1}^{m} v_{1j}} & \cdots & \frac{v_{1m}}{\sum_{j=1}^{m} v_{1j}} \\ \vdots & \ddots & \vdots \\ \frac{v_{m1}}{\sum_{j=1}^{m} v_{mj}} & \cdots & \frac{v_{mm}}{\sum_{j=1}^{m} v_{mj}} \end{bmatrix} \tag{6}$$

here, $n(V_{i+1})$ represents the normalized velocity. This normalized velocity serves as a guide map to update the positions of the particles in the search space. Every row of $n(V_{i+1})$ represents the probability of assigning all the m facilities to the location whose index is the same as that of the rows, i.e., the ith row of $n(V_{i+1})$ stores the probability of selection of all m facilities for the jth location. For each location, the particle updating is required to select a single facility from the set of m facilities (or from every row of the velocity matrix).

To solve a QAP, it should be ensured that the updated position corresponds to a feasible solution (or valid permutation). To find the new solution it may be possible that some of the locations are assigned with the same facilities, which represents an infeasible solution. For a feasible solution, each facility must be assigned to a unique location. To handle this situation, the positions of the particles are updated in three steps with the help of three source sets: first is the velocity, second is the current position of that particle, and third is the universal set $U = \{1, 2, 3, \ldots, m\}$ having all the m facilities.

The position update is illustrated by the following example: Let the updated velocity of any particle i be

$$V_{i+1} = \begin{bmatrix} 0.21 & 0.59 & 1.63 & 0.75 & 0.80 \\ 0.65 & 1.36 & 0.60 & 0.70 & 1.06 \\ 0.31 & 0.28 & 0.54 & 0.63 & 2.10 \\ 1.47 & 1.35 & 0.85 & 0.92 & 0.35 \\ 0.49 & 0.15 & 0.34 & 0.75 & 1.63 \end{bmatrix}$$

To update the particle position, first, we find the probability matrix of V_{i+1} according to Eq. 6, which is given follows:

$$n\left(V_{i+1}\right) = \begin{bmatrix} 0.05 & 0.15 & 0.41 & 0.19 & 0.20 \\ 0.15 & 0.31 & 0.14 & 0.16 & 0.24 \\ 0.08 & 0.07 & 0.14 & 0.16 & 0.54 \\ 0.30 & 0.27 & 0.17 & 0.19 & 0.07 \\ 0.14 & 0.04 & 0.10 & 0.22 & 0.48 \end{bmatrix}$$

The first source to update the position is the above normalized velocity $n\left(V_{i+1}\right)$. To update the position X_i of any particle i to a new position X_{i+1}, an element having the highest probability is selected from each row of $n\left(V_{i+1}\right)$. So from the first source set, the updated position is $X_{i+1} = \{3\ 2\ 5\ 1\ 5\}$.

In the second step, the feasibility of the updated particle is tested for the valid permutation and only the unique assignments are retained in the updated position X_{i+1}. As the fifth facility is assigned to the third and fifth locations, null ϕ is assigned to fifth location. Then the updated position is $X_{i+1} = \{3\ 2\ 5\ 1\ \phi\}$.

In the third step, the locations having empty assignments are identified. To complete the new position construction these identified locations are assigned with the respective facilities from their current position (second source set). To assign the fifth location in X_{i+1} the current position $X_i = \{2\ 5\ 4\ 3\ 1\}$ is checked. However, assigning the first facility to the fifth location from X_i turned into an infeasible solution ($\{3\ 2\ 5\ 1\ 1\}$) so it is not used. Hence, the fifth location is still to be assigned, for this, a universal set U is used as a source set and the null elements in X_{i+1} are assigned from the set $U' = \{a | a \in U \text{ and } a \notin X_{i+1}\}$ to respective dimensions (fifth here). Therefore, the final updated particle's position is $X_{i+1} = \{3\ 2\ 5\ 1\ 4\}$, which is a feasible solution to the QAPs.

The algorithmic framework of the proposed discrete AEFA for QAPs is presented in Algorithm 1.

Algorithm 1 Algorithmic framework of proposed discrete AEFA for QAPs

Initialize all the parameters of AEFA
Initialize the charged population $X = \{X_1, X_2, \ldots, X_N\}$ with random permutations
Initialize the velocity of size $m \times m$ having uniformly distributed random number between 0 and 1
Calculate the fitness, personal best of each particle, *best* and *worst* fitness over the entire charge population
if Stopping criteria is not met then
  for i=1 to N **do**
    Determine the fitness values fit_i
    Evaluate the force by applying the redefined subtraction or *setdiff* operation
    Calculate the acceleration
    Update the velocity according to the redefined velocity update rule given in Sect. 3.2
    Update the position according to redefined position update rule given in Sect. 3.3
  end for
  Evaluate fitness at updated position X_{i+1}, i.e., $fit(X_{i+1})$
  Determine the global *best* fitness
  if $fit(X_i) < fit(X_{i+1})$ **then**
    $X_{i+1} = X_i$
  end if
end if

4 Experimental Results and Discussions

4.1 Data and Experimental Environment

To check the efficiency of the proposed discrete AEFA for the quadratic assignment problems, different challenging QAPs *had*, *nug*, *chr*, *esc*, *lipa*, *scr*, *tai*, and *tho* are taken from the QAPLIB online repository [19]. Further, to check the effect of dimensionality on the performance of the algorithm both the low and high dimensions are selected. The experiments are performed on MATLAB 2019a with Intel Core i3 processor and 2GB RAM system. All the experiments are carried out for a fixed parameter setting of AEFA as suggested in [16] and given as follows: $K_0 = 500$, $\alpha = 30$, number of charged particles $N = 30$, and $maxitr = 1000$.

4.2 Performance Evaluation of DAEFA

Each of the benchmark problems is solved ten times using the proposed algorithm and the best results are recorded. The average percentage deviation (APD) and the best percentage deviation (BPD) from the best known optimal solution are evaluated for each test problem. The experimental results are presented in Table 1, where the time column corresponds to the CPU time in seconds. From the result table, it can be observed that the computational results are equivalent to the best known global optimal solution for 26 out of 30 problem instances, whereas for the remaining four

Table 1 Computational results obtained by proposed algorithm

Sr. no	Instance	APD	BPD	Time
1	had12	0.00	0.00	2.03
2	had14	0.00	0.00	3.02
3	had16	0.00	0.00	3.24
4	had18	0.00	0.00	3.45
5	had20	0.00	0.00	4.56
6	nug20	0.00	0.00	3.45
7	nug25	0.00	0.00	5.37
8	nug27	0.00	0.00	5.29
9	nug28	0.00	0.00	5.51
10	nug30	0.00	0.00	6.30
11	chr20a	0.00	0.00	3.05
12	chr25a	0.00	0.00	4.26
13	esc16a	0.00	0.00	2.36
14	esc32a	0.00	0.00	3.56
15	esc64a	0.00	0.00	20.05
16	esc128	0.00	0.00	64.07
17	tho30	0.00	0.00	5.30
18	tho40	0.00	0.00	7.60
19	tho150	0.0160	0.0074	120
20	lipa20a	0.00	0.00	3.50
21	lipa30a	0.00	0.00	5.02
22	lipa40a	0.00	0.00	8.30
23	lipa50a	0.00	0.00	10.45
24	lipa60a	0.00	0.00	18.36
25	lipa70a	0.0113	0.0040	24.50
26	lipa80a	0.4017	0.3831	45.30
27	lipa90a	0.7329	0.2867	95.36
28	scr12	0.00	0.00	2.49
29	scr15	0.00	0.00	3.54
30	scr20	0.00	0.00	4.09

problem instances namely $tho150$, $lipa70a$, $lipa80a$, and $lipa90a$ it is closer to the best known solution with a maximum average percentage deviation of 0.7329%. The computational results indicate that the proposed discrete framework is capable to solve all the selected benchmark QAP instances efficiently within a reasonable computational time.

4.3 Comparison with the Existing Optimizers

To validate the performance of the proposed algorithm, we compare it with the state-of-the-art algorithms ABC [6], best PSO variant UPSO [9], TLBO-RTS [11], ACO/GA/LS [20], and ITS [7]. For these state-of-the-art algorithms, the results are

Table 2 Comparison of results with existing optimizers

Instance	BKS	ABC	UPSO	TLBO-RTS	ACO/GA/LS	ITS	DAEFA
had12	1652	0.00	–	0.00	–	–	0.00
had14	2724	0.00	–	0.00	–	–	0.00
had16	3720	0.00	–	0.00	–	–	0.00
had18	5358	0.00	–	0.00	–	–	0.00
had20	6922	0.00	0.06	0.00	–	–	0.00
nug20	578	0.00	–	0.00	–	–	0.00
nug25	3744	0.00	–	0.00	–	–	0.00
nug27	5234	0.00	–	0.00	–	–	0.00
nug28	5166	0.00	–	0.00	–	–	0.00
nug30	6124	0.00	0.62	0.00	–	–	0.00
chr15a	9896	0.00	0.00	0.00	–	–	0.00
chr20a	2192	0.00	–	0.00	–	–	0.00
chr25a	3796	0.00	19.81	0.00	–	–	0.00
esc16a	68	0.00	–	0.00	–	–	0.00
esc32a	130	0.00	–	0.00	0.00	0.00	0.00
esc64a	116	0.00	0.00	0.00	0.00	0.00	0.00
esc128	64	0.00	–	0.00	0.00	0.01	0.00
tai20a	70382	0.00	0.80	0.00	–	0.00	0.00
tai25a	1167256	0.00	–	0.00	–	0.00	0.00
tai30a	1818146	0.00	2.40	0.00	0.341	0.00	0.00
tai35a	2422002	0.00	–	0.00	0.487	0.00	0.00
tai40a	3139370	0.00	3.29	0.074	0.593	0.220	0.00
tai50a	4938796	0.312	3.53	0.550	0.901	0.410	0.0609
tai60a	7205962	0.449	3.46	0.643	1.068	0.450	0.2195
tai80a	13499184	0.817	3.69	0.771	1.178	0.360	0.8180
tai100a	21052466	0.614	3.79	1.045	1.115	0.300	1.1638
lipa50a	62093	0.00	–	0.00	–	–	0.00
tho30	149936	0.00	–	0.00	–	–	0.00
tho40	240516	0.00	–	0.00	–	–	0.00
tho150	8133398	0.014	2.95	0.030	–	–	0.0113

taken from the literature. The comparison of the results is performed in terms of the average percentage deviation (APD) and the results are presented in Table 2, where the second column represents the best-known solution (BKS) for the corresponding problem. As seen in Table 2, the proposed discrete framework DAEFA performed significantly better/comparable to all the other existing algorithms for most of the problem instances. Only for problem instances $tai80a$ and $tai100a$, the proposed DAEFA obtained best results followed by the ABC algorithm whereas superior to all the others: UPSO, TLBO-RTS, ACO/GA/LS, and ITS. The comparison results suggest that the proposed discrete framework DAEFA is a powerful optimization tool for the QAPs and outperformed the existing algorithms.

5 Conclusion

In this paper, we developed a simple modified discrete artificial electric field algorithm (DAEFA) to make it suitable for the quadratic assignment problems. DAEFA obtained promising results closed to the known global optimal solutions with less than 1% of average percentage deviation (APD) and best percentage deviation (BPD). Also, the proposed algorithm has significantly better/comparable performance over the state-of-the-art algorithms. In future, the existing results may be improved and the proposed algorithm can be applied to solve the more challenging problem instances for high dimensionality and the other combinatorial problems of different domains. This may be done by implementing the parallel computation, designing the hybrids of DAEFA or tuning the parameters of DAEFA.

References

1. T.C. Koopmans, M. Beckmann, Assignment problems and the location of economic activities. Econ. J. Econ. Soc. 53–76 (1957)
2. P.M. Pardalos, F. Rendl, H. Wolkowitz, The quadratic assignment problem: a survey and recent developments. Quadratic assignment and related problem. DIMACS Ser Discrete Math Theor Comput Sci. **16**, 1–42 (1994)
3. R.K. Ahuja, J.B. Orlin, A. Tiwari, A greedy genetic algorithm for the quadratic assignment problem. Comput. Oper. Res. **27**(10), 917–934 (2000)
4. L.M. Gambardella, É.D. Taillard, M. Dorigo, Ant colonies for the quadratic assignment problem. J. Oper. Res. Soc. **50**(2), 167–176 (1999)
5. V. Nissen, Solving the quadratic assignment problem with clues from nature. IEEE Trans. Neural Netw. **5**(1), 66–72 (1994)
6. T. Dokeroglu, E. Sevinc, A. Cosar, Artificial bee colony optimization for the quadratic assignment problem. Appl. Soft Comput. **76**, 595–606 (2019)
7. A. Misevicius, An implementation of the iterated tabu search algorithm for the quadratic assignment problem. Oper. Res. Spect. **34**(3), 665–690 (2012)
8. T. James, C. Rego, F. Glover, Multistart tabu search and diversification strategies for the quadratic assignment problem. IEEE Trans. Syst. Man Cybern. Part A: Syst. Humans **39**(3), 579–596 (2009)
9. F. Hafiz, A. Abdennour, Particle Swarm algorithm variants for the quadratic assignment problems-A probabilistic learning approach. Expert Syst. Appl. **44**, 413–431 (2016)
10. A. Misevicius, A modified simulated annealing algorithm for the quadratic assignment problem. Informatica **14**(4), 497–514 (2003)
11. T. Dokeroglu, Hybrid teaching–learning-based optimization algorithms for the quadratic assignment problem. Comput. Ind. Eng. **85**, 86–101 (2015)
12. U. Benlic, J.K. Hao, Memetic search for the quadratic assignment problem. Expert Syst. Appl. **42**(1), 584–595 (2015)

13. Z. Drezner, Extensive experiments with hybrid genetic algorithms for the solution of the quadratic assignment problem. Comput. Oper. Res. **35**(3), 717–736 (2008)
14. W.L. Lim, A. Wibowo, M.I. Desa, H. Haron, A biogeography-based optimization algorithm hybridized with tabu search for the quadratic assignment problem. Comput. Intell. Neurosci. (2016)
15. M. Abdel-Basset, G. Manogaran, D. El-Shahat, S. Mirjalili, Integrating the whale algorithm with tabu search for quadratic assignment problem: a new approach for locating hospital departments. Appl. Soft Comput. **73**, 530–546 (2018)
16. A. Yadav, AEFA: artificial electric field algorithm for global optimization. Swarm Evol. Comput. **48**, 93–108 (2019)
17. A. Yadav, N. Kumar, Artificial electric field algorithm for engineering optimization problems. Expert Syst. Appl. **149**, 113308 (2020)
18. A. Yadav, Discrete artificial electric field algorithm for high-order graph matching. Appl. Soft Comput. **92**, 106260 (2020)
19. R.E. Burkard, S.E. Karisch, F. Rendl, QAPLIB—a quadratic assignment problem library. J. Glob. Optim. **10**(4), 391–403 (1997)
20. L.Y. Tseng, S.C. Liang, A hybrid metaheuristic for the quadratic assignment problem. Comput. Optim. Appl. **34**(1), 85–113 (2006)

Fuzzy-Based Kernelized Clustering Algorithms for Handling Big Data Using Apache Spark

Preeti Jha, Aruna Tiwari, Neha Bharill, Milind Ratnaparkhe, Neha Nagendra, and Mukkamalla Mounika

Abstract In this paper, we propose a novel Kernelized Scalable Random Sampling with Iterative Optimization Fuzzy c-Means (KSRSIO-FCM) and a Kernelized Scalable Literal Fuzzy c-Means (KSLFCM) clustering algorithms for big data framework. The evolution of kernelized clustering algorithms led us to deal with the nonlinear separable problems by applying kernel Radial Basis Functions (RBF) which map the input data space nonlinearly into a high-dimensional feature space. The experimental result shows that the KSRSIO-FCM algorithm achieves significant improvement in terms of F-score, Adjusted Rand Index (ARI), and Normalized Mutual Information (NMI) for Big Data. Experimentation is performed on well-known IRIS datasets to show the effectiveness of proposed KSRSIO-FCM in comparison with KSLFCM. The KSRSIO-FCM implemented on Apache Spark shows better potential for Big Data clustering.

P. Jha (✉) · A. Tiwari · N. Nagendra · M. Mounika
Department of Computer Science and Engineering, Indian Institute of Technology Indore, Indore, India
e-mail: phd1801201006@iiti.ac.in

A. Tiwari
e-mail: artiwari@iiti.ac.in

N. Nagendra
e-mail: nehanagendra02@gmail.com

M. Mounika
e-mail: mounikamukkamalla16@gmail.com

N. Bharill
Department of Computer Science and Engineering, Mahindra University, Ecole Centrale School of Engineering, Hyderabad, India
e-mail: neha.bharill@mechyd.ac.in

M. Ratnaparkhe
Biotechnology, ICAR-Indian Institute of Soybean Research Indore, Indore, India
e-mail: ratnaparkhe.milind@gmail.com

S. M. Nigdeli et al. (eds.), *Proceedings of 6th International Conference on Harmony Search, Soft Computing and Applications*, Advances in Intelligent Systems and Computing 1275, https://doi.org/10.1007/978-981-15-8603-3_37

1 Introduction

Clustering is an unsupervised learning technique, which is used on similar objects of data that can be grouped to form a subset of data; this results in finding patterns of datasets [1]. This technique can be applied to the area of data mining, machine learning, information retrieval, and cybersecurity. Clustering algorithms are majorly used for the following applications: pattern recognition, data mining, classification, image segmentation, data analysis, and modeling [2–4]. Clustering is broadly divided into two parts, i.e., hierarchical and partitioning clustering [5]. Hierarchical clustering finds the clusters by partitioning data in either top-down or bottom-up fashion in a recursive manner, whereas partitioning clustering divides a dataset into a number of disjoint clusters. Hard and fuzzy are the two methods of partitioning clustering [6]. In hard clustering, each data sample is assigned to one cluster. While in fuzzy clustering, a data sample can belong to more than one cluster with varying degrees of membership. Presently, there are several fuzzy clustering algorithms such as Fuzzy c-Means (FCM) [7] and kernelized Fuzzy c-Means (KFCM) [8] that are used for handling big data. Many variants of fuzzy clustering algorithms have been proposed by researchers such as single-pass FCM (SPFCM) [9], and an online FCM (OFCM) [10]. The SPFCM and OFCM algorithms partition the dataset into a number of subsets and then calculate the clusters for the entire dataset. Nowadays, applications are becoming extremely popular and easily accessible due to Internet advancements; a huge number of people are utilizing the Internet and generating enormous data.

As per the survey conducted by IDC [11], it has been concluded that social sites are flooded with a vast amount of data per day. For example, 6 billion searches are carried out on Google, 300 h of videos are uploaded to YouTube, 500 million of tweets per day are sent from Twitter, 53 million of likes come from Instagram, 4.75 billion of posts are uploaded on Facebook, and 4.3 billion messages are sent by Facebook messenger. Nowadays, data is increasing and transforming our society. The analytics breakthroughs brought an impressive and remarkable generation of biological data, which was a dream some years ago, that include sequencing of Protein, DNA, and RNA [12]. There is a need to perform the analysis on a vast amount of this biological data. To do this, there is a need for efficient data storage, searching, analysis, and feature extraction.

The kernel version of SPFCM is known as spKFCM, and the kernel version of OFCM is known as okFCM [13]. Both the algorithm are used to group huge data efficiently. The extension of FCM known as random sampling plus extension Fuzzy c-Means (rseFCM) [13] is also developed, but the overlapping of the cluster is the main issue of rseFCM. The overlapping is removed by Random Sampling with Iterative Optimization Fuzzy c-Means (RSIO-FCM) [14]. However, RSIO-FCM suffers from a sudden rise in several iterations during the clustering. To overcome the issues of RSIO-FCM, a Scalable Random Sampling with Iterative Optimization Fuzzy c-Means algorithm (SRSIO-FCM) [15] incremental fuzzy clustering has been developed. We have extended the SRSIO-FCM to a KSRSIO-FCM and the SLFCM to a KSLFCM. The Kernelized SLFCM algorithm has been derived from SLFCM by changing the vector norms, fuzzy cluster centers, and the partition matrix. The vector norm is defined by Radial Basis Function (RBF) instead of Euclidean distance as described in Sect. 4. The KSRSIO-FCM uses KSLFCM to compute membership matrix and cluster centers. The proposed kernel functions, i.e., KSLFCM and KSRSIO-FCM, are an improved version of traditional clustering SRSIO-FCM and SLFCM algorithms.

In today's scenario, many types of core computing frameworks have been invented for the implementation of Big Data precisely. Apache Spark is a well-known general-purpose distributed big data framework that keeps the benefits of map-reduce scalable and makes it more flexible. Spark handles the volume of data and provides efficient processing, a scalable framework with fault-tolerant capability. Spark is built on Hadoop's data volume model, i.e., Hadoop Distributed File System (HDFS). Apache Spark is excellent to deploy an application with a map-reduce technique [16], and thus, our aim for designing an implementation of a kernelized iterative clustering algorithm using the Apache Spark framework.

In this paper, we propose and implement a KSRSIO-FCM algorithm with the help of the proposed kernelized version of SLFCM to make it more efficient. KSLFCM algorithm is an extended version of SLFCM, which uses kernel functions to optimize the objective functions. We implemented the proposed algorithms on the Apache Spark cluster for handling big data to tackle challenges of fuzzy clustering. The KSRSIO-FCM algorithms partition the dataset into various subsets and then perform clustering on each subset. The clustering of the first subset is done by calculating the cluster center (V_1) and membership knowledge (M_1); then (V_1) is fed as an input for clustering of the second subset. Thereafter, the cluster center (V_2) and membership knowledge (M_2) are obtained. Finally, M_1 and M_2 are merged to obtain new cluster centers. These cluster centers are fed as an input to the third subset. This procedure repeats for all subsequent subsets. We have tested KSRSIO-FCM and KSLFCM algorithms on Apache Spark with the IRIS dataset as a benchmark dataset. Our proposed algorithm can handle huge datasets and achieve a practical result in the problem.

The next sections of this paper are standardized as follows: in Sect. 2, a review of fuzzy clustering approaches is discussed which forms the basis for our proposed algorithms. The overview of the Apache Spark cluster is detailed in Sect. 3. The proposed algorithm KSRSIO-FCM with KSLFCM implementation on the Apache Spark cluster is explained in Sect. 4. The results obtained from experiments conducted on benchmark datasets are reported in terms of F-score, ARI, and NMI in Sect. 5. Finally, the conclusions of our work are drawn in Sect. 6.

2 Preliminaries

The objective function of the Fuzzy c-Means (FCM) algorithm is stated as follows:

$$J_p(M, V') = \sum_{i=1}^{s}\sum_{j=1}^{c} m_{ij}^{p} \|x_i - v_j^{\prime}\|2, \; p > 1 \tag{1}$$

x_i is denoted as a data sample, v_j as a cluster center, p termed as fuzzification parameter, s termed as the total number of samples, c termed as number of clusters, and m_{ij} the membership degree of x_i regarding v_j.

The membership matrix m_{ij} corresponding to data sample x_i to each cluster v_j can be calculated as follows:

$$m_{ij}^{p} = \frac{\|x_i - v_j\|^{-1/(p-1)}}{\sum_{j=1}^{c} \|x_i - v_j\|^{-1/(p-1)}} \tag{2}$$

Cluster Center can be computed as follow:

$$v_j^{\prime} = \frac{\sum_{i=1}^{S} m_{ij}^{p} x_i}{\sum_{i=1}^{S} m_{ij}^{p}} \tag{3}$$

As mentioned earlier, Fuzzy c-Means can handle linear relations. To handle nonlinear relations, the concept of the kernel method is explained in Sect. 4.

2.1 *Scalable Literal Fuzzy c-Means (SLFCM)*

The SLFCM clustering algorithm is executed on Apache Spark to handle big data. In SLFCM [17], the computation of cluster membership knowledge is evaluated in parallel on slave nodes. Thus, it overcomes time complexity as compared to the linear execution of an algorithm on a standalone machine. The membership degrees are combined on the master node, and cluster center values are evaluated. This process is continued until the difference observed is not useful for the benefits of cluster centers. Each data sample is reserved in the form of an array of features in Resilient Distributed Datasets (RDDs) [18], which is a data structure to store objects precisely in memory.

2.2 *Scalable Random Sampling with Iterative Optimization Fuzzy c-Means (SRSIO-FCM)*

The SRSIO-FCM [15] is implemented to overcome the issue present in RSIO-FCM. The SRSIO-FCM algorithm executed on Apache Spark distributes huge data into various subsets to the slave node and processes each subset individually on the slave node in a parallel manner. The SRSIO-FCM helps in reducing runtime for the grouping of huge data, without negotiating the output of clustering quality. The RSIO-FCM does not save the membership matrix, and due to this the SRSIO-FCM algorithm runs faster.

The clustering of the initial subset is obtained through the usage of SLFCM. The cluster centers are declared, then its computation is performed, and the membership matrix for the initial subset is obtained by applying SLFCM. The output of initial cluster centers is fed to the second subset for clustering. The cluster centers of the two subsets are different from each other because the SRSIO-FCM doesn't feed the output of the cluster center from the second subset, as the input cluster centers to the third subset for clustering. Instead, it combines a membership degree of all subsequent subsets. SRSIO-FCM is a better algorithm than other iterative algorithms.

These two approaches are extended in Sect. 4, where we propose the KSLFCM and KSRSIO-FCM algorithms. Before presenting the proposed KSRSIO-FCM and KSLFCM algorithms, an overview of the Apache Spark cluster is discussed in the next section.

3 Apache Spark

The two approaches KSLFCM and KSRSIO-FCM can be made scalable using the Apache Spark framework for handling big data. Apache Spark is a scalable in-memory computation framework for big data processing. It allows subsets of the dataset to be processed in parallel across a cluster. Apache Spark is a high-speed cluster computing system with efficient and straightforward development APIs which allows the slave node to access the dataset iteratively and execute efficiently. The in-memory cluster computing technique of spark increases the processing speed of an application by executing a spark job on the Hadoop framework to share a cluster and dataset while satisfying consistent levels of service and response. Apache Spark works with YARN in Hadoop to access data from spark engines [19]. Spark builds with a stack of libraries, including SQL and Data Frames, MLlib, GraphX, and Spark Streaming. MLlib is a library that provides a machine learning algorithm for data science techniques.

Figure 1 shows the overview of Apache Spark clusters. The Apache Spark cluster consists of one master node and a number of worker nodes. The master is known as a driver which is used for task scheduling. Spark initiates a scheduling process in a hierarchical manner with jobs, steps, and tasks. Step is a subset of tasks partitioned

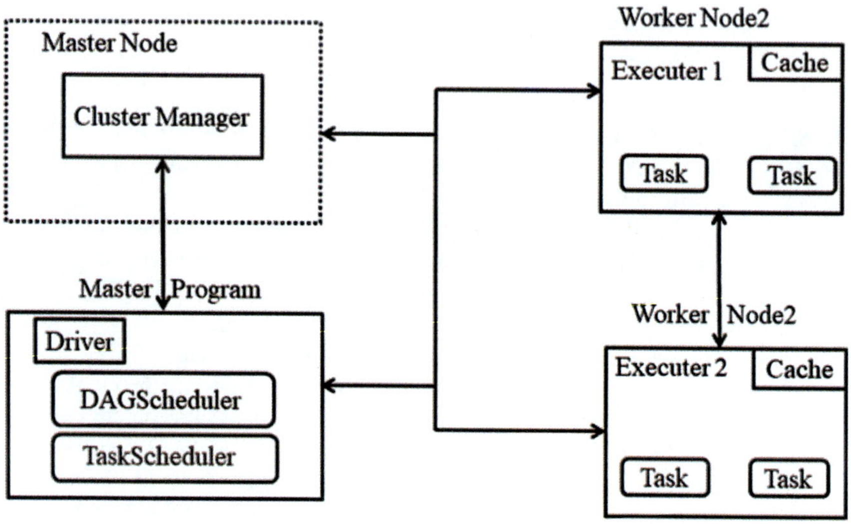

Fig. 1 Overview of Apache Spark cluster

from collective jobs, which is used to match the map and reduce phase. Apache Spark comprises DAG Scheduler and Task Scheduler. The DAG Scheduler calculates a directed acyclic graph (DAG) of steps for a job. It also keeps track of the record of RDD with each step outputs, whereas the Task Schedule submits tasks from each step to the cluster. Apache Spark provides different cluster modes to the user for the execution of their Apache Spark program by allowing the master to connect to a cluster manager, standalone in our case. There is an executor created for each program for each worker node. The executor is responsible for running the tasks and caching the data in memory or disk [20].

4 Proposed Work

In this paper, we proposed a kernelized version of SRSIO [15] termed as KSRSIO-FCM. KSRSIO-FCM designed to overcome challenges arises in fuzzy clustering for handling massive data. The two contributory approaches overcome the saving of substantial membership knowledge during the clustering of each block of datasets. Thus it overwhelms the space complexity. KSRSIO-FCM executed on Apache Spark is an optimization of the fuzzy-based clustering algorithm primarily; we produced a kernel version of SLFCM fuzzy clustering algorithms and named it as KSLFCM. Both the algorithms are discussed in the following subsections.

4.1 Kernelized Scalable Literal Fuzzy c-Means (KSLFCM)

KSLFCM is a kernelized version of SLFCM algorithm by applying the Radial Basis Function (RBF) kernel function. The RBF maps the input data space nonlinearly into a high-dimensional feature space. The kernel radial basis function is defined in Eq. (4). The membership degree is calculated using data samples and cluster center values, as stated in Eq. (5) in parallel on various slave nodes by reducing the run-time [21]. The membership degrees are then combined on master nodes, and cluster center values are calculated using Eq. (6). This procedure is repeated until the absolute difference of the membership matrix is less than the stopping criteria ($\in$). Apache Spark is used to implement the KSLFCM algorithm. Algorithm 1 describes the functioning of KSLFCM. The kernelized version of SLFCM [17] is defined as follows:

$$K\left(x_i, \mathrm{v_j}\right) = e^{\left(-\frac{\|x_i - \mathrm{v_j}\|^2}{2\sigma^2}\right)} \tag{4}$$

The membership degree is calculated using data samples and cluster center values as

$$m_{ij} = \frac{\left(1 - K\left(x_i, v_j\right)\right)^{-1/(p-1)}}{\sum_{j=1}^{c}\left(1 - K\left(x_i, v_j\right)\right)^{-1/(p-1)}} \tag{5}$$

The cluster centers can be computed as

$$v_j = \frac{\sum_{i=1}^{S} m_{ij}^{p} K\left(x_i, v_j\right) x_i}{\sum_{i=1}^{S} m_{ij}^{p} K\left(x_i, v_j\right)} \tag{6}$$

The choice of kernel parameter is the most critical task. In the proposed work, the Kernel parameter can be selected in the following way:

$$\sigma = \sqrt{\frac{\sum_{i=1}^{S}\left(d_i - \bar{d}\right)^2}{S - 1}} \tag{7}$$

where $d_i = \|x_i - \bar{x}\|$ and $\bar{d}$ is the average of all the distances of d_i.

The objective function is minimized using the proposed kernel-based fuzzy clustering algorithm explained in Algorithm 1.

In Algorithm 1, the membership degree is calculated separately for each data sample. In Line 2 of Algorithm 1, we have used map and reduce ByKey functions to obtain the parallel computation of the membership knowledge of all the data samples. Furthermore, Line 4 of Algorithm 1 is used to update the cluster center values from membership degrees of all data samples. Thus, Line 3 of Algorithm 1 is executed after the membership degrees of all locations have been computed. At the master node, the membership degree of all the data samples is merged and saved as a membership knowledge I', which is required in Eq. (6) to update the cluster center v_j. After that, we calculated the difference between the old value of the initialized cluster center and newly calculated cluster center values, this is given in Line 4 of Algorithm 1. Repeat this procedure until no change in the values of cluster centers is recognized.

Algorithm 1. *KSLFCM to iteratively Minimize* $J_p(M, V')$

Input: X, V, c, p, σ
Output: $\grave{I}, \grave{V}$

1. Declare cluster centers randomly
   $V = \{v_1, v_2, v_3 \ldots \ldots . v_c\}$
2. Calculate membership matrix and cluster center by using equation (3) & equation (4) respectively
   $\grave{I} = X.map(V).reduceByKey()$
3. If $||V' - V|| < \epsilon$ then stop,
4. Return $\grave{I}, \grave{V}$
5. Otherwise, continue with step 2.

The space requirement is reduced by avoiding the storage of the huge membership matrix. Instead of storing the membership matrix, we calculate the values of the parameter present in the numerator and the denominator of Eq. (5). For example, In Algorithm 1, the membership matrix M is needed to compute cluster center V' using Eq. (5).

4.2 *Kernelized Scalable Random Sampling with Iterative Optimization Fuzzy c-Means (KSRSIO-FCM)*

Then KSRSIO-FCM calculates the cluster centers and membership knowledge for initial subset X_1, represented by V' and $\grave{I}$, respectively, by applying KSLFCM. Then V' is fed as an input cluster center for clustering of second subset X. The KSRSIO-FCM performs clustering of X by applying KSLFCM and calculates the cluster centers and membership knowledge represented by I' and V'. However, KSRSIO-FCM does not feed V as an input for the clustering of the third subset. It instead merges the membership knowledge of all the processed subsets, i.e., it combines I' and I to find new cluster centers which are fed as input to the third subset.

Algorithm 2. *KSRSIO-FCM to Iteratively Minimize* $J_p(M, V`)$

Input: *X, V, c, p*
Output: $\grave{V}, \grave{I}$

1. Consider X as s_n sized from chosen subsets $X = \{x_1, x_2, \ldots \ldots . x_n\}$
2. Sample X_1 from X without replacement.
3. $\grave{V}, \grave{I} = KSLFCM(X_1, V, c, p)$
4. for $p = 2$ *to n do*
   - 4.1: $\grave{V}, I = KSLFCM\ (X_t, \grave{V}, c, p)$
   - 4.2: **Merge** the distribution of all blocks of prepared datasets $\grave{I} = \grave{I} \cup I$
   - 4.3: **Calculate** new cluster center $\grave{V}$ using $\grave{I}$

   end for
5. Return $\grave{V}, \grave{I}$

KSRSIO-FCM covers the entire high-dimensional feature space. Thus, the result produced is equivalent to that produced by clustering the whole data at once.

5 Experimental Results

In the experiments, the comparison of KSRSIO-FCM with KSLFCM is performed on the Apache Spark cluster. The performance of KSRSIO-FCM in comparison with KSLFCM using F-score, ARI, and NMI has been done.

5.1 Experimental Environment

The Apache Spark cluster is used to perform experiments evaluation. The Apache Spark cluster consists of five slave nodes and one master node. Each node has 16 GB RAM and 1 TB storage. HDFS is used across the cluster for data storage with spark standalone mode [22]. The algorithms were implemented in Python version 3.6.7, Apache Spark cluster 2.4.0 setup on Ubuntu 18.04 with Hadoop version 2.7.3.

5.2 Dataset

The IRIS dataset is present in the UCI machine learning repository. The IRIS dataset is a type of iris plant which contains three classes with 50 instances each. Predicted attributes of category for the iris plant information are as follows: width and length of sepals and petals are in cm. Iris Sentosa, Iris Versicolour, and Iris Virginica are the three types of classes of the IRIS dataset.

Table 1 Parameter specification for IRIS dataset

Parameters	KSLFCM	KSRSIO-FCM
P	1.75	1.75
∈	0.01	0.01
C	3	3

5.3 *Parameters Specification*

The value of the fuzzification parameter, i.e., $p = 1.75$, stopping criteria $\in = 0.01$, and c is a number of a cluster which is fixed as 3 since these parameter values work best for most of the datasets [23]. In our implementation of the proposed algorithm, parameter values used are listed in Table 1.

5.4 *Evaluation Criteria*

5.4.1 F-score

F-score [24] is used to calculate the accuracy of a clustering output. F-score for a cluster $C_y^{\backprime}$w.r.t. to a class C_x shows how better cluster $C_y^{\backprime}$ defines class C_x. F-score is defined below

$$p_{xy} = \frac{n_{xy}}{n_y} \tag{8}$$

$$r_{xy} = \frac{n_{xy}}{n_x} \tag{9}$$

$$f\left(C_x, C_y^{\backprime}\right) = \frac{2 * p_{xy} * r_{xy}}{p_{xy} + r_{xy}} \tag{10}$$

$$\text{F-score} = \sum_x \frac{n_{xy}}{n} \max_y f\left(C_x, C_y^{\backprime}\right) \tag{11}$$

where n_{xy} is the number of data samples in cluster C_x that is also present in class C_y. n_x and n_y are the number of data samples of class C_x and class C_y, respectively. p_{xy} is termed as precision and r_{xy} is recall. The total number of data samples is n.

5.4.2 Adjusted Rand Index (ARI)

The adjusted Rand index [25] is used to find the similarity between the clustering of two data samples. The ARI is calculated by setting the maximum element in each column to 1, and all others to 0. The ARI index is defined as follows:

$$ARI = \frac{\sum_{x,y}\binom{n_{xy}}{2} - [\sum_i \binom{n_{x.}}{2}\sum_j \binom{n_{.y}}{2}]/\binom{n}{2}}{\frac{1}{2}[\sum_x \binom{n_{x.}}{2} + \sum_y \binom{n_{.y}}{2}] - [\sum_x \binom{n_{x.}}{2}\sum_y \binom{n_{.y}}{2}]/\binom{n}{2}}. \tag{12}$$

5.4.3 Normalized Mutual Information (NMI)

NMI derives from entropy information theory, and it [26] is used for calculation of clustering quality, which measures the ratio of the mutual information of the clustering, ground truth, and their harmonic mean. NMI is defined as follows:

$$NMI = \frac{\sum_{x=1}^{t}\sum_{y=1}^{m} n_x^y \log(\frac{n \cdot n_x^y}{n_c \cdot n_p})}{\sqrt{(\sum_{x=1}^{t} n_x \log(\frac{n_x}{n}))(\sum_{y=1}^{m} n_y \log(\frac{n_y}{n}))}} \tag{13}$$

where n denotes the total number of data samples, n_x and n_y are the data samples in the xth cluster and the yth class, respectively, and n_y^x is the number of common data samples in class y and cluster x.

5.5 *Result*

This section discusses the experimental results to show the effectiveness of KSRSIO-FCM in comparison with KSLFCM on the benchmark IRIS dataset in terms of various measures such as F-score, ARI, and NMI. We compare the execution of KSRSIO-FCM with KSLFCM using the parameters as described in Table 2.

Table 2 F-score, ARI, and NMI for KSRSIO-FCM and KSLFCM with various block sizes on IRIS dataset

Algorithm	Block size (no. of subsets)	Train-test partition	F-score	ARI	NMI
KSRSIO-FCM	16.67% (6 subsets)	70–30	0.916	0.678	0.726
		80–20	0.679	0.673	0.773
	20.00% (5 subsets)	70–30	0.800	0.781	0.827
		80–20	0.750	0.738	0.806
	25.00% (4 subsets)	70–30	0.916	0.724	0.750
		80–20	0.679	0.673	0.773
	33.33% (3 subsets)	70–30	0.916	0.724	0.750
		80–20	0.750	0.738	0.806
	50.00% (2 subsets)	70–30	0.564	0.560	0.717
		80–20	0.895	0.741	0.757
KSLFCM	100% (1 subset)	70–30	0.792	0.739	0.803
		80–20	0.783	0.132	0.257

We have listed the value of F-score calculated for the benchmark IRIS dataset using KSLFCM and KSRSIO-FCM with different block sizes. The quality of clustering is measure by F-score and NMI. A higher value of F-score and NMI represents better clustering. Thus, we observe that KSRSIO-FCM functions, as well as KSLFCM in terms of F-score. Furthermore, KSRSIO-FCM shows relatively well for all the block sizes and in almost every case, 70–30 distribution has a better value compared to 80–20 distribution. KSLFCM and KSRSIO-FCM are very close for 70–30 train-test distribution whereas it is expressively low for KSLFCM as compared to KSRSIO-FCM in case of 80–20 distribution for the values of NMI. Also, the values of NMI for different block sizes are relatively equal. Thus, we conclude that KSRSIO-FCM as well as KSLFCM works for 70–30 distribution in terms of NMI but performs better than KSLFCM for 80–20 distribution. Then the ARI computation measures the clustering output with the ground truth labels. A higher value of ARI results in an improved clustering. The values of ARI for KSLFCM and KSRSIO-FCM are around for 70–30 distribution, whereas it is expressively low for KSLFCM compared to KSRSIO-FCM in case of 80–20 distribution. Additionally, there is no expressive variation in the value of ARI for different block sizes of KSRSIO-FCM. Thus, we can conclude that KSRSIO-FCM with the given block sizes performs as good as KSLFCM for 70–30 partition but performs better than KSLFCM for 80–20 partition.

6 Conclusion

In this paper, we proposed a new KSRSIO-FCM approach for Big Data using Apache Spark. We applied KSRSIO-FCM on the benchmark IRIS dataset to exhibit its utility. Experimental analysis conducted on the IRIS dataset shows that KSRSIO-FCM excels KSLFCM. KSRSIO-FCM compared with KSLFCM produces a proportionate clustering output in terms of F-score, ARI, and NMI. Moreover, we measure the performance of KSRSIO-FCM by varying the number of subsets using the parameter above. Thus, KSRSIO-FCM is convenient for achieving better clustering outputs on Big Data platforms efficiently and accurately.

References

1. A. Saxena, M. Prasad, A. Gupta, N. Bharill, O.P. Patel, A. Tiwari, M.J. Er, W. Ding, C.T. Lin, A review of clustering techniques and developments. Neuro-Comput. **267**, 664–681 (2017)
2. K. Hirota, W. Pedrycz, Fuzzy computing for data mining. Proc. IEEE **87**(9), 1575–1600 (1999)
3. N.S. Iyer, A. Kandel, M. Schneider, Feature-based fuzzy classification for interpretation of mammograms. Fuzzy Sets Syst. **114**(2), 271–280 (2000)
4. W. Phillips II., R. Velthuizen, S. Phuphanich, L. Hall, L. Clarke, M. Silbiger, Application of fuzzy c-means segmentation technique for tissue differentiation in mr images of a hemorrhagic glioblastoma multiforme. Magn. Reson. Imaging **13**(2), 277–290 (1995)
5. R.O. Duda, P.E. Hart, et al., *Pattern Classification and Scene Analysis*, vol. 3 (Wiley, New York, 1973)

6. Y. Leung, J.S. Zhang, Z.B. Xu, Clustering by scalespace filtering. IEEE Trans. Pattern Anal. Mach. Intell. **22**(12), 1396–1410 (2000)
7. J.C. Bezdek, *Pattern Recognition with Fuzzy Objective Function Algorithms* (Springer Science & Business Media, 2013)
8. Z. Zhang, T.C. Havens, Scalable approximation of kernel fuzzy c-means, in *2013 IEEE International Conference on Big Data* (IEEE, 2013), pp. 161–168
9. P. Hore, L.O. Hall, D.B. Goldgof, Single pass fuzzy c-means, in *2007 IEEE International Fuzzy Systems Conference* (IEEE, 2007), pp. 1–7
10. J.K. Parker, L.O. Hall, Accelerating fuzzy-c means using an estimated subsample size. IEEE Trans. Fuzzy Syst. **22**(5), 1229–1244 (2013)
11. J. Gantz, D. Reinsel, The digital universe in 2020: big data, bigger digital shadows, and biggest growth in the far east. IDC iView: IDC Analyze the Future 2007 (2012), pp. 1–16
12. G. Zhao, C. Ling, D. Sun, Sparksw: scalable distributed computing system for large-scale biological sequence alignment, in *2015 15th IEEE/ACM International Symposium on Cluster, Cloud and Grid Computing* (IEEE, 2015), pp. 845–852
13. T.C. Havens, J.C. Bezdek, C. Leckie, L.O. Hall, M. Palaniswami, Fuzzy c-means algorithms for very large data. IEEE Trans. Fuzzy Syst. **20**(6), 1130–1146 (2012)
14. N. Bharill, A. Tiwari, Handling big data with fuzzy based classification approach, in *Advance Trends in Soft Computing* (Springer, 2014), pp. 219–227
15. N. Bharill, A. Tiwari, A. Malviya, Fuzzy based scalable clustering algorithms for handling big data using apache spark. IEEE Trans. Big Data **2**(4), 339–352 (2016)
16. J. Dean, S. Ghemawat, Mapreduce: simplified data processing on large clusters. Commun. ACM **51**(1), 107–113 (2008)
17. N. Bharill, A. Tiwari, A. Malviya Fuzzy based clustering algorithms to handle big data with implementation on apache spark, in *2016 IEEE Second International Conference on Big Data Computing Service and Applications (BigDataService)* (IEEE, 2016), pp. 95–104
18. M. Zaharia, M. Chowdhury, T. Das, A. Dave, J. Ma, M. McCauley, M.J. Franklin, S. Shenker, I. Stoica, Resilient distributed datasets: a fault-tolerant abstraction for in-memory cluster computing, in *Proceedings of the 9th USENIX conference on Networked Systems Design and Implementation* (USENIX Association, 2012), pp. 2–2
19. D. Borthakur, et al., HDFS architecture guide. Hadoop Apache Project **53**, 1–13 (2008)
20. S. Tang, B. He, C. Yu, Y. Li, K. Li, A survey on spark ecosystem for big data processing (2018), arXiv preprint arXiv:181108834
21. J.F. Kolen, T. Hutcheson, Reducing the time complexity of the fuzzy c- means algorithm. IEEE Trans. Fuzzy Syst. **10**(2), 263–267 (2002)
22. V.K. Vavilapalli, A.C. Murthy, C. Douglas, S. Agarwal, M. Konar, R. Evans, T. Graves, J. Lowe, H. Shah, S. Seth, et al., Apache hadoop yarn: yet another resource negotiator, in *Proceedings of the 4th Annual Symposium on Cloud Computing* (ACM, 2013), p. 5
23. V. Schwämmle, O.N. Jensen, A simple and fast method to determine the parameters for fuzzy c–means cluster analysis. Bioinformatics **26**(22), 2841–2848 (2010)
24. Y. Wang, L. Chen, J.P. Mei, Incremental fuzzy clustering with multiple medoids for large data. IEEE Trans. Fuzzy Syst. **22**(6), 1557–1568 (2014)
25. K.Y. Yeung, W.L. Ruzzo, Details of the adjusted rand index and clustering algorithms, supplement to the paper an empirical study on principal component analysis for clustering gene expression data. Bioinformatics **17**(9), 763–774 (2001)
26. A. Strehl, J. Ghosh, Cluster ensembles—a knowledge reuse frame-work for combining multiple partitions. J. Mach. Learn. Res. **3**(Dec), 583–617 (2002)

Author Index

A
Akbay Arama, Zülal, 281, 295
Akın, Muhammed Selahaddin, 281
Alinejad, Babak, 45
Al-Shamiri, Abobakr Khalil, 11, 33
Alten, Oğuzhan, 281
Anita, 411
Ansari, Zahid Ahmed, 349
Aral, Sena, 187, 197
Arıkan, Eda, 269
Aydoğdu, İbrahim, 139, 207

B
Bekdaş, Gabriel, 45
Bekdaş, Gebrail, 53, 61, 73, 81, 93, 111, 127, 155, 179, 187, 197, 261, 317
Bharill, Neha, 423

C
Cakiroglu, Celal, 261
Chand Bansal, Jagdish, 401
Choi, Young Hwan, 1
Cosgun, Turgay, 219, 235

D
Demir, Emre, 139, 207
Dirir, Ahmed, 387

F
Farzam, Maziar Fahimi, 45

G
Gavgani, Seyyed Ali Mousavi, 45
Gençdal, Hazal Berrak, 295
Gifari, Zulkarnaen, 171
Gunes, Baris, 219, 235

H
Hassan, Ahmed, 377

J
Jha, Preeti, 423
Jung, Donghwi, 21
Ju, Young-Kyu, 171

K
Kale, Bahar Neslişah, 139
Karadağ, Bulut, 269
Kayabekir, Aylin Ece, 53, 81, 127, 155, 249
Kim, Joong Hoon, 1, 11, 21, 33, 171, 411
Kim, Taewook, 1
Kumar, Nitin, 411

L
Laghari, Mohammad Shakeel, 377, 387
Liu, Kun, 365
Li, Yilin, 307

M
Mangir, Atakan, 219, 235
Min, Kyoung Won, 21

S. M. Nigdeli et al. (eds.), *Proceedings of 6th International Conference on Harmony Search, Soft Computing and Applications*, Advances in Intelligent Systems and Computing 1275, https://doi.org/10.1007/978-981-15-8603-3

Mounika, Mukkamalla, 423
Mubuli, Apaer, 111

N
Nagendra, Neha, 423
Narendra, V. G., 325, 339
Nasir, Mohammad, 33
Nigdeli, Sinan Melih, 53, 61, 73, 81, 93, 111, 127, 155, 179, 187, 197, 317
Noman, Mubashir, 377, 387
Nuray, Said Enes, 281, 295

P
Pinto, Ancilla J., 325, 339
Prasad, G. Shiva, 339
Pu, Xiaojun, 307

Q
Qiao, Ying, 307

R
Rakıcı, Elmas, 179
Ratnaparkhe, Milind, 423

S
Sadollah, Ali, 11, 33
Sayin, Baris, 219, 235
Shafeeq, B. M. Ahamed, 349
Singh, Shitu, 401

T
Tiwari, Aruna, 423
Toklu, Yusuf Cengiz, 127

U
Ulusoy, Serdar, 61

W
Wang, Hongan, 307, 365
Wang, Zhonghui, 365
Wei, Wenting, 365
Wu, Yunkun, 365

Y
Yadav, Anupam, 411
Yılmaz, Nur, 187, 197
Yücel, Melda, 93, 127, 281, 295, 317

Z
Zhang, Zhi-Yu, 171
Zhu, Jiaqi, 365